*Current Topics
in Vector Research*

Current Topics in Vector Research

Edited by

Kerry F. Harris
Virus–Vector Laboratory, Department of Entomology, Texas A&M
University, College Station, Texas 77843, USA

Current Topics in Vector Research

Volume 3

Edited by Kerry F. Harris

With Contributions by
D.J. Gubler   C.Hiruki   B.H. Kay   K. Kiritani
S. Miyai   F. Nakasuji   Y.O. Paliwal
D.C. Ramsdell   Y. Robert   R. Stace-Smith
H.A. Standfast   D.S. Teakle

With 25 Illustrations

Springer Science+Business Media, LLC

Kerry F. Harris
Virus-Vector Laboratory
Department of Entomology
Texas A&M University
College Station, Texas 77843
USA

Volumes 1 and 2 of *Current Topics in Vector Research* were published by Praeger Publishers, New York, New York.

ISSN: 0737-8491

Typeset by David E. Seham Associates Inc., Metuchen, New Jersey.
ISBN 978-1-4612-9115-2   ISBN 978-1-4612-4688-6 (eBook)
DOI 10.1007/978-1-4612-4688-6

In Memoriam

We grieve the loss of our friend and fellow editorial board member, Dr. Harry Hoogstraal (1917–1986). Harry served as Head of the Department of Medical Zoology of the United States Naval Medical Research Unit Number Three in Cairo for 36 years. Early in his career, he became renowned as the world's foremost authority on the biology, distribution, systematics, behavior, and evolution of ticks, and their medical importance as vectors of viruses, rickettsiae, bacteria, and protozoans affecting human and animal health. His scientific genius and breadth of intellectual and cultural interests amply qualified Harry as a Renaissance man. Whereas he was one of the most highly honored and warmly regarded research biologists of our time, he was equally well known and regarded by others as a sculptor, collector of art, collector of rare cactuses, Egyptologist, and tropical fish aficionado. He was immensely important in our eyes and in the eyes of all his friends and colleagues; however, he was never self-important and always maintained a healthy sense of humor. One had only to be with Harry briefly to sense the presence of a kind and caring person. His reputation as a humanitarian will certainly live on in the hearts and minds of those who knew him and the many who were helped by him. Recently, I was pleasantly reminded of Harry's many quiet, exemplary qualities, both as man and scientist, while reading his tribute to Theobald Smith in the 1986 spring issue of the *Bulletin of the Entomological Society of America* (Volume 32, Number 1). I could not help but think that Theobald would have been honored too. Thanks for you, Harry. Knowing you has made our lives richer. We miss you.

Editor	Editorial Board Members	
Kerry F. Harris	Willy Burgdorfer	Edouard Kurstak
	Paul E.M. Fine	John J. McKelvey, Jr.
	Richard I.B. Francki	Robert F. Whitcomb
	Isaac Harpaz	Telford H. Work

Preface

Vector transmission of pathogens affecting human, animal, and plant health continues to plague mankind both in industrialized and Third World countries. The diseases caused by these pathogens cost billions of dollars annually in medical expenses and lost productivity. Some cause widespread destruction of food- and fiber-producing plants and animals, whereas others present direct and immediate threats to human life and further development in Third World countries.

During the past 15 years or so, we have witnessed an explosive increase in interest in how vectors acquire, carry, and subsequently inoculate disease agents to human, animal, and plant hosts. This interest transcends the boundaries of any one discipline and involves researchers from such varied fields as human and veterinary medicine, entomology, plant pathology, virology, physiology, microbiology, parasitology, biochemistry, molecular biology, genetic engineering, ultrastructure, biophysics, biosystematics, biogeography, ecology, behavioral sciences, and others. Accompanying and perhaps generating this renewed interest is the realization that fundamental knowledge of pathogen-vector-host interrelationships is a first and necessary step in our quest for efficient, safe methods of disease control.

With recent emphasis on a more organismic approach to vector research, today, as never before, researchers from the human, animal, and plant "camps" are working together in a common cause: control and eventual elimination of vector-associated diseases and human suffering. It is my hope that *Current Topics in Vector Research* both reflects and stimulates this newly emerging cooperativeness. It is the first book series to report vector research from all three health areas: human, animal, and plant. Its primary purpose is to encourage communication among vector specialists and generalists alike around the world by providing a common forum to share data, ideas, and technologies relative to all types of pathogen-vector-host transmission systems and disease control strategies and to discuss past, present, and likely future trends in research.

I thank the authors in Volume 3 for their outstanding contributions, and the members of the Editorial Board for their help and encouragement. We welcome and thank the staff of our new publishers, Springer-Verlag New York Inc., for their encouragement and technical assistance. We look forward to an enduring and mutually rewarding relationship. And, I again thank The Rockefeller Foundation and its Associate Director for Agricultural Sciences, now retired, Dr. John J. McKelvey, Jr., for their collaboration and support of our efforts from the very start. Indeed, more than any other factor, the interest and enthusiasm generated by several vector symposia at The Rockefeller Foundation's Bellagio Conference Center confirmed my belief that there was more than ample interest and progress in vector research worldwide and sufficient esprit de corps among researchers to warrant a series such as *Current Topics in Vector Research*.

Kerry F. Harris

Contents

Contributors

Duane J. Gubler
   Division of Vector-Borne Viral Diseases, Center of Infectious
   Diseases, Centers for Disease Control, Public Health Service,
   United States Department of Health and Human Services, San Juan,
   Puerto Rico 00936

Chuji Hiruki
   Department of Plant Science, University of Alberta, Edmonton,
   Alberta T6G 2P5, Canada

Brian H. Kay
   Vector Biology and Control Unit, Queensland Institute of Medical
   Research, Herston, Brisbane, Queensland, Australia 4006

Keizi Kiritani
   Insect Ecology Laboratory, Division of Entomology, National
   Institute of Agro-Environmental Sciences, Yatabe, Tsukuba, Ibaraki
   305, Japan

Shun'ichi Miyai
   Department of Environmental Planning, Division of Information
   Analysis, National Institute of Agro-Environmental Sciences,
   Yatabe, Tsukuba, Ibaraki 305, Japan

Fusao Nakasuji
   Entomological Laboratory, Faculty of Agriculture, Okayama
   University, Okayama 700, Japan

Yogesh C. Paliwal
   Chemistry and Biology Research Institute, Agriculture Canada,
   Ottawa, Ontario, K1A 0C6, Canada

Donald C. Ramsdell
Department of Botany and Plant Pathology, Michigan State
University, East Lansing, Michigan 48824, USA

Yvon Robert
Institut National de la Recherche Agronomique, Centre de
Recherches de Rennes, Laboratoire de Recherches de la Chaire de
Zoologie, Domaine de la Motte-au-Vicomte, B.P. 29-35650, Le Rheu,
France

Richard Stace-Smith
Department of Plant Pathology, Agriculture Canada, Research
Station, Vancouver, British Columbia V6T 1X2, Canada

Harry A. Standfast
CSIRO Division of Tropical Animal Science, Long Pocket
Laboratories, Indooroopilly, Brisbane, Queensland, Australia 4068

David S. Teakle
Department of Microbiology, University of Queensland, St. Lucia,
Brisbane, Queensland, Australia 4067

1
Ecology of Arboviruses and Their Vectors in Australia

Brian H. Kay and Harry A. Standfast

Introduction

Research on the vectors of arboviruses in Australia (Fig. 1.1) has been hampered by the vastness (7,686,855 km^2) of the continent in relation to its relatively sparse human population, currently about 16 million. In relation to the vectors of veterinary importance, the phrase "Australia rides on the sheep's back" implies national dependence on the pastoral industry and on a sheep and cattle herd of about 160 million. Research effort in both the medical and veterinary fields has fluctuated markedly during the last 30 years, with peaks of activity in the epidemic and the immediate postepidemic periods, falling to a low level in the interepidemic periods when politicians and administrators saw little merit in supporting studies of arthropod-borne viruses. Central to this theme are a small group of workers, stimulated by an even smaller series of events in Australian medical and veterinary history.

Research began auspiciously in 1906 when Thomas Bancroft suggested that *Stegomyia fasciata (Aedes aegypti* L.*)* was the vector of dengue (6), and was ably confirmed in 1918, by Cleland *et al.* (19). The main effort in isolation and identification of arboviruses began in 1957 when Doherty initiated the arbovirus program at the Queensland Institute of Medical Research (QIMR), and by 1977 the list of arboviruses known from Australia had increased from three (dengue, Murray Valley encephalitis, Bovine ephemeral fever)[1] to 41 (33). Contributions by teams led by Marshall (John Curtin School of Medical Research, JCSMR), Stanley (University of Western Australia, UWA), and St. George, Standfast and Dyce (Commonwealth Scientific and Industrial Research Organization, CSIRO) have

---

[1]Although epidemic polyarthritis was first recognized in 1928, Ross River virus was first isolated at QIMR in 1963 (31).

Brian H. Kay, Vector Biology and Control Unit, Queensland Institute of Medical Research, Herston, Brisbane, Queensland, Australia 4006.
Harry A. Standfast, CSIRO Division of Tropical Animal Science, Long Pocket Laboratories, Indooroopilly, Brisbane, Queensland, Australia 4068.

FIGURE 1.1. Australia, showing localities listed in the text together with 300- and 800-mm isohyets.

now increased this list to at least 77 (Table 1.1). Much of this work has been reviewed by Spradbrow (140), French (49), Doherty (29–31, 33), and Marshall and Miles (108) or has been referred to in the Arbovirus Research in Australia series (32, 142, 143). As most of this early work has received adequate coverage from the virologists' viewpoint, we shall focus on events since 1977 with greater entomological perspective than before and attempt to predict what may happen in the future.

Characters and Events

The regular occurrence of dengue epidemics especially in Queensland (from 1879), the Northern Territory (1914, 1920–1921, 1927–1930). Western Australia (13 years from 1909 to 1929), and New South Wales (1886, 1898, 1905, 1916, 1926, and 1942) (98) provided the incentive for studies by Hare, the Bancrofts, and Cleland and co-workers. The additional challenges of malaria from 1838 (8) and filariasis, which, according to Flynn in 1903, "threatens to become a national curse" (104), led to the establishment of

TABLE 1.1. Arboviruses of Australia, 1985.[a]

Family	Genus or group	Virus
Alphaviridae	*Alphavirus*	Barmah Forest (BF), Getah (GET), Ross River (RR), Sindbis (SIN)
Flaviviridae	*Flavivirus*	Alfuy (ALF), Edge Hill (EH), Gadgets Gully (GGY), Kokobera (KOK), Kunjin (KUN), Murray Valley encephalitis (MVE), Saumarez Reef (SRE), Stratford (STR), Dengue (DEN)
Bunyaviridae	*Bunyavirus*	
	Simbu group	Aino (AINO), Akabane (AKA), Douglas (DOU), Facey's Paddock), Peaton (PEA), Thimiri (THI), Tinaroo (TIN)
	Koongol group	Koongol (KOO), Wongal (WON)
	Mapputta group	Gan Gan (GG), Mapputta (MAP), Trubanaman (TRU)
	Upolu group	Upolu (UPO)
	Turlock group	Umbre (UMB)
	Ungrouped	Belmont (BEL), Kowanyama (KOW)
	Nairovirus	
	Dera Ghazi Khan group	Kao Shuan (KS)
	Sakhalin group	Taggert (TAG)
	Uukuvirus	
	Uukuniemi group	Precarious Point (PP)
Reoviridae	*Orbivirus*	
	Bluetongue group	Bluetongue serotypes 1, 15, 20, 21, 23
	Corriparta group	Corriparta (COR)
	Epizootic Hemorrhagic disease group	(CSIRO 157), (CSIRO 753), (CSIRO 775), (DDP 59), Ibaraki (IBA)
	Eubanangee group	Eubanangee (EUB), Tilligerry (TIL)
	Kemerovo group	Nugget (NUG)
	Palyam group	D'Aguilar (DAG), CSIRO Village (CVG), Bunyip Creek (BC), Marrakai (MAR)
	Wallal group	(Mudjinbarry), Wallal (WAL)
	Warrego group	Mitchell River (MR), Warrego (WAR)
	Ungrouped	Paroo River (PR)
Rhabdoviridae	Bovine ephemeral fever group	Adelaide River (ADE), Berrimah (BRH), Bovine ephemeral fever (BEF), Kimberley (KIM)
	Mossuril group	Charleville (CHV)
	Tibrogargan group	(Coastal Plains), Tibrogargan (TIB)
	Ungrouped	Almpiwar (ALM), (Humpty Doo), (Oak Vale), Kununurra (KNA), (Parry's Creek)
Undefined		(Beatrice Hill), (Lake Clarendon), (Leanyer), Ngaingan (NGA), (Parker's Farm), Picola (PIA), Termeil (TER), Wongorr (WGR), Yacaaba (YAC)
	Quaranfil group	Johnston Atoll (JA)

[a]Unregistered viruses in parentheses.

the Australian Institute of Tropical Medicine in Townsville during 1910, with Anton Breinl appointed as its first Director. Its establishment antedated by 6 years the Walter and Eliza Hall Institute in Melbourne (26), which, from 1951, was to exert a major influence on studies of the epidemiology of Murray Valley encephalitis virus (MVE), then known as Australian X disease. The Tropical Institute was transferred from Townsville to Sydney in 1930 and redesignated the School of Public Health and Tropical Medicine, from which D.J. Lee and others made a significant contribution to the taxonomy of mosquitoes and ceratopogonids.

The gap in research into fevers in northern Australia created by the transfer of the Townsville Institute to Sydney was filled almost two decades later by the Queensland Institute of Medical Research. Its first Director, Ian Mackerras, and four other staff members proudly occupied a temporary wartime hut on June 2, 1947 with a clear mandate from Burnet, the Director of the Hall Institute, to build QIMR into "a first class centre for the investigation of insect-borne diseases" (27). This building, constructed to last 5 years in 1940, was to be the Institute's home for 30 and was destined to become the epicenter of arbovirus research in Australia.

The release of myxoma virus for the control of rabbits between May and November 1950, and the coincidental reappearance of Murray Valley encephalitis (MVE) during January 1951, marked this period as being important and, to some extent, frustrating. The disturbing suggestion that myxoma virus may have been the etiological agent of MVE inspired self-inoculation by Fenner, Burnet, and Clunies-Ross to prove the opposite. Substantial studies of vector ecology of value to arbovirology were carried out to determine the mechanical vectors of myxoma virus (48).

At the Hall Institute, French isolated MVE virus from clinical material, while Anderson, aided by Victorian public health personnel, completed serological surveys of humans and domestic and wild animals. Anderson hypothesized that the vector of MVE was probably *Culex annulirostris* and that its epidemiology was linked to viral transport from northern Australia, or from countries to the north of Australia by migrating waterbirds in times of above average rainfall (2). Miles and collaborators at the Institute of Medical and Veterinary Research, Adelaide, were also successful in isolating the virus and arrived at a similar epidemiological conclusion. The validity of what is commonly known as the Miles–Anderson hypothesis will be discussed later. In 1952, Reeves (then of the Hooper Foundation, San Francisco) was invited to form a field team whose aim was to isolate MVE from mosquitoes. In this group were mosquito taxonomist E.N. Marks of the National Mosquito Control Committee, based at the University of Queensland, and M.J. Mackerras, Parasitologist (QIMR), who provided assistance and gained practical experience in the techniques of collection and preservation of arthropods, as practiced by United States workers (103). Successful isolation of MVE from mosquitoes eluded workers until 1960 (34).

At QIMR, I.M. and M.J. Mackerras began to describe the Tabanidae and Simuliidae of Australia while maintaining a modest working interest in MVE. In 1957, R.L. Doherty returned to QIMR after study leave at the Harvard School of Public Health and, influenced by the ecological approach of W.C. Reeves and others in the United States plus the Rockefeller Foundation work in Trinidad and Belem (28), began a successful program which lasted for 19 years. Based on serological survey data, Mitchell River Mission (Kowanyama), an aboriginal settlement on the Gulf of Carpentaria in northern Queensland, was selected for long-term study in relation to MVE; the initial field trip, from March 30 to April 13, 1960, opened a Pandora's box of 7 arboviruses new to Australia plus 4 prized isolations of MVE from mosquitoes (34). From 1960 to 1961, 25,901 mosquitoes were processed, returning 60 strains of 11 arboviruses. In subsequent years, only the February 1976 Charleville (southwestern Queensland) trip, 27,079 mosquitoes for 74 isolations (42, 65), outstripped this productivity. One wonders the outcome had this initial visit not been successful. Nevertheless, the basis for establishment of another arbovirus outpost had been made and in 1966, the Rockefeller Foundation funded the QIMR field station.

In 1968, the third recognized Bovine ephemeral fever epidemic probably had greater significance than the first in 1936, which had resulted in its isolation from "buffy coat" (leucocyte-platelet fraction) of cattle blood (102). Although I.M. and M.J. Mackerras, then CSIRO scientists, failed to demonstrate cyclical transmission by *Stomoxys calcitrans* L., *Aedes vigilax* (Skuse), *Ae. alternans* (Westwood), and *Cx. annulirostris,* they were attracted to the possibility that the Ceratopogonidae might be the vectors. The third epidemic in 1968 paved the way for Doherty to expand collections into attempted isolation of arboviruses from biting midges, phlebotomine sandflies, tabanids, and simuliids. This saw the valuable collaboration of A.L. Dyce (CSIRO), the development of new sorting and handling techniques (46), and isolation of an interesting group of viruses, mainly from ceratopogonids. Two Simbu group viruses, Akabane and Aino (formerly known as Samford) (37), were subsequently linked with teratogenic disease in domestic animals. These initial findings provided sufficient stimulus for the inauguration of a strong CSIRO program directed toward arthropod-borne viruses affecting livestock, which has discovered 26 new viruses, highlighted, perhaps, by the first isolation of bluetongue virus in Australia (146).

Isolation of arboviruses Upolu and Johnston Atoll from the tick *Ornithodoros capensis* Neumann followed the visit of H.N. Johnson from the Yale Arbovirus Research Unit to a coral cay on the Great Barrier Reef. Since then, seven other viruses have been isolated, including Saumarez Reef and Gadgets Gully, flaviviruses related to the tick-borne encephalitis group and therefore of possible medical importance (152,145).

The 1974 epidemic of MVE stimulated research and resulted in the first isolations from *Cx. annulirostris* during an epidemic (42, 111). By this time, QIMR was moving toward definitive virological and entomological study of vectors of selected viruses; the isolations were but spinoff from other studies. The Doherty era, affectionately known as "the stamp-collecting era," was nearing an end. Two years prior to the MVE epidemic, the damming of the Ord Valley in northwestern Australia had provided N.F. Stanley (UWA) with a productive study area and it appeared that an endemic focus of MVE had been discovered. Regular wet-season collections had indicated that approximately 1 in 650 *Cx. annulirostris* were infected with MVE (95), with a bonus of 8 isolates during the dry-transitional seasons (June–November).

In 1979, Ross River virus gained international status when it caused about 50,000 clinical cases in Fiji and then was transported to other Pacific islands (108). Stanley described this event as "Australia's gift to the Pacific" (159), but in retrospect, it boosted arbovirology in Australia as well as in Oceania.

Ironically, this section ends as it began, with dengue fever, which reappeared in north Queensland in 1981, after an apparent absence of 26 years (70). Although the epidemic was mainly confined to north Queensland, extensive *Ae. aegypti* surveys were conducted not only through this state, but also in New South Wales (136), Northern Territory, and Western Australia. Breeding of *Ae. aegypti* was still associated with rainwater tanks, drums, tires, and other backyard rubbish in many country towns as it had been during the early epidemics, although the popularity of indoor potted plants had caused major problems by furnishing breeding sites in the larger coastal cities. This epidemic provided a reminder that northern Australia is again receptive to dengue fever and the sometimes fatal dengue hemorrhagic fever/dengue shock syndrome, first described in north Queensland by Hare in 1898 (60).

Virus Isolations

Insects cannot be confirmed as vectors on the basis of virus isolation from unengorged insects. This merely indicates that they can be infected. It does not guarantee that the insect will shed virus in its saliva, will live long enough to transmit, or be present in sufficient numbers to maintain the virus in the community. However, 99, 105, and 159 isolations, respectively, of Sindbis, MVE, and Kunjin from *Cx. annulirostris* and 32 isolations of Corriparta virus from *Aedeomyia catasticta* Knab are suggestive of an ecological association (Table 1.2). The small number of isolates of Sindbis, Kunjin, Alfuy, and even MVE, all reported to have avian hosts (29), from less well-sampled species, e.g., *Cx. squamosus* (Taylor), *Cx. pullus* Theobald, and *Cx. pipiens australicus* Dobrotworsky and Drummond, is compatible with their predilection for bird feeding (76, 82).

That *Cx. annulirostris* is widely accepted as the major mosquito vector of arboviruses (638 isolates) in Australia is not just because the sample

TABLE 1.2. Summary of arbovirus isolations from selected mosquito species up to 1985, based on published and some unpublished data.

Arthropod	Number processed[a]	Virus (number of isolates)
Culex annulirostris	454,437	SIN (99), RR (41), BF (4), MVE (105), KUN (159), KOK (31), EH (5), KOO (23), WON (40), COR (10), WAR (3), Facey's Paddock (2), GG (3), WGR (12), Parker's Farm (2), UMB (1), BEL (1), KOW (2), PR (9), TER (2), PIA (2), OR540 (6), Parry's (3), KIM (1), OR379 (2), OR512 (2), EUB (4), Bunyavirus unidentified (64), ungrouped (2)
Aedes vigilax	59,481	SIN (1), EH (12), STRAT (1), KOK (1), RR (10), GG (4), YAC (1), TER (1)
Aedes normanensis	46,136	SIN (11), MVE (1), RR (8), KOK (1), EH (13), Facey's Paddock (2), GG (6), BF (2), UMB (1), TER (1), WGR (1)
Ae. pseudonormanensis	317	—
Ae. tremulus	327	KUN (1)
Ae. funereus	4,410	—
Ae. camptorhynchus	330	TER (2)
Ae. vittiger	2,791	SIN (1), Parker's Farm (1)
Ae. eidsvoldensis	1,877	SIN (2), GG (2)
Ae. lineatopennis	4,169	WGR (1)
Ae. procax	63	—
Ae. theobaldi	433	(RR)[b]
Cx. quinquefasciatus	59,099	SIN (2), KUN (1), MVE (1)
Cx. sitiens	10,204	—
Cx. australicus	1,174	MVE (1), (SIN, KUN)[b]
Cx. edwardsi	1,829	SIN (1), COR (3), Oak Vale (9), unidentified (3)
Cx. squamosus	1,011	KUN (3), SIN (1)
Cx. starckeae	3,100	SIN (1)
Cx. bitaeniorhynchus	5,166	GET (1), MVE (1)
Cx. pullus	5,434	KUN (2), SIN (1), ALF (3)
Aedeomyia catasticta	7,892	ALF (1), COR (32), KNA (1), OR869 (4)
Mansonia uniformis	11,566	EH (1), RR (1)
Ma. septempunctata	2,509	SIN (1)
Coquillettidia crassipes/ xanthogaster	17,954	WON (1)
Cq. linealis	1,489	RR (1)
Anopheles bancroftii	27,161	KOO (2), KUN (1), BEF (1)
An. annulipes	39,761	Parker's Farm (3), KOW (2), MAP (3), TRUB (2), TIL (1)
An. amictus	16,783	GET (1), KOW (1), RR (3)[b], EH (1), MAP (2)
An. farauti	4,638	KOO (1), KUN (1), EUB (1)
An. meraukensis	2,486	EH (1), MAP (2), Leanyer (1), WAR (1)

[a]Only identified species listed. Viruses isolated by Woodroofe (see 32) included but details of 13,506 mosquitoes processed absent.
[b]Marshall (107) and Marshall and Miles (108) stated that these viruses also had been recovered, but details as yet unpublished.

Table 1.3. Supplementary list of arbovirus isolations from Australia, 1977–1985.[a]

Genus/group	Virus	Year	Source of isolation		
			Place	Species	Reference
Alphavirus	Barmah Forest[b]	1974, 1976	Charleville, Queensland	*Culex annulirostris, Aedes normanensis*	42
		1975	Beatrice Hill, Northern Territory	*Culicoides marksi, Cx. annulirostris*	158
	Ross River	1974	Murray Valley, Victoria	*Cx. annulirostris*	111
		1972	Termeil forest, New South Wales	*Ae. vigilax, Aedes* spp.	109
		1977	Derby, Western Australia	*Cx. annulirostris*	165
		1977	Oakey, Queensland	Horse	126
	Sindbis	1976	Kowanyama, Queensland	*Cx. quinquefasciatus, Cx. pullus*	42, 64
		1972, 1974, 1976	Ord Valley, Western Australia	*Cx. annulirostris, Ae. normanensis*	97
		1984	Peachester, Queensland	*Cx. edwardsi*	e
Flavivirus	Edge Hill	1970	Nelson Bay, New South Wales	*Ae. vigilax*	109
	Gadgets Gully	1975–1977, 1979	Macquarie Island	*Ixodes uriae*	152
	Kokobera	1976	Kowanyama, Queensland	*Cx. annulirostris*	42, 64
		1976	Charleville, Queensland	*Cx. annulirostris, Ae. normanensis*	42, 64
	Kunjin	1974, 1975	Ord Valley, Western Australia	*Cx. annulirostris*	97
		1974	Murray Valley, Victoria	*Cx. annulirostris*	111
	Murray Valley encephalitis	1974	Murray Valley, Victoria	*Cx. australicus*	111
		1975, 1977, 1978	Ord Valley, Western Australia	*Cx. annulirostris, Cx. quinquefasciatus*	95
		1977	Derby, Western Australia	*Cx. annulirostris*	165
	Saumarez Reef	1974	Coral Cays, Queensland	*Ornithodoros capensis*	145
		1974	Northern Tasmania	*I. eudyptidis*	145
	Dengue (Type 1)	1981, 1982	Thursday Island, Queensland	Human	70
		1982	Cairns, Queensland	Human	70

Virus	Year	Location	Host/vector	Reference
Bunyavirus				
Simbu group				
Aino	1979	Peachester, Queensland	Cattle, *C. brevitarsis*	22
Akabane	1979	Peachester, Queensland	Cattle, *C. brevitarsis*	22
Douglas (Facey's Paddock)	1978	Douglas Station, Queensland	Cattle	21, 147
	1974, 1976	Charleville, Queensland	*Cx. annulirostris, Ae. normanensis*	42
Peaton	1974–1976	Beatrice Hill, Northern Territory	*Culicoides* spp.	158
	1976	Peachester, Queensland	*C. brevitarsis*	147, 149
	1977	Grafton, New South Wales	Cattle	147, 149
Thimiri	1974	Beatrice Hill, Northern Territory	*C. histrio*	154
Tinaroo	1978	Kairi, Queensland	*C. brevitarsis*	21, 143
Wongal	1974	Murray Valley, Victoria	*Cx. annulirostris*	111
Koongol group				
Gan Gan	1977	Derby, Western Australia	*Cx. annulirostris*	165
Mapputta group	1970	Nelson Bay, New South Wales	*Ae. vigilax*	109
Mapputta	1974–1976	Beatrice Hill, Northern Territory	*Anopheles annulipes, An. amictus*	158
Turlock group				
Umbre[b]	1974	Charleville, Queensland	*Cx. annulirostris, Ae. normanensis*	42
Ungrouped				
Belmont	1974–1976	Beatrice Hill, Northern Territory	*C. marksi*	158
Kowanyama	1974	Beatrice Hill, Northern Territory	*Culicoides* spp.	158
		Murray Valley, Victoria	*Cx. annulirostris*	111
Nairovirus				
Sakhalin group				
Taggert	1976–1977, 1979	Macquarie Island	*I. uriae*	152
Uukuvirus				
Precarious Point	1975–1977	Macquarie Island	*I. uriae*	152
Orbivirus				
Bluetongue group				
Bluetongue #20	1975	Beatrice Hill, Northern Territory	*Culicoides* spp.	146
Bluetongue #21	1979	Victoria River, Northern Territory	Cattle	148
Bluetongue #1	1979	Beatrice Hill, Northern Territory	Cattle	148
	1979	Tortilla, Northern Territory	Cattle	148

TABLE 1.3. *Continued*

Genus/group	Virus	Year	Source of isolation – Place	Source of isolation – Species	Reference
		1979	Beatrice Hill, Northern Territory	*C. fulvus*[c]	156
		1983	Peachester, Queensland	*C. brevitarsis*	144
	Bluetongue #15	1982	Adelaide River, Northern Territory	Cattle	*g*
Corriparta group	Bluetongue #23	1982	Beatrice Hill, Northern Territory	Cattle	*f*
	Corriparta	1974–1976	Beatrice Hill, Northern Territory	*Cx. annulirostris*	158
		1973–1976	Ord Valley, Western Australia	*Cx. annulirostris, Aedeomyia catasticta*	97
	(CSIRO157)	1984	Peachester, Queensland	*Cx. edwardsi*	*e*
Epizootic hemorrhagic disease (EHD) group		1981	Peachester, Queensland	*C. brevitarsis*	150
		1979	Beatrice Hill, Northern Territory	Cattle	150
	(CSIRO753)	1981	Peachester, Queensland	*C. brevitarsis*	150
		1981	Peachester, Queensland	Cattle	150
	(CSIRO775)	1981	Kairi, Queensland	Cattle	150
	(DPP59)	1981	Northern Territory	Cattle	150
	Ibaraki	1980	Beatrice Hill, Northern Territory	Cattle	150
Eubenangee group	Eubenangee	1974–1976	Beatrice Hill, Northern Territory	*An. farauti, Cx. annulirostris, C. marksi*	158
	Tilligerry	1971	Nelson Bay, New South Wales	*An. annulipes*	158
Kemerovo group	Nugget	1977, 1979	Macquarie Island	*I. uriae*	109
Palyam group	CSIRO Village	1974–1976	Beatrice Hill, Northern Territory	*Culicoides* spp.	152
	Bunyip Creek	1974–1976	Beatrice Hill, Northern Territory	*C. oxystoma*	158
	Marrakai	1974–1976	Beatrice Hill, Northern Territory	*C. oxystoma*	158
Wallal group	(Mudjinbarry)	1971	Mudjinbarry, Northern Territory	*C. marksi*	41
	Wallal	1974–1976	Beatrice Hill, Northern Territory	*C. marksi*	158
Warrego group	Warrego	1974–1976	Beatrice Hill, Northern Territory	*C. marksi, Culicoides* spp.., *Cx. annulirostris*	158
Ungrouped	Paroo River	1974	Murray Valley, Victoria	*Cx. annulirostris*	109

Rhabdovirus

BEF group				
Adelaide River	1981	Tortilla Flat, Northern Territory	Cattle	55
Berrimah	1981	Berrimah, Northern Territory	Cattle	54
Kimberley	1973	Ord Valley, Western Australia	Cx. annulirostris	97
	1980	Tortilla Flat, Northern Territory	Cattle	23
Tibrogargan group				
(Coastal Plains)	1981	Beatrice Hill, Northern Territory	Cattle	e
Tibrogargan	1976	Peachester, Queensland	C. brevitarsis	24
Ungrouped				
(Humpty Doo)	1974–1976	Beatrice Hill, Northern Territory	C. marksi, Lasiohelea spp.	158
Kununurra	1973	Ord Valley, Western Australia	Ad. catasticta	97
(Parry's Creek)	1973, 1974, 1976	Ord Valley, Western Australia	Cx. annulirostris	97
(Oak Vale)	1984	Peachester, Queensland	Cx. edwardsi	f
Undefined				
(Beatrice Hill)	1974–1976	Beatrice Hill, Northern Territory	C. perigrinus	158
Lake Clarendon	1980	Lake Clarendon, Queensland	A. robertsi	151
(Leanyer)[d]	1974	Darwin, Northern Territory	An. meraukensis	40
	1975	Beatrice Hill, Northern Territory	C. marksi	158
(Parker's Farm)	1976	Charleville, Queensland	Cx. annulirostris	42
	1974–1976	Beatrice Hill, Northern Territory	C. marksi	158
Picola	1974	Murray Valley, Victoria	Cx. annulirostris	111
Termeil	1972	Termeil forest, New South Wales	Ae. camptorhynchus, Ae. vigilax, Cx. annulirostris	109
Yacaaba	1970	Nelson Bay, New South Wales	Ae. vigilax	109

[a]Unregistered viruses in parentheses.
[b]Barmah Forest and Umbre listed as Ch 16313 and Ch 19546, respectively (42).
[c]Listed as *Culicoides* (*Avaritia*)#5 (156).
[d]Listed as NT 16701 (33).
[e]G. Gard and D. Cybinski (unpublished data).
[f]H.A. Standfast and T.D. St. George (unpublished data).
[g]G. Gard (unpublished data).

size was eight times larger than for any other species, but also because of its climatic and behavioral adaptability and versatility. The high virus isolation rate from *Ae. normanensis* (Taylor) (1 per 981) suggests this species should be investigated further, although care with methods, timing, and possibly site selection undoubtedly influenced these returns. In contrast, the poor isolation rate from the 59,099 *Cx. quinquefasciatus* Say processed (1 per 14,775) corroborates vector competence findings and suggests that this species is unlikely to be important in Australia. Table 1.2 also indicates that many species have been poorly sampled and efforts to process substantial numbers of (1) the ephemeral inland temporary ground pool species, e.g., *Ae. vittiger* (Skuse), *Ae. theobaldi eidsvoldensis* Mackerras, *Ae. procax* Skuse, and *Ae. theobaldi* (Taylor), (2) coastal saltmarsh *Ae. camptorhynchus* Thompson and even *Ae. vigilax* and *Ae. funereus* (Theobald), and (3) *Cx. australicus* should be intensified. Further data on these species may aid in either the understanding of virus survival in semiarid or temperate zones or the causes of human morbidity on the heavily populated coast.

With respect to biting midges, *Culicoides marksi* Lee and Reye has been a most productive species, but because it feeds mainly on marsupials (75, 121), a link with diseases of introduced livestock seems unlikely. On the other hand, continued processing of those species closely associated with cattle, sheep, and horses is still warranted and evident from the number of new arboviruses isolated over recent years (Table 1.3). This list supplements those of Doherty (30, 31, 33). Analysis of isolates from the *Culicoides* processed (Table 1.4) demonstrates a strong association with

TABLE 1.4. Summary of arbovirus isolations from *Culicoides* species up to 1985, based on published and unpublished data.[a]

Arthropod	Number processed	Viruses (number of isolates)
Culicoides actoni	147	WAR (1)
C. austropalpalis	18,233	KNA (1), WAR (1), Facey's Paddock (1)
C. brevitarsis	198,880	AKA (81), AINO (29), BEF (1), BLU 1 (3), BC (31), CVG (87), CSIRO 157 (13), CSIRO 753 (3), CSIRO 775 (2), DAG (40), DOU (4), KIM (2), PEA (18), TIB (7), TIN (5), Palyam group (10)
C. bundyensis	6,468	BEL (1)
C. dycei	6,884	WAL (4), WAR (1)
C. fulvus	158	BLU (1)
C. histrio	1,005	THI (1)
C. marksi	102,575	BF (1), BEL (1), EUB (1), Leanyer (4), Mudjinberry (1), Parker's Farm (1), Humpty Doo (1), WAL (40), WAR (30)
C. oxystoma	45,833	BC (2)
C. pallidothorax	80	WGR (1)
C. peregrinus	17,381	Beatrice Hill (1)
C. wadai	3,587	AKA (1)

[a]Only identified species included.

bunyaviruses (particularly Simbu group), orbiviruses, and rhabdoviruses and the importance of *C. brevitarsis,* which has yielded 336 isolates of 16 different viruses. However, the ongoing taxonomic elucidation of particularly the subgenus *Avaritia,* but also other *Culicoides,* suggests two things: (1) early *C. brevitarsis* isolates may not always have been derived from uniform pools and (2) other newly recognized species, e.g., *C. fulvus* Sen and Das Gupta and *C. wadai* Kitaoka, require more virological investigation. A similar situation may exist in the case of *Culex edwardsi* Barraud, a species which could be confused with *Cx. annulirostris.* Recently 1829 *Cx. edwardsi* yielded 16 isolates of at least three different viruses (H.A. Standfast and T.D. St. George, unpublished data).

In 1984, Standfast and colleagues (158) pointed out that "there is no more reason to expect a virus to be confined to one insect family than there is to expect it to be restricted to one vertebrate family." Whereas only 35 of the 409 arboviruses listed in the *International Catalogue* (7) were isolated from insects representing more than one family of Diptera, nine of 22 viruses isolated from Beatrice Hill from 1974 to 1976 (Barmah Forest, Belmont, Facey's Paddock, Bovine ephemeral fever, Eubenangee, Warrego, Leanyer, Parker's Farm, Wongorr) have this property.

Isolations of Barmah Forest virus from both mosquitoes and biting midges (three isolates from *C. marksi*) is of interest, as, to our knowledge, this is the first alphavirus from species other than culicines or anophelines. The virus was originally classified as a bunyavirus-like virus, but on subsequent analysis of structure, virus-induced proteins, and in infected cells, Dalgarno and co-workers (25) concluded that it was an atypical alphavirus.

Lvov (99) discussed the possible role of migratory sea birds in relation to the possible occurrence of Tyuleniy, Kemerovo, Uukuniemi, and Sakhalin (or related viruses) at opposite ends of the Pacific basin. Occasional collections, mainly from *Ixodes uriae* White from Macquarie Island, but also from *Ixodes eudyptidis* Maskell from Tasmania, have now substantiated Lvov's suggestion with isolations of Saumarez Reef, Nugget, Precarious Point, and Taggert, respectively. The latter three viruses were isolated from *I. uriae* also known to occur in the subarctic (as *I. putus*) and possibly have discontinuous distributions. However, additional isolations of Saumarez Reef virus from *Ornithodoros capensis* Neumann collected on two reefs at 20 and 22°S latitude (145) suggest that they may not, and further collections may end considerable speculation on transfer pathways.

Isolations of Johnston Atoll and Upolu viruses from the widely distributed *O. capensis* in tern colonies on the Great Barrier Reef illustrated the utilization of a simple bird–tick ecosystem in Australia, but also created zoogeographic interest, as Johnston Atoll virus had previously been isolated from bird colonies in the northern Pacific (36). Collections of *Argas robertsi* Hoogstraal, Kaiser, and Kohls from mainland colonies of ciconiiforms, including the ubiquitous cattle egret *Bubulcus ibis coromandus,* have resulted in the isolation of Kao Shuan, previously known from Taiwan and Java (38), and Lake Clarendon virus, which to date has no shared antigenic relationships with any other known viruses (151). Hence, the viruses isolated offer an evolutionary range from those occurring elsewhere (Kao Shuan, Johnston Atoll), to those with antigenic relationships elsewhere (Upolu, Saumarez Reef, Nugget, Taggert, Gadgets Gulley, Precarious Point), to Lake Clarendon virus, which, so far, has proved distinct.

The precise number of immature or adult ticks processed is difficult to ascertain, but is approximately 5600, yielding 116 strains of nine viruses. However, the majority of those processed have been *I. uriae, O. capensis,* and *A. robertsi* in three different habitats occupied by birds. The late H. Hoogstraal's cryptic comments (personal communication) probably summarized the general thesis that other species require evaluation: "If any of your virological colleagues would care to check these ticks [*Aponomma* and *Amblyomma*] for viruses he might make history."

Vectors and Their Competence

The taxonomy of Australian mosquitoes is well known, but there remains a paucity of ecological information with respect to even those important species from which arboviruses have been frequently isolated. In 1974, this prompted Doherty (31) to state, "Certainly there is at present less information available about the bionomics of *Culex annulirostris,* the likely major vector of Murray Valley encephalitis virus in Australia, than had been accumulated 28 years ago about *Aedes aegypti,* the vector of dengue." In the same review, Doherty pointed out that "Colonization of *Culex annulirostris* is needed to allow more detailed studies of virus–vector relationships but this goal has so far eluded investigators."

These shortcomings have been rectified to some extent with colonization of strains of *Cx. annulirostris* from the Murray Valley (114) and Brisbane (119) which has facilitated detailed investigation of larval bionomics (119), and vector competence (73, 77, 81) and provided donor mosquitoes for the experimental infection of a wide range of domestic and wild vertebrates (83, 84). In 1974, the MVE epidemic stimulated field study of *Cx. annulirostris* in the Murray Valley (107, 112, 113, 115), whereas definitive investigations of seasonal abundance (65, 96, 164), age structure (66), biting habits (69), host preference and feeding patterns (76, 82), resting habits (68), and breeding physiology (85, 119) were ongoing in Queensland and elsewhere.

Earlier investigations of *Cx. annulirostris* have been summarized by Kay and colleagues (64, 78). Contemporary studies have supported the view that this species is a most adaptable mosquito, capable of colonizing rainwater pools within a day of their formation at Mildura, Victoria, and reaches maximum larval densities within 8 days, thus utilizing them prior to maximum buildup of predators (113). However, although temperatures of 25°C or slightly higher were optimal for development and survival of immatures in the Murray Valley (115) and in Brisbane (119), their upper and lower temperature thresholds were different, indicating not only climatic adaptation, but also the potential danger in extrapolating biological data from one part of Australia to another. Because water temperatures of 25–30°C are conducive to maximum growth potential of *Cx. annulirostris* from Brisbane, the risk of arbovirus transmission may actually be reduced in extremely hot summer conditions. Further studies of this nature have been done in tropical Australia by D. Rae (unpublished data).

To date, autogenous egg development has been associated with four Australian mosquito species, including *Ae. vigilax,* considered to be the major coastal vector of the alphavirus Ross River (78, 139). Autogeny has yet to be detected in field populations of *Cx. annulirostris,* although 5–9% of colonies from Mildura and Brisbane possess this trait (85). It may bear little relevance to arbovirus transmission, although it may provide this species with another survival strategy under conditions of high larval nutrition.

Studies of the host preference of *Cx. annulirostris* using animal-baited traps at Kowanyama and Charleville indicated that dogs, cattle, and pigs were more attractive than, in descending order of attractiveness, kangaroos, humans, and domestic chickens (76). Subsequent studies of host-feeding patterns in the Northern Territory (121) and at the above Queensland sites (76, 82) support this view, and, in fact, human feeding in both rural and urban situations seldom exceeded 5% of the total blood meals reacting by precipitin test. The study at Charleville supports the generally accepted view that *Cx. annulirostris* has mammalian predilections, but also indicate that the feeding patterns of this species are strongly influenced by host availability and abundance. Although this suggests that *Cx. annulirostris* populations may need to reach high densities prior to epidemic spillover to man, it also demonstrates that this mosquito is well suited for the maintenance of arboviruses in a wide range of vertebrate hosts.

From 1972 to 1973, 33 viruses were fed orally to *Cx. annulirostris* in blood to demonstrate its susceptibility to 18 (73). All 12 arboviruses previously isolated from field collections of this species replicated after feeding, thus providing some support for incrimination of potential vectors from virus isolation programs. At least 27 different arboviruses have now been isolated from *Cx. annulirostris* (Table 1.2), including Kowanyama and Eubenangee viruses to which *Cx. annulirostris* was also susceptible in this study. Four of the 18 viruses remain to be isolated from naturally infected *Cx. annulirostris:* Getah, Alfuy, Bovine ephemeral fever, and Mapputta. Four other species, *Ae. vigilax* (73), *Ae. funereus* (74), *Ae. aegypti* (77), and *Cx. quinquefasciatus* (17), were also evaluated for their susceptibility to a broad range of arboviruses. Although their narrower distribution and behavior patterns may preclude association with some of the arboviruses to which they could be infected, these studies also demonstrated the surprisingly broad and narrow susceptibilities of *Ae. vigilax* and *Cx. quinquefasciatus,* respectively, and the moderate susceptibilities of *Ae. funereus* and *Ae. aegypti.*

Apart from the experimental studies of Mackerras, *et al.* in 1940 (102) and those of Standfast, vector competence investigations with mosquitoes have concentrated on the medically important arboviruses. Following the 1951 epidemic of MVE, McLean (116, 117) established that *Cx. annulirostris* was an efficient vector, and additional evaluations of Australian and exotic species (1, 74, 77, 131) plus *Cx. sitiens* Wiedemann and *Anopheles farauti* Laveran (B.H. Kay, unpublished data) increased the list of susceptible species to 19. Seven of the Australian species have been shown to transmit MVE experimentally: *Cx. annulirostris, Cx. quinquefasciatus, Ae. rubrithorax* (Macquart), *Ae. aegypti, Ae. vigilax, Ae. vittiger,* and *Ae. lineatopennis* Ludlow. That *An. farauti* s.l. could be infected orally, with virus reaching the salivary glands in 1 of 12 tested, is of interest, as neither *An. annulipes* nor *An. quadrimaculatus,* Say, tested by McLean, became infected. However, the closely related Kunjin has previously been isolated from *An. farauti* (35). Thus, species of four genera, *Culex, Aedes, Mansonia,* and *Anopheles,* are known to be at least susceptible to MVE infection.

Because of known geographical and temporal differences in vector competence (59), studies based on single samples are of limited value. Positive susceptibility data may be useful but negative data, may be misleading. These considerations led to broader investigations in which ten populations of *Cx. annulirostris* showed considerable heterogeneity in their responses to oral infection and transmission of MVE and Kunjin (81) and in which seven populations of *Cx. quinquefasciatus* were shown to be either poorly susceptible or refractory to MVE, Kunjin, and Ross River (79). The contrast in the vector competence of these two species was as marked as their virus isolation histories (Table 1.2). Liehne and co-workers (95) have suggested that their isolation of MVE in the Ord River Valley was probably a reflection of the intense MVE activity of March–April 1974 rather than a true indication of vector status. We would support this suggestion on the basis of vector competence data. Furthermore, it is doubtful, given present susceptibility, that *Cx. quinquefasciatus* warrants serious control efforts in the face of outbreaks of any of these three viruses.

Currently, the effect of larval diet and temperature is being evaluated with *Cx. annulirostris* and MVE (B.H. Kay, unpublished data). Whereas Grimstad and Haramis (57) have demonstrated differences in transmission efficiency of La Crosse virus fed to *Ae. triseriatus* of different nutritional status, *Cx. annulirostris* so far have failed to exhibit any differences in competence, although low-diet mosquitoes feed and oviposit less readily. Partially fed, but nevertheless nulliparous *Cx. annulirostris* have been shown to contain MVE virus and to be capable of transmission. Recent studies have concentrated on vector competence at temperatures similar to that of the Murray Valley during spring. After rearing at 27°C, infected *Cx. annulirostris* adults held at 20°C and 13–24°C not only had lower titers of MVE than those held at 27°C, but also transmitted with lowered efficiency. This result was analogous to evaluations of *Cx. tarsalis* (Coquillett) with western equine encephalomyelitis virus, in which elevated temperature selected for a viral modulating trait, resulting in decreased transmission (J.L. Hardy and L.D. Kramer, personal communication). Whether such modulation actually decreases the number of virions or simply causes increased production of defective ones is an interesting question, as is whether such modulation is reversible when ambient conditions become more favorable. However, the results with *Cx. annulirostris* were quite different when larvae were reared at the adult holding temperature and there seemed to be no such loss in transmission efficiency. This phenomenon is currently being investigated.

The range of mosquitoes susceptible to Ross River virus is even greater than for MVE. Ross River virus has been associated with five genera, *Culex, Aedes, Mansonia, Coquillettidia,* and *Anopheles* (80). To date 22 of 25 species have shown some susceptibility, with the inclusion of recent work of Ballard (5) adding *Ae. camptorhynchus, Ae. alboannulatus,* and *Ae. rubrithorax* and B.H. Kay (unpublished data) adding *Ae. vittiger* (positive), *Ae. alternans* (negative), and *An. farauti* (negative). However, regional differences in vector competence to Ross River virus are also apparent with *Ae. australis* and *Ae. notoscriptus* from New Zealand and Australia (4, and B.H. Kay, unpublished data). Whereas the ecology of what is believed to be the major inland vector, *Cx. annulirostris,* has been fairly well worked, relatively little is known of the strategies and associations of the chief coastal vector, *Ae. vigilax.*

Sinclair (139) and Kay *et al.* (78) have reviewed knowledge of the saltmarsh mosquito, *Aedes vigilax,* which, besides its surprisingly broad susceptibility to arboviruses (67, 73), is the chief coastal vector of canine filariasis, *Dirofilaria immitis* (Leidy). *Aedes vigilax* is an important pest, attacking man and animals mainly during crepuscular periods, but also by day, and dispersing widely up to 64 km from its breeding sites. Because (1) the greatest centers of human population reside on the eastern seaboard and (2) tourist development is burgeoning in Queensland, efforts are increasing to find an acceptable control strategy.

In Queensland, tourism is currently threatened by the resurgence of *Ae. aegypti* caused by the gradual decrease in mosquito control activities since 1955, when a major epidemic of dengue occurred in northern and central Queensland (132). Periodic surveys organized from 1965 by E.N. Marks indicated a decline of this species, but nevertheless indicated its presence especially in coastal northeastern Queensland (70). In 1981, there was some reluctance to accept that dengue transmission was occurring regularly in north Queensland. After all, dengue was considered as a problem associated with developing countries. Queenslanders felt more comfortable that Texas had also joined the Third World with a small outbreak during 1980 (58). Recent surveys indicate that *Ae. aegypti* is well established throughout Queensland almost to the southern border, sometimes in high densities (Fig. 1.2). It is notable that *Ae. aegypti* is also capable of transmitting the three other medically important pathogens, MVE, Kunjin, and Ross River virus (77). Recently C.J. Leake (unpub-

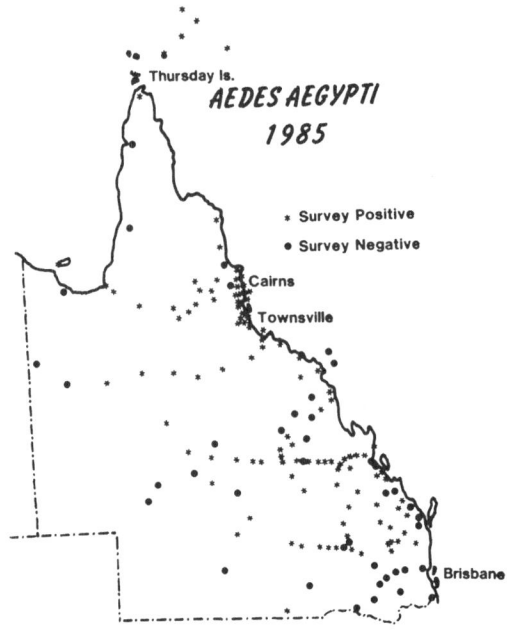

FIGURE 1.2. Known distribution of *Aedes aegypti* in Queensland, Australia, 1985.

lished data) concluded that *Ae. (Stegomyia) katherinensis* was unlikely to be an important vector of dengue type 2 in Australia within the range of its limited distribution.

Aedes (Finlaya) notoscriptus is a common peridomestic species which bites man. It is now known that subgenera other than *Stegomyia* may be susceptible to dengue infection, e.g., *Ae. (Finlaya) niveus* group in Malaysia (133). Sporadic cases of dengue in localities where *Ae. aegypti* is thought to be absent, e.g., Brisbane, suggests that an evaluation of the competence of *Ae. notoscriptus* is warranted. The susceptibility of this species to yellow fever should also be reevaluated, as the negative finding of Russell and colleagues (137) was based on ten mosquitoes.

But what of the other Australia species about which little is known? The entomological monographs by Lee *et al.* (92–94) of the Culicidae of the Australasian region are extremely valuable, but perhaps their exhaustive reference lists give a false impression of a wealth of information. As with the *Cx. annulirostris* group mosquitoes (106), the taxonomy of some Australian anophelines is under review with the discovery of sibling species of *An. farauti* (105) and *An. annulipes* by C. Green (unpublished data). Few species, from which arboviruses have been isolated, have been subjects of definitive ecological study. Although Russell (134, 135) studied environmental factors influencing the development and fecundity of *An. annulipes* and *An. hilli,* most data on anophelines have been derived from World War II studies.

Surveys of the Murray Valley for suspected MVE vectors (129) showed that besides *Cx. annulirostris, Ae. theobaldi* and *Cq. linealis* were abundant and worthy of consideration. Since then, Ross River virus has been isolated from *Cq. linealis* at Nelson Bay near Newcastle, New South Wales (51). *Aedes normanensis,* the source of 46 isolates of virus, including Ross River and MVE (Table 1.2) and other common species of *Ochlerotatus* offer an outstanding and relevant challenge to the understanding of both arboviruses and vectors. This has been emphasized even more with the relatively recent recognition of transovarial transmission and its implications for virus survival in nature.

Whereas the culicids are more or less regarded as medical pests, biting midges are firmly established as vectors affecting livestock. When Mackerras *et al.* (102) suggested in 1940 that ceratopogonids might be the vectors of Bovine ephemeral fever, "it was practically impossible to identify specimens *[Culicoides]* collected in the field . . . not a single life history was known and very little was known of the distribution or habits of any Australian species" (90). Prior to 1971, phlebotomine sand flies were considered "a quite rare component of the Australian biting fly fauna" (43). Although the taxonomy of the tabanids and simuliids had been reasonably evaluated (100, 101), Hunter and Moorhouse (61) were probably first to publish a definitive paper on the bionomics of simuliids. In 1963, Lee and co-workers (91) summarized all known data on ceratopogonids, phlebotomines, and simuliids in relation to disease in domesticated animals in Australia and concluded that both these latter groups had been almost completely neglected. Judging by the numbers of *Lasiohelea* (6853), phlebotomines (4032), simuliids (683), and tabanids (1661) processed for virus isolation to date, this may still be true. However, the predominantly reptile-feeding phlebotomines are unlikely to be important vectors of viruses in livestock or in humans.

In contrast, ceratopogonids and, in particular, *Culicoides* have been the subject of intense taxonomic and viral study since 1968. This is particularly true of the subgenus *Avaritia,* some of which breed in dung pats and so maintain a close association with cattle.

Culicoides brevitarsis has become the most notable of the nine Australian members of this subgenus (44), not only because bluetongue type 1 has been isolated from it, but also because of its association with 15 other viruses (Table 1.4). Bovine ephemeral fever has been shown to replicate in *C. brevitarsis* as well as *C. marksi* Lee and Reye and *Cx. annulirostris* (155).

Three members of this subgenus, *C. brevitarsis, C. brevipalpis* Delfinado, and *C. wadai* Kitaoka, are thought to have entered northern or northwestern Australia as aerial plankton since the 1820's. *Culicoides brevitarsis* now extends well into temperate New South Wales, whereas *C. brevipalpis* is restricted to a small area in the north of the Northern Territory (44, 122). There is evidence, however, of a recent ongoing southern extension of *C. wadai.* In 1982 its southern limit was thought to be Mackay, Queensland, but since then regular trapping by CSIRO workers has provided evidence of its establishment in southeastern Queensland

(157) and by summer 1983–1984 had extended into northern New South Wales. The implications of this extension on *C. brevitarsis* densities and on subsequent virus transmission to cattle remain interesting topics for study.

Field studies of *C. brevitarsis* at Beatrice Hill (121) and at Kowanyama (75) have established that this species feeds on marsupials rather than on bovids. Standfast and Dyce (153) recorded attack rates of greater than 3272 *C. brevitarsis* per hour on cattle in southeastern Queensland and Riek (130) noted maximum attack of horses between 1800 and 2200 hr (causing "Queensland itch"). Campbell and Kettle (16) quantitated biting activity with respect to meteorological conditions, host color, and distribution on the host. It is generally accepted that peak activity of *C. brevitarsis* occurs prior to and just after sunset and sunrise. In southeastern Queensland, swarming occurred about an hour before sunset (15); biting in cattle, particularly on the ridgeline at the tail, increased rapidly from zero 30 min before sunset to a peak 30 min after sunset, decreasing to zero over the next 5–6 hr of darkness.

Oviposition on dung pats occurred throughout the diel, but was maximal between 1400 and 1800 hr (14). Eggs were deposited all over the upper surface of pats for 7 days after their defecation: at 22–29°C, averages of 53–150 adults/pat emerged 11–24 days following oviposition. Dung pats over 7 days old were thought to be unsuitable for oviposition because of the development of a thick crust. At 24–26°C, an average of 31 eggs/female matured between 30 and 50 hr; no autogenous egg development occurred (13). These studies, and those of seasonal abundance (45, 72, 122, 156), have provided a firm basis toward understanding why this species is important.

Vector competence evaluations of ten *Culicoides* species with respect to bluetongue (122, 156) have been both rewarding and baffling. After resolution of problems associated with feeding and maintenance, *C. brevitarsis*, *C. fulvus*, *C. actoni* Smith, *C. perigrinus* Kieffer, *C. oxystoma* Kieffer (= *C. schultzei*), and *C. wadai* showed some susceptibility to bluetongue serotype 20, but *C. austropalpalis* Lee and Reye, *C. brevipalpis*, *C. bundyensis* Lee and Reye, and *C. marksi* did not. All six susceptible species have distributions extending outside Australia. *Culicoides fulvus* and *C. actoni* were the only species that transmitted the virus to sheep. The negative result for the 1761 *C. marksi* fed was particularly pleasing, as this species is widely distributed throughout Australia and reaches high densities in sheep grazing areas. In contrast, those species found positive are widespread in northern areas stocked with cattle receiving rainfall in excess of 1000 mm/year (45).

Subsequent trials with *C. brevitarsis* and bluetongue serotypes 20 and 21 (CSIRO 154) and 1 (CSIRO 156) confirmed the low infection rate achieved previously with *C. brevitarsis*, from <0.1 to 0.9%. On this basis, from data of host viremia and susceptibility and from allowance for 0.85 daily survival, a hypothetical model for bluetongue transmission required 35,670 engorged *C. brevitarsis* per single transmission. At present, the only attempt to gauge survival of *C. brevitarsis* in nature has been at Ocean Shores, New South Wales (72), where survival rates were of the above order during 1972–1973. Given future adjustments to some of the parameters in this model, it is unlikely that *C. brevitarsis* is a major vector of bluetongue despite two isolations from southeastern Queensland (144), where high attack rates do occur in the dairying areas. However, even in areas where attack rates are very high, T.D. St. George (personal communication) has noted that vertebrate infection rates are low. On the basis of WHO criteria for rating vectors, *C. fulvus* was considered the most likely vector of bluetongue serotype 20 in the north of the Northern Territory, although there was not enough evidence to exclude several other species (156). Clearly, a similar study of potential vectors is warranted for southeastern Queensland, where many of the above species, including *C. fulvus,* do not occur.

Whereas *C. brevitarsis* has been associated with arboviruses of introduced livestock, antibodies to the nine arboviruses isolated from the indigenous *C. marksi* and *C. dycei* have been found mainly in native marsupials and rarely in introduced animals. Both of these species are known to breed along the margins of freshwater streams, swamps, and wallows and feed mainly on both native and introduced mammals during crepuscular periods (75, 91, 121). *Culicoides marksi* is especially common inland of the Great Dividing Range, whereas *C. dycei* is reasonably common from northern Queensland to southern New South Wales. For example, both species were commonly collected at Charleville and Kowanyawa for virus isolation (31), but were uncommon in coastal New South Wales (72).

Apart from some of the estuarine *Culicoides* species that pose substantial pest problems in tourist areas, little definitive knowledge of the bionomics of the freshwater species exists. Over the last 10 years, Kettle and Elson (88) have described and provided keys for the immature stages of 11 of about 70 species (many undescribed) occurring in Australia. It is hoped that the arbovirus isolations and investigations of diseases of livestock will provide an enduring stimulus to research on the ceratopogonids. Such research is basic to development of suitable control strategies.

Vertebrate Hosts

Most of the suspected hosts of the 77 arboviruses of Australia have been listed as such on the basis of virus isolation and, in some cases, limited antibody surveys, and for this reason tabulation seems premature. In relation to the timing and extent of arbovirus disseminations, Doherty (33) listed host density, distribution, movement, age structure, and survival as being important ecological considerations, but commented that significant work in this field had yet to be done.

From 1979, Kay and co-workers (83, 84) utilized their Brisbane *Cx. annulirostris* colony as both donor and recipients of MVE virus in an investigation of potential vertebrate hosts. The 14 species selected were chosen on the basis of their abundance, distribution, habits, and attractiveness to *Cx. annulirostris* (the major vector) and titer, duration of viremia, and serological response assayed. This work complemented that of Boyle *et al.* (10, 11), who infected four ciconiiform species with either MVE, Kunjin, or Japanese encephalitis.

Because (A) the hypothesis relating to MVE epidemics related to waterbirds (2) and (B) mammals are often either expensive and or difficult to handle and house, an obvious gap existed. This was emphasized further from field investigations of *Cx. annulirostris,* which clearly demonstrated that this species fed mainly on mammals (76, 82, 121). Of the 204 individuals inoculated, feral rabbits and Grey Kangaroos *(Macropus giganteus)* proved to be the most competent hosts, with Sulphur-crested Cockatoos *(Cacatua galerita),* Galahs *(Eulophus roseicapillus),* Little Corellas *(Cacatua sanginea),* Black Ducks *(Anas superciliosa),* domestic chickens, feral pigs, and dogs responding with low to moderate viremias. Other species did not respond or, at the most, contained trace amounts of virus, e.g., sheep, cattle, horses, Agile Wallabies *(Macropus agilis),* and wild mice. Trace viremias were often detected in recipient *Cx. annulirostris* placed on infected hosts, but not in the suckling mouse assay system used.

The intraspecific response of these infected animals often proved to vary as much as the interspecific response. This also proved the case with the HI antibody response, which was either undetectable or of low titer and transient in sheep and Agile Wallabies, but sometimes also in cattle, horses, pigs, domestic chickens, corellas, and Black Ducks. These demonstrated two points: (1) that many species of vertebrates are capable amplifiers of MVE and (2) that caution is required in interpretation of serological surveys, especially in the absence of *a priori* knowledge of the response of the species involved.

Marshall and Miles (108) have listed vertebrate hosts of Ross River virus. Further experimental infections (B.H. Kay, unpublished data) have established that Agile Wallabies and Grey Kangaroos are good hosts, whereas Merino Sheep, Black Ducks, and horses may respond more moderately. As with MVE infection, some pigs, rabbits, and domestic chickens failed to give an HI antibody response following infection. It would seem, therefore, from experimental infection, field isolation, and antibody survey that Ross River virus with its wide range of insect vectors also has a wide range of vertebrate hosts: mainly mammals, particularly native ones. Despite evidence from laboratory infections, the association of nervous disease (53) or contracted tendons (126) in horses with this virus remains to be proved.

Far less is known of the hosts of Kunjin virus, thought to have a primary cycle in *Cx. annulirostris* and birds (29). Its association as a causal agent of human encephalitis was not established until 1974. Only a small number of vertebrates have been inoculated experimentally, but pigs, Rufous Night Herons, and Intermediate and Little Egrets develop low to moderate viremias, while calves do not (9, 10, 141). Three of five calves inoculated either intravenously, intradermally, or subcutaneously developed mild clinical signs, and nonpurulent encephalitis was diagnosed in all brains examined. During summer 1984–1985 in Victoria, Kunjin virus was isolated from the brain of a moribund horse (I.D. Marshall, personal communication).

In his 1972 review, Spradbrow (140) listed Bovine ephemeral fever as the only known disease of animals produced by an arbovirus in Australia. Since then, the Simbu group viruses have been the subject of considerable effort. Akabane causes congenital deformities and Aino, Peaton, and Tinaroo have been implicated as possible causal agent of deformities in cattle and/or sheep (20). The Australian bluetongue isolates have not warranted the concern engendered by the initial isolation (56), as this isolate and two others have been shown to produce no clinical response in cattle and only a mild to moderate response in experimentally inoculated sheep (148). In Japan, Getah has been linked with encephalitis in horses (63) and Ibaraki has been established as a pathogen of cattle (62). Of the five EHD group viruses isolated from Australia, none has been shown to be a cause of disease, although CSIRO 157, 753, and DPP59 produced fever, malaise, and hyperemia of skin and the mucous membranes of sheep (150).

Although there has been considerable gain in understanding the host relationship of arboviruses, the data gathered would only partially satisfy the plea made by Spradbrow (140) for more research.

Strategies for Survival

Australia is a continent of tremendous contrast. Almost half is arid or semiarid (often referred to as "the dead heart"). Tropical northern areas, e.g., Kowanyama, Beatrice Hill, and the Ord Valley, experience a well-defined wet season, usually from December to March, during which approximately 85% of an annual rainfall of >1000 mm occurs. With the onset of the dry season, lush tall grasses are reduced to brown stubble; abundant water progressively disappears, and by November is restricted to chains of waterholes which are important to animal survival. A swamp today may be parched clay tomorrow. Animal populations concentrate around permanent water during the harsh dry season or move nomadically according to availability of food and water resources. Nix (124) has related the seasonal movement of birds to the plant growth indices in Australia and found that, in general, southern movement occurs after the end of the wet season.

The extensive riverine plains of New South Wales, one of the major waterbird breeding areas in Australia (50), either receive both summer and winter (northern NSW) or winter (southern NSW, Murray Valley of Victoria) rainfall. As only 8% of Australian birds can be regarded as migratory in the true sense (87), the major influx to these regions during spring (September–November) occurs as a result of sporadic nomadic movements. In these areas, low winter temperatures have a profound influence on both plant and animal growth indices.

From first principles and from what has been published elsewhere in the world, it was expected (31) that "arbovirus activity would be concentrated in the tropical high rainfall areas, and would occur there throughout the year, spreading in favourable wet seasons to inland and southern areas where aridity or temperature make the environment less favourable." However, on the basis of virus isolations and antibody surveys, the tropical rainforests of northeastern Queensland showed less activity than localities such as Kowanyama in the drier northwest. Although 17 years of study at Kowanyama demonstrated that arboviruses survived through adverse dry season conditions, it was also abundantly clear that this site did not represent a permanent focus of MVE infection. In view of the transience of water and wildlife at any particular point during the dry season, the hypothesis of Miles and Dane (118) relating to shifting temporary foci of infection seems to make the greatest sense in terms of arbovirus survival. To define such foci over hundreds of thousands of square kilometers would be akin to looking for a needle in a haystack. To complicate matters further, several virus survival mechanisms may operate at the same or at different times. That arbovirus activity is reduced in the dry season would suggest that, in fact, they do.

There seems little point to dwell on what Marshall (107) described as "the dogma of pre-epidemic introduction of arboviruses from tropical to temperate areas." In relation to MVE, major bird departures after the end of the tropical wet season, followed by random dispersal in a hedge-hopping manner over several months, is not a tenable means of seeding temperate zones with MVE in spring. Nor is direct bird migration at the end of the dry season when arbovirus activity and mosquito populations are at a low ebb (42, 65). Sellers' contention (138) that the southern extension of the intertropical convergence zone was responsible for an 1800-km transportation of infected migratory *Ae. vigilax* to the Murray Valley could only be described as ridiculous. Certainly, long-range movement of mosquitoes up to 500 km are known (3), and numerous hematophagous arthropods, including mosquitoes, *Culicoides,* and *Phlebotomus* have been collected in Australia from kite trap collections 100–300 m above ground level by R.A. Farrow and B.H. Kay (unpublished data). However, prior to the 1951 and 1974 epidemics, the synoptic pattern was such that prolonged windborne southern dispersal was not possible. In contrast, the progressive southern dissemination of Bovine ephemeral fever from September 1967 to March 1968 by airborne vectors seems plausible (123, 155).

Several field studies (52, 107, 120) have now provided sufficient evidence to suggest that Ross River and Sindbis (at least) may survive quite well in temperate areas: MVE and Kunjin are both known to be active occasionally in interepidemic periods. Although there is compelling evidence, based on waterbird breeding and the incidence of MVE infection during epidemic years (12, 110), to still consider that possibly pelecaniforms and ciconiiforms play a key role in the rural dissemination of MVE, there now exists a large body of information to indicate that the maintenance role through mammalian infection has previously been understated. Recent mathematical modeling of rural amplification cycles by Kay *et al.* (86) also suggests that initiation of rural amplification precedes clinical expression in the human population by at least 80 days. This time scale creates a dilemma, because *Cx. annulirostris* populations are not abundant enough during spring (112, 115) to be solely involved as the initiating vector, at least prior to the 1951 and 1974 epidemics. If *Cx. annulirostris* is not always involved, then what is?

Marshall, McDonald, and others (107, 112) have suggested possible initiation by *Aedes* spp. or *Cx. australicus,* common during winter and spring. As transovarial transmission of both MVE and Kunjin viruses have been demonstrated in laboratory trials, albeit with species irrelevant to this ecological jigsaw (71, 160), transovarial initiation seems probable. This does not imply, however, that other survival strategies are not employed. Although long-term viral persistence has been demonstrated for several viruses, including western equine encephalomyelitis, Japanese encephalitis, Russian Spring Summer encephalitis, and St. Louis encephalitis in a variety of cold- and warm-blooded vertebrates (see 128), there is little evidence to suggest that this may be important to MVE survival in temperate areas. This possibility has been invoked, however, to explain rises in antibody of two patients convalescing after MVE infections in 1974 (39). Kay and co-workers (84) recently demonstrated concurrent HI antibody and high-titered viremias in rabbits.

Little is known of the range of survival options available to other Australian arboviruses, although one may speculate on how many employ transovarial transmission. With regard to overseas studies, dengue has been isolated from field-collected larvae and male *Ae. aegypti* in Burma (89), whereas transovarial transmission of several Bunyaviruses (California encephalitis group) and Rhabdoviruses is well established (161, 162). Experiments with transovarial transmission of alphaviruses, such as western, eastern, and Venezuelan equine encephalomyelitis, have returned mixed results (163). In Australia, Kay (67) reported infection of *Ae. vigilax* progeny with Ross River, and Overden and Mahon (125) have isolated Sindbis from the progeny of *Ae. australis,* both following intrathoracic inoculation. Confirmation of these findings by oral infection is required.

The high isolation rate of Ross River and Sindbis viruses from *Ae. normanensis* from semiarid Charleville during February 1976 (42) may also suggest a mechanism involving vertical transmission. Regardless of the outcome with alphaviruses, it is certain that some Australian flaviviruses utilize this form of survival. However, for many viruses in Australia, cyclic transmission may simply be all that is required for survival.

Future Impressions

While the period up to 1977 has been referred to as the "stamp collecting" era, it is obvious that we have many more viruses to collect before the set is complete. Each change in direction of research has resulted in the isolation of new viruses. The move to study insects affecting livestock prompted by the 1968 outbreak of Bovine ephemeral fever resulted in the isolation of 26 viruses either new to Australia or new to science. Undoubtedly a study of insects associated with native mammals would be equally productive, while no attempt has been made to study in any detail the *Culicoides* of coastal Australia as hosts of viruses, and studies of ticks associated with the large reptile fauna are nonexistent.

The tropical north of Australia is a sparsely populated region which until recently was little more than a vast cattle ranch, but this is rapidly changing with the large-scale exploitation of the mineral wealth and increased agricultural activity. After cattle were introduced into northern Australia and were present in substantial numbers, we saw the appearance of a number of *Culicoides* that attack livestock and had their origins in the Oriental Region. These insects are efficient vectors of viruses that affect livestock, viruses that are also widely distributed in the Oriental Region. Will we now see a parallel in human disease in northern Australia?

To support the new industries and new population centers in the north, a controlled water regime to irrigate crops and provide water for livestock and man will be needed. Situations analogous to the western equine encephalomyelitis problem that developed in the Central Valley of California (127) may arise. The Ord Valley irrigation scheme of Western Australia provides some evidence for this, although, unfortunately, no preflooding data are available. Baseline data for two major dams in northern Queensland are currently being gathered in the hope that engineers and administrators can be influenced with respect to the biological implications of design. The future expansion of rice monoculture, mainly in Queensland, could see exacerbation of problems with Murray Valley encephalitis, Kunjin, and Ross River viruses, transmitted by *Cx. annulirostris,* which thrives in ricefields, as does its Asian counterpart *Cx. tritaeniorhynchus* Giles. During summer 1983–1984 a major epidemic of Ross River virus occurred in the rice-growing areas of New South Wales.

Public complacency, a burgeoning tourist industry, and a tropical climate could ensure regular epidemics of dengue or even dengue hemorrhagic fever, introduced through new international airports at Cairns and Townsville, north Queensland. If dengue hemorrhagic fever can reach Cuba, as it did in 1981 (18), then why not northern Queensland? To complicate the problem further, the southern extension of *Ae. albopictus* through Papua New Guinea, Solomon, and Santa Cruz islands since 1972 (47), plus illicit traffic across the Torres Straits, suggest that Australia may have a secondary dengue vector before too long.

The current trend toward expedient problem-solving will be maintained, often following leads gleaned from more elaborate investigations overseas. The future for research into arboviruses affecting livestock seems on the surface more certain, as fewer basic studies have been completed. For example, definition of the vectors of bluetongue, bovine ephemeral fever, Akabane, and other potentially important Simbu group viruses has not been completed. Studies of the bionomics of dung-breeding *Culicoides* (*Avaritia*) spp. are just past the anecdotal stage. However, disease control in the veterinary field relies heavily on the use of vaccines and many see the development of a vaccine the prime task of the scientist and consider basic studies of disease and vector ecology an irrelevant waste of resources.

There is no doubt that the reactions of politicians and administrators will still be the major motivating force behind research funding. Sadly, argument for priority in studies designed to understand the ecology of viruses and their vectors prior to development of large-scale problems is difficult to sustain. Australia is a land of boom or bust, not just because of its stop–go ecosystems controlled principally by rainfall, but also in terms of funding, dominated by epidemics or epizootics (or the lack of them). However, this situation is not unique to Australia.

Acknowledgments. We wish to thank Dr. B.M. Gorman, QIMR, and T.D. St. George, CSIRO Division of Tropical Animal Science, for their constructive comments on this manuscript; permission to quote unpublished studies mentioned throughout the text is gratefully acknowledged.

References

1. Altman, R.M., 1963, The behavior of Murray Valley encephalitis virus in *Culex tritaeniorhynchus* Giles and *Culex pipiens quinquefasciatus* Say, *Am. J. Trop. Med. Hyg.* **12**:425–434.
2. Anderson, S.G., 1954, Murray Valley encephalitis and Australian X disease, *J. Hyg. (Camb).* **52**:447–468.
3. Asahina, S., 1970, Transoceanic flight of mosquitoes on the northwest Pacific, *Jpn. J. Med. Sci. Biol.* **23**:255–258.
4. Austin, F.J., and Maguire, T., 1982, Ross River virus in the Southwest Pacific in: J.S. Mackenzie (ed.), *Viral Diseases in South-east Asia and the Western Pacific*, Academic Press, London, pp. 528–531.
5. Ballard, W.J., 1982, Ross River virus (RRV) vector competence of four mosquito species collected on the south coast of New South Wales, B.Sc. (Hons.) Thesis, Australian National University. 107 pp.
6. Bancroft, T.L., 1906, On the etiology of dengue fever, *Austr. Med. Gaz.* **25**:17–18.
7. Berge, T.O. (ed.), 1975, *International Catalogue of Arboviruses Including Certain Other Viruses of Vertebrates,* 2nd ed., U.S. Department of Health, Education and Welfare, Public Health Service, Public. No. (CDC)75-8301. 789 pp.
8. Black, R.H., 1972, Malaria in Australia, Service Publications School Public Health and Tropical Medicine No. 9.
9. Boyle, D.B., 1979, Comparative studies of two related Australian flaviviruses, Murray Valley encephalitis and Kunjin, Ph.D. Thesis, Australian National University. 161 pp.
10. Boyle, D.B., Dickerman, R.W., and Marshall, I.D., 1983a, Primary viraemia responses of herons to experimental infection with Murray Valley enceph-

alitis, Kunjin and Japanese encephalitis viruses, *Aust. J. Exp. Biol. Med. Sci.* **61**:655–664.

11. Boyle, D.B., Marshall, I.D., and Dickerman, R.W., 1983b, Primary antibody responses of herons to experimental infection with Murray Valley encephalitis and Kunjin viruses, *Aust. J. Exp. Biol. Med. Sci.* **61**:665–674.

12. Briggs, S.V., 1982, Possible medical impacts of waterbirds on humans—three examples, in: M.E. Fowler (ed.), *Wildlife Diseases of the Pacific Basin and Other Countries,* Fruitridge, Sacramento, California, pp. 25–32.

13. Campbell, M.M., and Kettle, D.S., 1975, Oogenesis in *Culicoides brevitarsis* Kieffer (Diptera: Ceratopogonidae) and the development of a plastron-like layer on the egg, *Aust. J. Zool.* **23**:203–218.

14. Campbell, M.M., and Kettle, D.S., 1976, Number of adult *Culicoides brevitarsis* Kieffer (Diptera: Ceratopogonidae) emerging from bovine dung exposed under different conditions in the field, *Aust. J. Zool.* **24**:75–85.

15. Campbell, M.M., and Kettle, D.S., 1979a, Swarming of *Culicoides brevitarsis* Kieffer (Diptera: Ceratopogonidae) with reference to markers, swarm size, proximity to cattle and weather, *Aust. J. Zool.* **27**:17–30.

16. Campbell, M.M., and Kettle, D.S., 1979b, Abundance and temporal and spatial distribution of *Culicoides brevitarsis* Kieffer (Diptera: Ceratopogonidae) on cattle in south-east Queensland, *Aust. J. Zool.* **27**:251–260.

17. Carley, J.G., Standfast, H.A., and Kay, B.H., 1973, Multiplication of viruses isolated from arthropods and vertebrates in Australia in experimentally infected mosquitoes, *J. Med. Entomol.* **10**:244–249.

18. Centers for Disease Control, 1981, *Morbidity Mortality Weekly Report* **30**:317.

19. Cleland, J.B., Bradley, B., and McDonald, W., 1918, Dengue fever in Australia, *J. Hyg.* **16**:317–376.

20. Coverdale, R., Cybinski, D.H., and St. George, T.D., 1978, Congenital abnormalities in calves associated with Akabane virus and Aino virus, *Aust. Vet. J.* **54**:151–152.

21. Cybinski, D.H., 1984, Douglas and Tinaroo viruses: two Simbu group arboviruses infecting *Culicoides brevitarsis* and livestock in Australia, *Aust. J. Biol. Sci.* **37**:91–97.

22. Cybinski, D.H., and Zakrzewski, H., 1983a, A dual infection of a bull with Akabane and Aino viruses, *Aust. Vet. J.* **60**:283.

23. Cybinski, D.H., and Zakrzewski, H., 1983b, The isolation and preliminary characterization of a rhabdovirus in Australia related to bovine ephemeral fever virus, *Vet. Microbiol.* **8**:221–235.

24. Cybinski, D.H., St. George, T.D., Standfast, H.A., and McGregor, A., 1980, Isolation of Tibrogargan virus, a new Australian rhabdovirus from *Culicoides brevitarsis, Vet. Microbiol.* **5**:301–308.

25. Dalgarno, L., Marshall, I.D., Hardy, C.M., Short, N.J., and Irving, A.M., 1982, Studies on Barmah Forest virus, in: T.D. St. George and B.H. Kay (eds.), *Arbovirus Research in Australia, Proceedings 3rd Symposium CSIRO–QIMR,* pp. 97–101.

26. Derrick, E.H., 1957, The challenge of north Queensland fevers, *Aust. Ann. Med.* **6**:173–188.

27. Derrick, E.H., 1972, The birth of the Queensland Institute of Medical Research, *Med. J. Aust.* **2**:952–959.

28. Doherty, R.L., 1964, Viruses, mosquitoes and epidemics, *Queensl. Health* **1**:1–9.

29. Doherty, R.L., 1972a, Arboviruses of Australia, *Aust. Vet. J.* **48**:172–180.
30. Doherty, R.L., 1972b, Arthropod-borne viruses in Australia and their relation to infection and disease, M.D. Thesis, University of Queensland, 170 pp.
31. Doherty, R.L., 1974, Arthropod-borne viruses in Australia and their relation to infection and disease, *Progr. Med. Virol.* **17**:136–192.
32. Doherty, R.L. (ed.), 1976, *Arbovirus Research in Australia, Proceedings 1st Symposium CSIRO–QIMR.* 107 pp.
33. Doherty, R.L., 1977, Arthropod-borne viruses in Australia, 1973–1976, *Aust. J. Exp. Biol. Med. Sci.* **55**:103–130.
34. Doherty, R.L., Carley, J.G., Mackerras, M.J., and Marks, E.N., 1963, Studies of arthropod-borne virus infections in Queensland. III. Isolation and characterization of virus strains from wild-caught mosquitoes in north Queensland, *J. Exp. Biol. Med. Sci.* **41**:17–39.
35. Doherty, R.L., Whitehead, R.H., Wetters, E.J., and Gorman, B.M., 1968a, Studies of the epidemiology of arthropod-borne virus infections at Mitchell River Mission, Cape York Peninsula, North Queensland. II. Arbovirus infections of mosquitoes, man and domestic fowls, 1963–1966, *Trans. R. Soc. Trop. Med. Hyg.* **62**:430–438.
36. Doherty, R.L., Whitehead, R.H., Wetters, E.J., and Johnson, H.N., 1968b, Isolation of viruses from *Ornithodoros capensis* Neumann from a tern colony on the Great Barrier Reef, North Queensland, *Aust. J. Sci.* **31**:363–364.
37. Doherty, R.L., Carley, J.G., Standfast, H.A., Dyce, A.L., and Snowdon, W.A., 1972, Virus strains isolated from arthropods during an epizootic of bovine ephemeral fever in Queensland, *Aust. Vet. J.* **48**:81–86.
38. Doherty, R.L., Carley, J.G., Filippich, C., and Kay, B.H., 1976a, Isolation of virus strains related to Kao Shuan virus from *Argas robertsi* in Northern Territory, Australia, *Search* **7**:484–486.
39. Doherty, R.L., Carley, J.G., Filippich, C., White, J., and Gust, I.D., 1976b, Murray Valley encephalitis in Australia, 1974: Antibody responses in cases and community, *Aust. N. Z. J. Med.* **6**:446–453.
40. Doherty, R.L., Carley, J.G., Filippich, C., Kay, B.H., Gorman, B.M., and Rajapaksa, N., 1977, Isolation of Sindbis (alphavirus) and Leanyer viruses from mosquitoes collected in the Northern Territory of Australia, 1974, *Aust. J. Exp. Biol. Med. Sci.* **55**:485–489.
41. Doherty, R.L., Standfast, H.A., Dyce, A.L., Carley, J.G., Gorman, B.M., Filippich, C., and Kay, B.H., 1978, Mudjinbarry virus, an orbivirus related to Wallal virus isolated from midges from the Northern Territory of Australia, *Aust. J. Biol. Sci.* **31**:97–103.
42. Doherty, R.L., Carley, J.G., Kay, B.H., Filippich, C., Marks, E.N., and Frazier, C.L., 1979, Isolation of virus strains from mosquitoes collected in Queensland, 1972–1976, *Aust. J. Exp. Biol. Med. Sci.* **57**:509–520.
43. Dyce, A.L., 1971, Widespread distribution of *Phlebotomus* Rondani and Berte (Diptera: Psychodidae) in Australia, *J. Aust. Entomol. Soc.* **10**:105–108.
44. Dyce, A.L., 1982, Distribution of *Culicoides* (*Avaritia*) spp. (Diptera: Ceratopogonidae) west of the Pacific Ocean, in: T.D. St. George and B.H. Kay (eds.), *Arbovirus Research in Australia, Proceedings 3rd Symposium CSIRO–QIMR*, pp. 35–43.
45. Dyce, A.L., and Standfast, H.A., 1979, Distribution and dynamics of suspected vectors of bluetongue virus serotype 20 in Australia, in: T.D. St. George and E.L. French (eds.), *Arbovirus Research in Australia, Proceedings 2nd Symposium CSIRO–QIMR*, pp. 29–351.

46. Dyce, A.L., Standfast, H.A., and Kay, B.H., 1972, Collection and preparation of biting midges (Fam. Ceratopogonidae) and other small Diptera for virus isolation, *J. Aust. Entomol. Soc.* **11**:91–96.
47. Elliott, S.A., 1980, *Aedes albopictus* in the Solomon and Santa Cruz Islands, South Pacific, *Trans. R. Soc. Trop. Med. Hyg.* **74**:747–748.
48. Fenner, , F., and Ratcliffe, F.M., 1965, *Myxomatosis,* Cambridge University Press, Cambridge. 379 pp.
49. French, E.L., 1973, A review of arthropod-borne virus infections affecting man and animals in Australia, *Aust. J. Exp. Biol. Med. Sci.* **51**:131–158.
50. Fullagar, P., and Davey, C., 1983, Herons, egrets, ibises and spoonbills, *Parks Wildlife Wetlands* **1983**:39–42.
51. Gard, G., Marshall, I.D., and Woodroofe, G.M., 1973, Annual recurrent epidemic polyarthritis and Ross River virus activity in a coastal area of New South Wales. II. Mosquitoes, viruses, and wildlife, *Am. J. Trop. Med. Hyg.* **22**:551–560.
52. Gard, G.P., Giles, J.R., Dwyer-Gray, R.J., and Woodroofe, G.M., 1976, Serological evidence of inter-epidemic infection of feral pigs in New South Wales with Murray Valley encephalitis virus, *Aust. J. Exp. Biol. Med. Sci.* **54**:297–302.
53. Gard, G.P., Marshall, I.D., Walker, K.H., Acland, H.M., and de Sarem, W.G., 1977, Association of Australian arboviruses with nervous disease in horses, *Aust. Vet. J.* **53**:61–66.
54. Gard, G.P., Cybinski, D.H., and St. George, T.D., 1983, The isolation in Australia of a new virus related to bovine ephemeral fever virus, *Aust. Vet. J.* **60**:89.
55. Gard, G.P., Cybinski, D.H., and Zakrzewski, H., 1984, The isolation of a fourth bovine ephemeral fever group virus, *Aust. Vet. J.* **61**:332.
56. Gee, R.W., 1979, Bluetongue-effects on trade, in: T.D. St. George and E.L. French (eds.), *Arbovirus Research in Australia, Proceedings 2nd Symposium CSIRO–QIMR,* pp. 35–41.
57. Grimstad, P.R., and Haramis, L.D., 1984, *Aedes triseriatus* (Diptera: Culicidae) and La Crosse virus. III. Enhanced oral transmission by nutrition-deprived mosquitoes, *J. Med. Entomol.* **21**:249–256.
58. Hafkin, B., Kaplin, J.E., Reed, C., Elliott, L.B., Fontaine, R., Sather, G.E., and Kappus, K., 1982, Reintroduction of dengue fever into the continental United States. I. Dengue surveillance in Texas. 1980, *Am. J. Trop. Med. Hyg.* **31**:1222–1228.
59. Hardy, J.L., Houk, E.G., Kramer, L.D., and Reeves, W.C., 1983, Intrinsic factors affecting vector competence of mosquitoes for arboviruses, *Annu. Rev. Entomol.* **28**:229–262.
60. Hare, F.E., 1898, The 1897 epidemic of dengue in north Queensland, *Aust. Med. Gaz.* **17**:98–107.
61. Hunter, D.M., and Moorhouse, D.E., 1976, Comparative bionomics of adult *Austrosimulium pestilens* Mackerras and Mackerras and *A. bancrofti* (Taylor) (Diptera, Simuliidae), *Bull. Entomol. Res.* **66**:453–467.
62. Inaba, Y., 1975, Ibaraki disease and its relationship to Bluetongue, *Aust. Vet. J.* **51**:178–185.
63. Kamada, M., Ando, Y., Fukunaga, Y., Kumonomido, T., Imagawa, H., Wada, R., and Akiyama, Y., 1980, Equine Getah virus infection: Isolation of the virus from racehorses during an epizootic in Japan, *Am. J. Trop. Med. Hyg.* **29**:984–988.

64. Kay, B.H., 1978, Aspects of the vector potential of *Culex annulirostris* Skuse 1889 and other mosquitoes (Diptera: Culicidae) in Queensland, with particular reference to arbovirus transmission at Kowanyama and Charleville, Ph.D. Thesis, University of Queensland. 200 pp.

65. Kay, B.H., 1979, Seasonal abundance of *Culex annulirostris* and other mosquitoes at Kowanyama, north Queensland, and Charleville, south west Queensland, *Aust. J. Exp. Biol. Med. Sci.* **57**:497–508.

66. Kay, B.H., 1979, Age structure of populations of *Culex annulirostris* (Diptera: Culicidae) at Kowanyama and Charleville, Queensland, *J. Med. Entomol.* **16**:309–316.

67. Kay, B.H., 1982, Three modes of transmission of Ross River virus by *Aedes vigilax* (Skuse), *Aust. Exp. Biol. Med. Sci.* **60**:339–344.

68. Kay, B.H., 1983, Collection of resting adult mosquitoes at Kowanyama northern Queensland and Charleville, south-western Queensland, *J. Aust. Entomol. Soc.* **22**:19–24.

69. Kay, B.H., 1985, Man–vector contact at Kowanyama, northern Queensland, Australia, *J. Am. Mosq. Control Assoc.* **1**:191–194.

70. Kay, B.H., Barker-Hudson, P., Stallman, N.D., Wiemers, M.A., Marks, E.N., Holt, P.J., Muscio, M., and Gorman, B.M., 1984, Dengue fever reappearance in north Queensland after 26 years, *Med. J. Aust.* **140**:264–268.

71. Kay, B.H., and Carley, J.G., 1980, Transovarial transmission of Murray Valley encephalitis virus by *Aedes aegypti* (L.), *Aust. J. Exp. Biol. Med. Sci.* **58**:501–504.

72. Kay, B.H., and Lennon, T., 1982, Seasonal prevalence and bionomics of biting midges (Ceratopogonidae) at Ocean Shores, New South Wales, *J. Aust. Entomol. Soc.* **21**:207–216.

73. Kay, B.H., Carley, J.G., and Filippich, C., 1975, The multiplication of Queensland and New Guinean arboviruses in *Culex annulirostris* Skuse and *Aedes vigilax* (Skuse) (Diptera: Culicidae), *J. Med. Entomol.* **12**:279–283.

74. Kay, B.H., Carley, J.G., and Filippich, C., 1977, The multiplication of Queensland and New Guinean arboviruses in *Aedes funereus* (Theobald) (Diptera: Culicidae), *J. Med. Entomol.* **13**:451–453.

75. Kay, B.H., Boreham, P.F.L., Dyce, A.L., and Standfast, H.A., 1978, Blood feeding of biting midges (Diptera: Ceratopogonidae) at Kowanyama, Cape York Peninsula, north Queensland, *J. Aust. Entomol. Soc.* **17**:145–149.

76. Kay, B.H., Boreham, P.F.L., and Williams, G.M., 1979a, Host preferences and feeding patterns of mosquitoes (Diptera: Culicidae) at Kowanyama, Cape York Peninsula, northern Queensland, *Bull. Entomol. Res.* **69**:441–457.

77. Kay, B.H., Carley, J.G., Fanning, I.D., and Filippich, C., 1979b, Quantitative studies of the vector competence of *Aedes aegypti* (Linn.), *Culex annulirostris* Skuse and other mosquitoes (Diptera: Culicidae) with Murray Valley encephalitis and other Queensland arboviruses, *J. Med. Entomol.* **16**:59–66.

78. Kay, B.H., Sinclair, P., and Marks, E.N., 1981, Mosquitoes: Their interrelationships with man, in: R.L. Kitching and R.E. Jones (eds.), *The Ecology of Pests. Some Australian Case Histories*, CSIRO, Melbourne, Australia, pp. 157–174.

79. Kay, B.H., Fanning, I.D., and Carley, J.G., 1982a, Vector competence of *Culex pipiens quinquefasciatus* for Murray Valley encephalitis, Kunjin, and Ross River viruses from Australia, *Am. J. Trop. Med. Hyg.* **31**:844–848.

80. Kay, B.H., Miles, J.A.R., Gubler, D.J., and Mitchell, C.J., 1982b, Vectors of Ross River virus: An overview, in: J.S. Mackenzie (ed.), *Viral Diseases*

in South-East Asia and the Western Pacific, Academic Press, London pp. 532–536.

81. Kay, B.H., Fanning, I.D., and Carley, J.G., 1984, The vector competence of Australian *Culex annulirostris* with Murray Valley encephalitis and Kunjin viruses, *Aust. J. Exp. Biol. Med. Sci.* **62:**641–650.

82. Kay, B.H., Boreham, P.F.L., and Fanning, I.D., 1985a, Host feeding patterns of *Culex annulirostris* and other mosquitoes (Diptera: Culicidae) at Charleville, southwestern Queensland, Australia, *J. Med. Entomol.* **22:**529–535.

83. Kay, B.H., Hall, R.A., Fanning, I.D., and Young, P.L., 1985b, Experimental infection with Murray Valley encephalitis virus: Galahs, Sulphur-crested cockatoos. Corellas, Black ducks and wild mice, *J. Exp. Biol. Med. Sci.* **63:**599–606.

84. Kay, B.H., Young, P.L., Hall, R.A., and Fanning, I.D., 1985c, Experimental infection with Murray Valley encephalitis virus. Pigs, cattle, sheep, dogs, rabbits, macropods and chickens, *Aust. J. Exp. Biol. Med. Sci.* **63:**109–126.

85. Kay, B.H., Edman, J.D., and Mottram, P., 1986, Autogeny in *Culex annulirostris* from Australia. *J. Am. Mosq. Control Assoc.* **2:**11–13.

86. Kay, B.H., Saul, A.J., and McCullagh, A., 1987, A mathematical model for the rural amplification of Murray Valley encephalitis virus in southern Australia *Am. J. Epidem.* **125** (April).

87. Keast, A., 1959, Australian birds: Their zoogeography and adaptations to an arid continent, in: A. Keast and C.S. Christian (eds.), *Biogeography and Ecology in Australia,* Junk, The Hague, pp. 89–114.

88. Kettle, D.S., and Elson, M.M., 1976, The immature stages of some Australian *Culicoides* Latreille (Diptera: Ceratopogonidae), *J. Aust. Entomol. Soc.* **15:**303–332.

89. Khin, M.M., and Than, K.A., 1983, Transovarial transmission of dengue 2 virus by *Aedes aegypti* in nature, *Am. J. Trop. Med. Hyg.* **32:**590–594.

90. Lee, D.J., 1947, Australian Ceratopogonidae (Diptera, Nematocera). I. Relation to disease, biology, general characters and generic classification of the family, with a note on the genus *Ceratopogon, Proc. Linn. Soc. N.S.W.* **72:**313–331.

91. Lee, D.J., Reye, E.J., and Dyce, A.L., 1963, "Sandflies" as possible vectors of disease in domesticated animals in Australia, *Proc. Linn. Soc. N.S.W.* **87:**364–376.

92. Lee, D.J., Hicks, M.M., Griffiths, M., Russell, R.C., and Marks, E.N., 1980, The Culicidae of the Australian region, *School Public Health Tropical Medicine Monograph Series No. 2* **1:**1–286.

93. Lee, D.J., Hicks, M.M., Griffiths, M., Russell, R.C., and Marks, E.N., 1982, The Culicidae of the Australian region, School Public Health Tropical Medicine Monograph Series No. 2 **2:**1–248.

94. Lee, D.J., Hicks, M.M., Griffiths, M., Russell, R.C., and Marks, E.N., 1984, The Culicidae of the Australian region, School Public Health Trop. Med. Monogr. Ser. No. 2 **3:**1–157.

95. Liehne, P., 1980, The ecology of Australian encephalitis in north west Australia, *W. Aust. Health Surv.* **38:**19–23.

96. Liehne, P.F.L., Stanley, N.F., Alpers, M.P., and Liehne, C.G., 1976, Ord River arboviruses—The study site and mosquitoes, *Aust. J. Exp. Biol. Med. Sci.* **54:**487–497.

97. Liehne, P.F.L., Anderson, S., Stanley, N.F., Liehne, C.G., Wright, A.E.,

Chan, K.H, Leivers, S., Britten, D.K., and Hamilton, N.P., 1981, Isolation of Murray Valley encephalitis virus and other arboviruses in the Ord River Valley 1972–1976, *Aust. J. Exp. Biol. Med. Sci.* **59**:347–356.

98. Lumley, G.F., and Taylor, F.H., 1943, Dengue, Service Publications School Public Health Tropical Medicine No. 3. 171 pp.

99. Lvov, D.K., 1973, Arbovirus infections in Australia and Oceania, *Med. Parazitol.* **42**:342–348.

100. Mackerras, I.M., 1962, Speciation in Australian Tabanidae, in *The Evolution of Living Organisms, Symposium Royal Society of Victoria, December 1959, pp. 328–358.*

101. Mackerras, I.M., and Mackerras, M.J., 1952, Notes on Australian Simuliidae (Diptera). III, *Proc. Linn. Soc. N.S.W.* **77**:104–113.

102. Mackerras, I.M., Mackerras, M.J., and Burnet, F.M. 1940, Experimental studies of ephemeral fever in Australian cattle, *CSIRO Bull.* **136**:1–116.

103. Mackerras, I.M., Mackerras, M.J., and Domrow, R., 1952, Murray Valley encephalitis, *Rep. Queensl. Inst. Med. Res.* **7**:12–13.

104. Mackerras, M.J., 1958, The decline of filariasis in Queensland, *Med. J. Aust.* **1**:702–710.

105. Mahon, R.J., 1983, Identification of three sibling species of *Anopheles farauti* Laveran by the banding pattern of their polytene chromosomes, *J. Aust. Entomol. Soc.* **22**:31–34.

106. Marks, E.N., 1982, Recent taxonomic studies of *Culex annulirostris* and allied species and their possible significance for arbovirus research, in: T.D. St. George and B.H. Kay, eds., *Arbovirus Research in Australia, Proceedings 3rd Symposium CSIRO–QIMR*, pp. 146–151.

107. Marshall, I.D., 1979, Epidemiology of Murray Valley encephalitis in eastern Australia—patterns of arbovirus activity and strategies of arbovirus survival, in: T.D. St. George and E.L. French (eds.), *Arbovirus Research in Australia, Proceedings 2nd Symposium CSIRO–QIMR*, pp. 47–53.

108. Marshall, I.D., and Miles, J.A.R., 1984, Ross River virus and epidemic polyarthritis, in: K.F. Harris (ed.) *Current Topics in Vector Research*, Vol. 2, Praeger, New York, pp. 31–56.

109. Marshall, I.D., Woodroofe, G.M., and Gard, G.P., 1980, Arboviruses of coastal south-eastern Australia, *Aust. J. Exp. Biol. Med. Sci.* **58**:91–102.

110. Marshall, I.D., Brown, B.K., Keith, K., Gard, G.P., and Thibos, E., 1982a, Variation in arbovirus infection rates in species of birds sampled in a serological survey during an encephalitis epidemic in the Murray Valley of south-eastern Australia, February 1974, *Aust. J. Exp. Biol. Med. Sci.* **60**:471–478.

111. Marshall, I.D., Woodroofe, G.M., and Hirsch, S., 1982b, Viruses recovered from mosquitoes and wildlife serum collected in the Murray Valley of south-eastern Australia, February 1974, during an epidemic of encephalitis, *Aust. J. Exp. Biol. Med. Sci.* **60**:457–470.

112. McDonald, G., 1980, Population studies of *Culex annulirostris* Skuse and other mosquitoes (Diptera: Culicidae) at Mildura in the Murray Valley of southern Australia, *J. Aust. Entomol. Soc.* **19**:37–40.

113. McDonald, G., and Buchanan, G.A., 1981, The mosquito and predatory insect fauna inhabiting fresh-water ponds, with particular reference to *Culex annulirostris* Skuse (Diptera: Culicidae), *Aust. J. Ecol.* **6**:21–27.

114. McDonald, G., Smith, I.R., and Sheldon, G.P., 1977, Laboratory rearing of *Culex annulirostris* Skuse (Diptera: Culicidae), *J. Aust. Entomol. Soc.* **16**:353–358.

115. McDonald, G., McLaren, I.W., Shelden, G.P., and Smith, I.R., 1980, The effect of temperature on the population growth potential of *Culex annulirostris* Skuse (Diptera: Culicidae), *Aust. J. Ecol.* **5**:379–384.

116. McLean, D.M., 1953, Transmission of Murray Valley encephalitis by mosquitoes, *Aust. J. Exp. Biol.* **31**:481–490.

117. McLean, D.M., 1957, Vectors of Murray Valley encephalitis, *J. Infect. Dis.* **100**:223–227.

118. Miles, J.A.R., and Dane, D.M.S., 1956, Further observations relating to Murray Valley encephalitis in the northern Territory of Australia, *Med. J. Aust.* **1**:389–393.

119. Mottram, P., Kay, B.H., and Kettle, D.S., 1986, The effect of temperature on eggs and immature stages of *Culex annulirostris* Skuse (Diptera: Culicidae), *J. Aust. Entomol. Soc.* **25**:131–135.

120. Mudge, P.R., McColl, D., and Sutton, D., 1981, Ross River virus in Tasmania, *Aust. Med. J.* **2**:256.

121. Muller, M.J., Murray, M.D., and Edwards, J.A., 1981, Blood-sucking midges and mosquitoes feeding on mammals at Beatrice Hill, N.T., *Aust. J. Zool.* **29**:573–588.

122. Muller, M.J., Standfast, H.A., St. George, T.D., and Cybinski, D.H., 1982, *Culicoides brevitarsis* (Diptera: Ceratopogonidae) as a vector of arboviruses in Australia, in: T.D. St. George and B.H. Kay (eds.), *Arbovirus Research in Australia, Proceedings 3rd Symposium CSIRO–QIMR*, pp. 43–49.

123. Murray, M.D., 1970, The spread of ephemeral fever of cattle during the 1967–68 epizootic in Australia, *Aust. Vet. J.* **46**:77–82.

124. Nix, H.A., 1976, Environmental control of breeding, post-breeding dispersal and migration of birds in the Australian region, *Proc. Int. Ornithol. Congr.* **16**:273–305.

125. Ovenden, J.R., and Mahon, R.J., 1984, Venereal transmission of Sindbis virus between individuals of *Aedes australis* (Diptera: Culicidae), *J. Med. Entomol.* **21**:292–295.

126. Pascoe, R.R.R., St. George, T.D., and Cybinski, D.H., 1978, The isolation of a Ross River virus from a horse, *Aust. Vet. J.* **54**:600.

127. Reeves, W.C., 1965, Developing balanced programs in the University of California for mosquito control—Medical aspects, *Proc. Calif. Mosq. Control Assoc.* **33**:46–49.

128. Reeves, W.C., 1974, Over-wintering of arboviruses. *Progr. Med. Virol.* **17**:193–220.

129. Reeves, W.C., French, E.L., Marks, E.N., and Kent, N.E., 1954, Murray Valley encephalitis: A survey of suspected mosquito vectors, *Am. J. Trop. Med. Hyg.* **3**:147–159.

130. Riek, R.F., 1954, Studies on allergic dermatitis (Queensland itch) of the horse: The aetiology of the disease, *Aust. J. Agric. Res.* **5**:109–129.

131. Roseboom, L.E., and McLean, D.M., 1956, Transmission of the virus of Murray Valley encephalitis by *Culex tarsalis* Coquillett, *Aedes polynesiensis* Marks and *A. pseudoscutellaris* Theobald, *Am. J. Hyg.* **63**:136–139.

132. Rowan, L.C., 1956, An epidemic of dengue-like fever, Townsville, 1954: Clinical features with a review of the literature, *Med. J. Aust.* **1**:651–655.

133. Rudnick, A., 1978, Ecology of dengue virus, *Asian J. Infect. Dis.* **2**:156–160.

134. Russell, R.C., 1979a, A study of the influence of some experimental factors

on the development of *Anopheles annulipes* Walker and *Anopheles amictus hilli* Woodhill and Lee (Diptera: Culicidae). Part 1. Influence of salinity, temperature and larval density on the development of the immature stages, *Gen. Appl. Entomol.* **11**:32–41.

135. Russell, R.C., 1979b, A study of the influence of some experimental factors on the development of *Anopheles annulipes* Walker and *Anopheles amictus hili* Woodhill and Lee (Diptera: Culicidae). Part 2. Influence of salinity, temperature and larval densities during the development of the immature stages on adult fecundity, *Gen. Appl. Entomol.* **11**:42–45.

136. Russell, R.C., and Bryan, J.H., 1985, A survey of domestic container-breeding mosquitoes in New South Wales for the presence of *Aedes aegypti* L., the vector of dengue fever, *J. Aust. Entomol. Soc.* **24**:193–194.

137. Russell, R.C., Mukwaya, L.G., and Lule, M., 1977, Laboratory studies on the transmission of Yellow Fever virus by *Aedes (Finlaya) notoscriptus* (Dipt., Culicidae), *Aust. J. Exp. Biol. Med. Sci.* **55**:649–651.

138. Sellers, R.F., 1980, Weather, host and vector—Their interplay in the spread of insect-borne animal virus diseases, *J. Hyg. (Camb.)* **85**:65–102.

139. Sinclair, P., 1976, Notes on the biology of the salt-marsh mosquito *Aedes vigilax* (Skuse) in south-east Queensland, *Queensl. Nat.* **21**:134–139.

140. Spradbrow, P.B., 1972, Arbovirus infections of domestic animals in Australia, *Aust. Vet. J.* **48**:181–185.

141. Spradbrow, P.B., and Clark, L., 1966, Experimental infection of calves with a group B arbovirus (Kunjin virus), *Aust. Vet. J.* **42**:65–69.

142. St. George, T.D., and French, E.L. (eds.), 1979, *Arbovirus Research in Australia, Proceedings 2nd Symposium CSIRO–QIMR.* 169 pp.

143. St. George, T.D., and Kay, B.H. (eds.), 1982, *Arbovirus Research in Australia, Proceedings 3rd Symposium CSIRO–QIMR.* 249 pp.

144. St. George, T.D., and Muller, M.J., 1984, The isolation of a bluetongue virus from *Culicoides brevitarsis, Aust. Vet. J.* **61**:95.

145. St. George, T.D., Standfast, H.A., Doherty, R.L., Carley, J.G., Filippich, C., and Brandsma, J., 1977, The isolation of Saumarez Reef virus, a new flavivirus, from bird ticks *Ornithodoros capensis* and *Ixodes eudyptidis* in Australia, *Aust. J. Exp. Biol. Med. Sci.* **55**:493–499.

146. St. George, T.D., Standfast, H.A., Cybinski, D.H., Dyce, A.L., Muller, M.J., Doherty, R.L., and Carley, J.G., 1978, The isolation of a bluetongue virus from *Culicoides* collected in the Northern Territory of Australia, *Aust. Vet. J.* **54**:153–154.

147. St. George, T.D., Cybinski, D.H., Filippich, C., and Carley, J.G., 1979, The isolation of three Simbu group viruses new to Australia, *Aust. J. Exp. Biol. Med. Sci.* **57**:581–582.

148. St. George, T.D., Cybinski, D.H., Della-Porta, A.J., McPhee, D.A., Work, M.C., and Bainbridge, M.H., 1980a, The isolation of two bluetongue viruses from healthy cattle in Australia, *Aust. Vet. J.* **56**:562–563.

149. St. George, T.D., Standfast, H.A., Cybinski, D.H., Filippich, C., and Carley, J.G., 1980b, Peaton virus: A new Simbu Group arbovirus isolated from cattle and *Culicoides brevitarsis* in Australia, *Aust. J. Biol. Sci.* **33**:235–243.

150. St. George, T.D., Cybinski, D.H., Standfast, H.A., Gard, G.P., and Della-Porta, A.J., 1983, The isolation of five different viruses of the epizootic haemorrhagic disease of deer serogroup, *Aust. Vet. J.* **60**:216–217.

151. St. George, T.D., Cybinski, D.H., Main, A.J., McKilligan, N., and Kemp, D.H., 1984, Isolation of a new arbovirus from the tick *Argas robertsi* from a cattle egret *(Bubulcus ibis coromandus)* colony in Australia, *Aust. J. Biol. Sci.* **37**:85–89.

152. St. George, T.D., Doherty, R.L., Carley, J.G., Filippich, C., Brescia, A., Casals, J., Kemp, D.H., and Brothers, N., 1985, The isolation of arboviruses including a new flavivirus and a new bunyavirus from *Ixodes (Ceratixodes) uriae* (Ixodoidea: Ixodidae) collected at Macquarie Island, Australia, 1975–1979, *Am. J. Trop. Med. Hyg.* **34**:406–412.

153. Standfast, H.A., and Dyce, A.L., 1968, Attacks on cattle by mosquitoes and biting midges, *Aust. Vet. J.* **44**:585–586.

154. Standfast, H.A., and Dyce, A.L., 1982, Isolation of Thimiri virus from *Culicoides histrio* (Diptera: Ceratopogonidae) collected in northern Australia, *J. Med. Entomol.* **19**:212.

155. Standfast, H., Murray, M.D., Dyce, A.L., and St. George, T.D., 1973, Report on ephemeral fever in Australia, *Bull. Off. Int. Epizoot.* **79**:615–625.

156. Standfast, H.A., Dyce, A.L., St. George, T.D., Cybinski, D.H., and Muller, M.J., 1979, Vectors of a bluetongue virus in Australia, in: T.D. St. George and E.L. French (eds.), *Arbovirus Research in Australia, Proceedings 2nd Symposium CSIRO–QIMR*, pp. 20–28.

157. Standfast, H.A., Muller, M.J., and Dyce, A.L., 1983, A recent southern extension of the range of *Culicoides wadai* to south-east Queensland, *Aust. Vet. J.* **60**:383–384.

158. Standfast, H.A., Dyce, A.L., St. George, T.D., Muller, M.J., Doherty, R.L., Carley, J.G., and Filippich, C., 1984, Isolation of arboviruses from insects collected at Beatrice Hill, Northern Territory of Australia, 1974–1976, *Aust. J. Biol. Sci.* **37**:351–366.

159. Stanley, N.F., 1982, Human arbovirus infections in Australia, in: T.D. St. George and B.H. Kay (eds.), *Arbovirus Research in Australia, Proceedings 3rd Symposium CSIRO–QIMR*, pp. 216–226.

160. Tesh, R.B., 1980, Experimental studies of the transovarial transmission of Kunjin and San Angelo viruses in mosquitoes, *Am. J. Trop. Med. Hyg.* **29**:657–666.

161. Tesh, R.B., and Modi, G.B., 1983, Growth and transovarial transmission of Chandipura virus (Rhabdoviridae: Vesiculovirus) in *Phlebotomus papatasi*, *Am. J. Trop. Med. Hyg.* **32**:621–623.

162. Turell, M.J., Reeves, W.C., and Hardy, J.L., 1982, Transovarial and transstadial transmission of California encephalitis virus in *Aedes dorsalis* and *Aedes melanimon*, *Am. J. Trop. Med. Hyg.* **31**:1021–1029.

163. Watts, D.M., and Eldridge, B.F., 1975, Transovarial transmission of arboviruses by mosquitoes: A review, *Med. Biol.* **53**:271–278.

164. Wright, A.E., 1981, Ord River arboviruses—Mosquito captures during 1976–77, *J. Aust. Entomol. Soc.* **20**:47–57.

165. Wright, A.E., Anderson, S., Stanley, N.F., Liehne, P.F.S., and Britten, D.K., 1981, A preliminary investigation of the ecology of arboviruses in the Derby area of the Kimberley region, Western Australia, *Aust. J. Exp. Biol. Med. Sci.* **59**:357–367.

2
Current Research on Dengue

Duane J. Gubler

Introduction

Dengue is the most important arbovirus disease of humans, in terms of both morbidity and mortality (1). Since the end of World War II, the incidence of dengue disease has increased greatly. Coincident with that increase has been the emergence and spread of a severe and fatal form of the disease, dengue hemorrhagic fever/dengue shock syndrome (DHF/DSS), as a major public health problem in many areas of the tropics. Today DHF is a leading cause of hospitalization and death among children in many countries of Southeast Asia (2), and in recent years it has become increasingly important in the Pacific Islands and the Americas.

Three factors are primarily responsible for the increased epidemic dengue activity. First, there has been a continuous trend toward urbanization in many areas of the tropics, especially in Asia, since World War II (3). Second, there has been a near complete breakdown of effective mosquito control in these same urban centers, making conditions ideal for transmission of domestic mosquito-borne diseases. Last, with the advent of the jet airplane in the early 1960s, there has been a consistent increase in air travel by man, and this has provided the ideal mechanism for transportation and spread of dengue viruses between population centers of the world. Thus, viremic humans are primarily responsible for introduction of new dengue virus strains and serotypes into areas made permissive by lack of mosquito control. It is unlikely that this set of circumstances, which is so favorable to the spread of epidemic dengue, will change in the near future. Consequently, the prospects are good for continued and frequent epidemic dengue and DHF in most populated areas of the tropics.

The increased importance of dengue as a public health and economic problem in recent years has led to an expansion of research on this disease. Major advances have been made in our knowledge of the epidemiology

Duane J. Gubler, San Juan Laboratories, Division of Vector-Borne Viral Diseases, Center for Infectious Diseases, Centers for Disease Control, Public Health Service, United States Department of Health and Human Services, San Juan, Puerto Rico 00936.

of dengue infection, in diagnostic methodology, and in molecular virology. There is still considerable controversy, however, about what the important risk factors are in the pathogenesis of DHF. In the area of prevention and control, most effort has been directed toward developing vaccines as well as more effective surveillance that can provide early warning for epidemic transmission. Space limitations prevent a comprehensive review of research on dengue viruses in this chapter. Rather, the purpose is to highlight briefly areas that are important to our understanding of dengue viruses and to development of more effective methods for diagnosis, prevention, and control.

Epidemiology

Natural History of Dengue Viruses

Early work in the Philippines demonstrated that nonhuman primates were infected with dengue viruses in nature, leading to speculation that the natural maintenance cycle of dengue viruses involved nonhuman primates and forest mosquitoes in the jungles of Asia (4, 5). Subsequently, extensive field work was done in Malaysia over a 20-year period, eventually documenting that dengue viruses were maintained in the forests of Malaysia in a cycle involving canopy-dwelling *Aedes (Finlaya) niveus* complex mosquitoes and monkeys (5–7). Evidence for this jungle cycle was as follows: (1) A high percentage of wild monkeys (68%) had flavivirus antibodies, primarily against dengue and zika viruses, (2) natural dengue virus infection was demonstrated in sentinel monkeys in the forest, both by isolation of virus (dengue 1, 2, and 4), and by seroconversion (dengue 1, 2, and 3), (3) dengue 4 virus was isolated from a pool of *Ae. (F.) niveus* collected from the canopy of the forest. It was concluded that dengue in peninsular Malaysia exists in a silent enzootic jungle cycle involving canopy-dwelling mosquitoes of the *Ae. (F.) niveus* complex and monkeys, in a rural endemic cycle involving *Ae. (Stegomyia) albopictus* and humans, and in an urban cycle involving *Ae. (S.) aegypti* and humans.

Further evidence of a forest maintenance cycle for dengue viruses has been obtained by French workers in Vietnam and Africa. In Vietnam, dengue virus of an unknown serotype was isolated from *Ae. (F.) niveus niveus* collected in a forest area (8). In West Africa, the evidence is stronger. Over 300 dengue 2 viruses were isolated from wild-caught mosquitoes in 1980 and 1981 (1, 9, 10). The isolates were from five species of mosquitoes, *Ae. (Stegomyia) africanus, Ae. (S.) leuteocephalus, Ae. (S.) opok, Ae. (Diceromyia) taylori,* and *Ae. (D.) furcifer.* Furthermore, two of the isolates were from pools of male mosquitoes, suggesting that transovarial transmission may play a role in the natural maintenance of dengue viruses in certain situations. This aspect of dengue ecology is discussed in more detail below.

In the Americas, evidence of a forest maintenance cycle for dengue viruses is more circumstantial because field studies have not yet been carried out. Nevertheless, dengue 2 neutralizing antibody has been detected in Ayoreo Indians living in the remote Rincón del Tigre area of Bolivia. These persons had not traveled outside the area, and *Ae. aegypti* was not present in that part of the country (11). These infections may be part of a forest cycle similar to that observed in Asia and West Africa. Other, more circumstantial evidence suggests that *Ae. (Gymnometopa) mediovittatus* may be involved in a natural maintenance cycle for dengue on some Caribbean islands (12). This species is a forest mosquito that has moved into the peridomestic environment and shares many larval habitats with *Ae. aegypti. Ae. mediovittatus* is a very common mosquito in rural and suburban communities of Puerto Rico where dengue virus transmission has been maintained continuously for over 11 years (San Juan Laboratories, CDC, unpublished data). Field studies have shown that *Ae. mediovittatus* feed avidly on humans, and their biting activity cycle is similar to that of *Ae. aegypti,* with peaks of activity in the morning right after daybreak and late in the afternoon. Laboratory studies have shown that *Ae. mediovittatus* has a significantly higher susceptibility to oral infection with dengue viruses than *Ae. aegypti,* and that they can transmit the virus both horizontally by bite to vertebrate hosts, and vertically, through the eggs to their progeny (12).

The role of transovarial transmission in the maintenance cycles of flaviviruses has been reinforced in recent years by laboratory and field studies that have shown that Koutango, dengue, yellow fever, Japanese encephalitis, Kunjin, Murray Valley encephalitis, and St. Louis encephalitis viruses can all be transmitted transovarially by certain species of mosquitoes (13–19). Of special interest is the fact that transovarial transmission of dengue 2 and dengue 4 has been documented in nature. The first evidence of vertical transmission in nature was obtained by French workers in West Africa. Dengue 2 was isolated from a pool of male *Aedes (furcifer) taylori* collected in the forest of Ivory Coast (9). According to Rosen *et al.* (19), these same French workers also isolated dengue 2 from another pool of *Ae. taylori* collected in Senegal. In Burma, dengue 2 was isolated from 5 of 199 pools of *Ae. aegypti* larvae (13,930 mosquitoes) collected from natural breeding containers in Rangoon (20). Two of the isolates were from pools of male mosquitoes that had been reared to adult stage with the sexes separated before testing. In Trinidad, dengue 4 was isolated from 1 of 158 pools (10,957 mosquitoes) of adult *Ae. aegypti* collected as eggs (21). By contrast, no dengue virus was isolated from over 8000 larvae (80 pools) collected from breeding containers in six locations of Jakarta, Indonesia in 1977 (D.J. Gubler, unpublished data) or from over 5000 larvae collected in Bangkok, Thailand (22).

In the laboratory, it has been shown that although transovarial transmission of dengue viruses occurs in *Ae. aegypti,* the filial infection rate is very low compared to that of other species of *Aedes* (19). Also, this species has a lower oral susceptibility to dengue viruses than other species, including *Ae. albopictus* and *Ae. mediovittatus* (12, 23–26). Collectively, the data suggest that *Ae. aegypti* is a less efficient host for dengue viruses than certain species of forest *Aedes* (19).

There is no doubt that *Ae. aegypti* is the most important epidemic vector of dengue and DHF, primarily because of this species' highly domesticated habitats and close association with humans. Because of its lower susceptibility to oral infection with dengue viruses, however, *Ae. aegypti* must feed on persons with high viremia to become infected. It has been documented that considerable variation exists among strains of dengue viruses in their ability to produce viremia (27–30), and there is some evidence that high human viremia may be associated with epidemic strains of dengue virus, while lower viremia is associated with endemic dengue virus strains (29, 30). Furthermore, even with the same strain of virus, there appears to be considerable variation in the duration and magnitude of viremia among individuals. With the low oral susceptibility to dengue virus infection of *Ae. aegypti*, it is likely that only those viruses associated with high human viremia would be transmitted by this species, while those viruses causing low human viremia would probably not be transmitted. It can be speculated that this virus–vector relationship may be a major factor in selecting and propagating epidemic strains of dengue viruses in urban situations. By contrast, viruses in rural areas or in forests could be maintained by more efficient *Aedes* spp. in a cycle combining transovarial transmission with periodic amplification in humans or monkeys.

Rosen *et al.* have suggested that the four dengue virus serotypes may have evolved in association with different mosquito species in different geographic areas (19). Thus, the natural maintenance cycles of dengue viruses may be associated with the forests of tropical Asia, Africa, or perhaps America in a cycle similar to that presented in Figure 2.1. Important questions remain to be answered: (1) do rural and/or forest maintenance cycles exist in the Americas, and (2) what role do variation in vector competence and natural maintenance cycles for dengue viruses play in the distribution and spread of epidemic dengue and DHF?

Risk Factors Associated with Dengue Hemorrhagic Fever

Controversy still exists concerning the risk factors associated with DHF/DSS, even though considerable research has been done in this area. One important question that remains to be answered is why some countries, such as Thailand and Indonesia, have major epidemics of DHF every few years, while other areas, such as India and Sri Lanka, which also have high endemicity for multiple dengue virus serotypes, have only sporadic cases of DHF. Evidence, primarily from Thailand, suggests that secondary infection with dengue 2 is the most important risk factor for DHF/DSS (31–33). The work leading to this conclusion has been summarized (32, 33) and is the basis for the secondary infection hypothesis for DHF. Briefly,

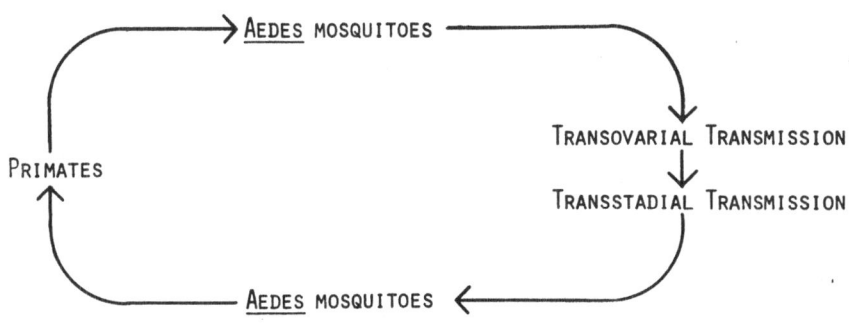

39

2

AFRICA : AEDES (DICEROMYIA)
         AEDES (STEGOMYIA)

ASIA   : AEDES (FINLAYA)
         AEDES (STEGOMYIA)

AMERICAS: AEDES SP.

ʀE 2.1. Proposed sylvatic maintenance cycle for dengue viruses.

. is hypothesized that mononuclear phagocytes are primary sites of dengue virus replication, that nonneutralizing heterologous dengue antibodies enhance monocyte infection by forming complexes with the infecting virus, and that the infected monocytes are responsible for release of factors that increase vascular permeability. Enhancement of dengue infection in these cells has been demonstrated *in vitro* with a number of flavivirus antibodies as well as with heterologous dengue antibodies. It should be noted, however, that as yet there is no solid evidence that this phenomenon influences dengue virus infection or replication in the human host. In Thailand, as well as in other parts of Southeast Asia, the Pacific, and the Americas, documented severe and fatal hemorrhagic disease may be associated with primary infections of all dengue serotypes (27, 29, 34, 35 and San Juan Laboratories, unpublished data), thus suggesting that virulence characteristics of the virus may also be an important risk factor for DHF (29, 30, 36). Finally, evidence is accumulating that suggests that host genetic factors also play a role in the severity of disease associated with dengue infection. Thus, there are many questions to be answered regarding risk factors associated with DHF that are very important in terms of epidemiology and future plans for prevention and control.

In 1979, the World Health Organization (WHO) sponsored a study group meeting to design prospective epidemiologic studies in high-risk DHF areas (Thailand and Indonesia) and low-risk DHF areas (Sri Lanka) with the objective of identifying risk factors associated with epidemic dengue and DHF. Two of these studies in Thailand and Sri Lanka are on-going and have produced some interesting results. In Thailand, the first year of the study supported conclusions from earlier work suggesting that secondary infection with dengue 2 was the most important risk factor for DHF (31). By contrast, a dengue epidemic in Colombo, Sri Lanka in 1982 was clinically mild and of the classical type, with no documented DHF cases despite the fact that the majority of confirmed cases in that outbreak were dengue 2 (37). This is reminiscent of the Tonga epidemic of dengue 2 in 1974, in which disease was clinically very mild (29). In Indonesia, from 1975 through 1978, most of the severe and fatal hemorrhagic disease was associated with dengue 3, not dengue 2 (38–40).

The studies in Asia are continuing and have now been expanded to include Puerto Rico, an area in the Caribbean that has experienced repeated epidemics of dengue without DHF (San Juan Laboratories, unpublished data). In the latter study, emphasis is also being directed toward the possibility that there are endemic and epidemic strains of dengue viruses. Limited field data already exist in support of this hypothesis (29, 30), and are consistent with the concept that forest and peridomestic species of *Aedes* may be primarily responsible for maintenance of endemic virus strains, while *Ae. aegypti* selects out and transmits epidemic virus strains.

Molecular Epidemiology

Dengue virus strain variation has been investigated by comparing the oligonucleotide "fingerprints" produced following RNase T1 digestion of genomic RNA and separation of the labeled oligonucleotides by two-dimension gel electrophoresis (41, 43). Oligonucleotide fingerprints of each of the four dengue virus serotypes are unique (41). Analysis of multiple isolates of dengue 1 and 2 viruses during epidemic conditions, during interepidemic years, and from different geographic areas has indicated that strains of virus from each major geographic region are similar to each other and distinct from viruses isolated from other areas (42–43). Comparison of fingerprints has facilitated separation of dengue 1 virus isolates into three distinct genotypic varieties representing Africa, Pacific/Southeast Asia, and the Caribbean. On the basis of fingerprint similarity, dengue 2 viruses have been divided into six genotypic varieties representing the Caribbean Basin/South Pacific/Central America/Mexico, Jamaica/Trinidad/Haiti, the Philippines, Thailand/Burma/Vietnam, West Africa/Upper Volta, and Kenya/Egypt (42, and D.W. Trent, personal communication). In both studies, the virus strains analyzed do not represent all areas of the world where dengue 1 and 2 virus infections occur and, therefore, it is likely that additional genotypic varieties exist. The data suggest that the evolution of dengue virus within a specific geographic region is much slower than that observed for other non-insect-transmitted RNA viruses (44–50), and, therefore, it may be postulated that the mosquito vector plays an important role in selecting viruses within populations conferring genetic stability on the population of virus transmitted. A second observation relating severity of disease with virus genotype suggests there is no direct relationship between fingerprint type and disease severity; mild to severe disease with hemorrhagic symptoms and shock have been associated with viruses having almost identical fingerprint patterns (D.W. Trent, personal communication).

The epidemiology and spread of dengue viruses are now better understood because of genetic analysis of virus strains by oligonucleotide mapping. Introduction of different strains of virus into a new geographic area can be rapidly detected and the spread of disease caused by the newly introduced virus documented. This capability is of considerable value where genetically distinct, but antigenically similar strains can be distinguished, and their transmission separately followed. Because viral antigens tend to be more conserved than the overall genetic sequence, fingerprinting is more sensitive than detecting antigenic divergence. However, the availability of monoclonal antibodies promises to improve greatly the precision and discrimination of viral serodiagnosis (51, 52). Fingerprinting of the genome is direct in that it does not require new reagents to detect the presence of a new viral genotype. Furthermore, fingerprinting analysis presents an average representation of the entire genome and is not restricted to those genomic sequences specifying viral surface polypeptides.

In the last few years, our knowledge of the molecular biology of viruses has been greatly expanded by the development of techniques for cloning and sequencing viral nucleic acids. This new knowledge has elucidated the organization of the viral genome, replication and translation strategies, the nature of the viral proteins, and the biochemical processing of viral proteins that are involved in virion morphogenesis. Despite the wealth of knowledge about the alphaviruses (53–55), the molecular structure of the flavivirus genome has yet to be determined, although it is now known that the coding sequences for the structural proteins are located at the 5' end of the viral genome in the order 5'-capsid-membrane-envelope glycoprotein-3' (J.H. Strauss, personal communication). It is anticipated that molecular cloning and nucleotide sequence data will soon be available for several of the dengue viruses. Genome sequence data must be collected that will permit construction of hybridization probes for the analysis of genetic variation and rapid identification of virus genotypic variants or types. The larger oligonucleotides of some of the dengue 2 geographic genotypic variants have already been sequenced and synthetic oligonucleotide hybridization probes complementary to the unique oligonucleotides have been constructed (J. Kerschner, personal communication). These probes, specific for each of the topotypes, will facilitate rapid and specific genetic characterization of new dengue 2 isolates and provide the information needed to facilitate primer extension nucleotide sequencing in regions of the genome where genetic variation related to antigenic structure and virulence occurs.

Vector Competence

Another important aspect of dengue ecology involves genetic heterogeneity in the mosquito vector populations. It has been documented that geographic strains of both *Ae. albopictus* and *Ae. aegypti* vary considerably in their susceptibility to oral infection with dengue viruses, and thus in their ability to transmit the infection (23, 24). These studies also showed that susceptibility to oral infection was dose-related and, therefore, a mosquito strain could be infected by increasing the amount of virus ingested, regardless of how resistant they were to dengue virus infection. It was also observed that in both mosquito species, the factors controlling susceptibility to oral infection were the same for all four dengue serotypes, that susceptibility was controlled by a midgut or mesenteronal barrier, and that there was no evidence of dissemination or salivary gland barriers. Finally, susceptibility to oral infection with dengue viruses was genetically controlled in both species (23, 24).

The data suggest that selective pressures in the environment may cause changes in susceptibility of the mosquito population and that vector competence may be an important risk factor for epidemic dengue transmission. (56).

Investigators in several laboratories are studying natural genetic variation among strains of *Ae. aegypti* by isozyme analysis, using gel electrophoresis (57–61). Since susceptibility to dengue viruses in *Ae. aegypti* is genetically controlled, attempts have been made to characterize strains of this mosquito with high and low dengue susceptibility by isozyme analysis using gel electrophoresis in an effort to identify markers for susceptibility and refractoriness. Preliminary results, however, have failed to show any conclusive relationship between dengue susceptibility and either enzyme banding patterns among geographic strains or morphotype of *Ae. aegypti* (San Juan Laboratories, unpublished data). These studies have shown that this mosquito species can be separated into geographic groups based on genetic relationships, an observation that has obvious importance in determining the origin of mosquito strains that have reinfested an area after eradication. Whether this type of genetic analysis of mosquito populations will increase our understanding of mosquito vector competence and the role it plays in the distribution and spread of epidemic dengue must be the subject of further study.

Clinical Studies

Evidence, mainly from Thailand, suggests that the primary pathophysiologic change that occurs in DHF/DSS is increased vascular permeability leading to a loss of plasma from the vascular compartment, hypovolemic shock, and death if not corrected (32, 33, 62). Although this leaky capillary syndrome has also been described in the majority of confirmed DHF cases in other areas of Asia, patients with a more severe bleeding tendency have been observed in Indonesia (40). Thus, only 53% of virologically confirmed fatal cases had a confirmed leaky capillary syndrome with hemoconcentration, while an additional 10% had some evidence of hemoconcentration, but with massive gastrointestinal (GI) hemorrhage. Thirty percent of these cases in Indonesia, however, died of shock due to blood loss from GI hemorrhage. This type of hemorrhagic disease is very rarely observed in Thailand, but appears to be more common in recent epidemics in the Americas (63) (San Juan Laboratories, unpublished data). Thus, considerable controversy still exists over what actually constitutes a case of DHF. One group states that only those patients with documented hemoconcentration and thrombocytopenia should be called DHF and all others, regardless of the extent of hemorrhagic manifestations and outcome of the illness, be called dengue with hemorrhagic manifestations (2, 32, 33). Others have called for a redefinition of the disease (36, 40).

More detailed studies are needed to clarify the spectrum of illness associated with dengue infection. The disease should be characterized in different parts of the world, in association with different human populations and strains and serotypes of dengue viruses. This is especially important in the American region, where hemorrhagic disease has not been common. Unanswered questions requiring answers include whether there are two or more pathogenetic mechanisms involved in causing severe and fatal dengue disease, or whether dengue disease simply represents a spectrum ranging from inapparent or mild febrile illness on one end to severe hemorrhagic disease with shock and death on the other. We also need to know whether the severity of illness varies with infection by different dengue virus strains and serotypes, and whether some persons (or populations) are genetically more susceptible to the severe and fatal forms of hemorrhagic disease. Last, we need to know the pathogenetic mechanism(s) that results in severe neurologic disorders associated with dengue infections, and whether encephalitic signs observed in patients with dengue infection are associated with certain virus strains or serotypes, or with persons or populations who are genetically predisposed to neurologic disorders.

Laboratory Diagnosis

Virus Isolation and Identification

Dengue viruses have been among the most difficult of the arboviruses to isolate and propagate in the laboratory. None of the standard laboratory animals nor mammalian cell cultures are sufficiently susceptible to dengue virus infection to use for routine dengue virus isolation. It was not until the development of the mosquito inoculation technique that a highly sensitive virus isolation system became available and for the first time allowed routine isolation of dengue viruses from clinical specimens (64). The technique is easy, relatively rapid, and economical. It has the disadvantage, however, of being labor-intensive and requires rearing and maintenance of mosquitoes, the expertise for which is not available in most virology laboratories.

In recent years, several mosquito tissue culture cell lines have been developed that are much more susceptible to dengue virus infection than mammalian cell cultures (65–67). Three cell lines, the C6/36 clone of *Aedes albopictus* cells, the AP-61 cell line from *Ae. pseudoscutellaris*, and the TRA-284 cell line from *Toxorhynchites amboinensis,* have been the most widely used (65–70). Comparative studies carried out with these three cell lines by two different groups, however, have shown that the TRA-284 cells have a higher susceptibility to dengue viruses than either the AP-61 or C6/36 cells (71, 72). Furthermore, the TRA-284 cells have been adapted to serum-free medium (TRA-284-SF) without a reduction in their susceptibility to dengue viruses (67). Because good-quality bovine serum is very expensive and frequently unavailable in dengue endemic areas, and because the TRA-284-SF are more sensitive, they are the culture system of choice for isolation of dengue viruses (71).

Dengue virus infection in cell cultures can be detected by immunofluorescence or by cytopathic effects (CPE). Reports of CPE in mosquito cells have been variable, but most frequently described in AP-61 and C6/36 cell cultures. In our experience, however, development of CPE is very unreliable, especially in C6/36 cells. CPE can be induced in infected C6/36 cells, however, by increasing incubation temperature to 36°C and using medium with pH 6.8 (73). Even in AP-61 cells, which are most consistent in developing CPE, it is likely that slow-growing viruses that infect only a few cells do not produce CPE. In these cultures, even after 10 days' incubation, it is not uncommon to have less than 5% of the cells infected with virus (69, 71). This type of infection, however, is easily deteced by either direct or indirect immunofluorescence. (69).

Identification of dengue viruses has been greatly facilitated by the recent development of hybridomas that produce serotype-specific monoclonal antibodies (74, 75). The monoclonal antibodies have been most frequently used in an indirect fluorescent antibody test (IFAT) (52, 69, 75), but can also be used in an enzyme-linked immunosorbent assay (ELISA) (76). They have been used successfully to identify dengue viruses in a variety of mammalian and mosquito cell cultures as well as in mosquito brain tissue (52).

In general, the monoclonal antibodies have been very specific. However, caution should be exercised, because not all dengue viruses, in particular dengue 1 and 2, react well with the monoclonal antibodies (52, 76). Preliminary data suggest that strain variation among viruses may influence reactivity, thus leading to inaccurate identification. On the other hand, it should be possible to produce monoclonal antibodies that can be used to distinguish different virus strains (76).

The serotype-specific monoclonal antibodies used in conjunction with sensitive mosquito cell cultures have provided, for the first time, a rapid, simple, economical, and accurate method for routine isolation and identification of dengue viruses (69). The use of these methods by laboratories in dengue endemic areas should facilitate development of more effective virologic surveillance systems that will be necessary to prevent epidemic dengue and DHF.

Serology

Three basic tests are routinely used for serologic diagnosis of dengue infections in most laboratories. These are the hemagglutination-inhibition (HI), complement fixation (CF), and neutralization (NT) tests. Unfortunately, no new diagnostic tests have been developed that have the sensitivity and specificity to replace any of these.

Research on dengue diagnosis in recent years has emphasized rapid diagnostic tests that would help clinicians establish etiology in severely ill patients. Most emphasis has been given to immunofluorescent assays, radioimmunoassays, and enzyme immunoassays, but to date, no test has been developed that is rapid, specific, and sensitive enough to be useful on a routine basis. Antibody detection assays show the least promise for rapid diagnosis because in primary infections (those persons experiencing their first flavivirus infection) antibody is relatively slow in developing.

The antibody-capture ELISA probably shows the most promise, especially when used for IgM-class antibody (MAC-ELISA). This test was developed for Japanese encephalitis virus (77) and has been adapted for dengue (78, 79, and San Juan Laboratories, unpublished data). However, the results obtained in Puerto Rico have been rather disappointing, in that the test is not as specific for the dengue viruses as was expected. Furthermore, the test has only limited use for rapid diagnosis of dengue, because detectable levels of anti-dengue IgM are not present in all patients before day 5 of illness. When used in conjunction with an IgG-capture ELISA, however, primary and secondary dengue infections can generally be differentiated with confidence.

Like the HI, the MAC-ELISA is not specific enough to identify the infecting virus serotype, even though monotypic reactions may be occasionally observed in both primary and secondary infections. Because anti-dengue IgM antibody persists for only 60 to 90 days in most patients, however, the MAC-ELISA has become a very useful test for surveillance for dengue and DHF. Moreover, it is nearly as sensitive as the HI test, and can be used in seroepidemiologic studies as well (San Juan Laboratories, unpublished data).

Several studies have suggested that peripheral blood leucocytes (PBLs) may be a primary site of virus replication (80–83). If this is the case, detection of infected PBLs by immunofluorescent assay might provide a method for rapid diagnosis of dengue infection. A study carried out on 19 patients in Puerto Rico, however, suggested that PBLs were not a primary site of dengue virus replication in patients with dengue fever (84). Thus, virus was more frequently isolated from plasma than PBLs, and dengue antigen was detected by DFA in the PBLs of only one patient.

Prevention and Control

There have been no new breakthroughs in the control of *Aedes aegypti* for over 15 years. Moreover, the economic problems of many tropical countries of the world have led to a near breakdown in routine mosquito control for dengue as well as for other diseases. While epidemic control is still considered high priority in most endemic areas, little effort has been made to develop programs designed to actually prevent epidemic transmission. The normal course of events has been to wait until an epidemic occurs, and then implement control measures. This approach generally has little effect on the course of an epidemic because ineffective surveillance programs usually do not detect increased dengue activity until peak transmission has already been reached.

Mosquito Control

A major problem with using mosquito control for prevention of epidemic dengue is that to be effective, it must be continuous, even during periods of low or no dengue transmission. During interepidemic periods, most people, including control agencies, lose interest in mosquito control and as a result, large mosquito population densities are allowed to build up. Equally unfortunate is that surveillance programs are generally too insensitive to detect increased dengue activity until the epidemic has reached near peak transmission. By that time, it is too late to effectively intervene and control the epidemic.

Because of the problems of relying on routine mosquito control for epidemic prevention, an attempt is being made in Puerto Rico to develop an early warning surveillance program that will provide a predictive capability for epidemic dengue and DHF (San Juan Laboratories, unpublished data). The system is absed on virologic surveillance using viral syndrome cases from selected cities on the island, and is designed to detect, without too much delay, the introduction of new dengue serotypes. Closely tied into the surveillance system is the development of a rapid-response emergency vector control unit that can respond to new virus introductions and control incipient epidemics before they spread. The vector control program is integrated and will use a variety of approaches, including community education, environmental sanitation and chemicals for larval control, and space sprays for adult control.

Dengue Vaccines

Considerable effort has been put into dengue vaccine development in the past several years. Unfortunately, little progress has s been made. Most attention has been directed toward dengue 2 virus because that serotype is considered by some to be the most important in causing severe and fatal disease (31). A live dengue 2 candidate vaccine (PR-159/S-1) was developed from a Puerto Rico virus by attenuation in fetal rhesus monkey lung cells (DBS-FRhL-2) which is temperature-sensitive, produces uniform small plaques in cell culture, and has decreased mouse neurovirulence (85–87). Human trials have shown that only about 61% of recipients without previous flavivirus infection develop anti-dengue antibodies, compared to 90% in persons with previous yellow fever vaccination (88). Clinically, recipients have presented with a variety of symptoms, including low-grade fever, chills, abdominal pain, headache, night sweats, nausea, and anorexia (88–89). The PR-159/S-1 vaccine virus infection has been shown to produce viremia in human recipients, a potentially beneficial factor considering the fact that the viruses isolated from vaccine recipients have shown no evidence of reversion to wild-type growth characteristics. (90, 91).

More recently, a dengue 4 candidate vaccine, attenuated by passage in primary dog kidney cells, was prepared in fetal rhesus lung cells and tested in five human volunteers. Only two of the recipients developed neutralizing antibodies, and both of these persons had detectable viremia and symptoms compatible with mild dengue, including a rash. The virus recovered from the two recipients showed characteristics of the parental virus, suggesting reversion to the wild type (92–96).

Attenuation of dengue 1 virus has also been attempted, but results with monkeys indicated lack of immunogenicity. Current work on development of dengue vaccines is ongoing in at least three laboratories, in the United States, Thailand, and China. Unfortunately, progress is slow and it is unlikely that a safe, immunogenic dengue vaccine will be available for general use in the near future.

Conclusions

The increasing public health importance of dengue viruses in many parts of the tropics has led to expanded research support in recent years. Important breakthroughs have come in laboratory diagnosis of dengue viruses, primarily in isolation, assay, and identification. A major problem remaining is development of a rapid diagnostic method that is reasonably economical and simple, but which is also sensitive and specific enough to be useful in a clinical laboratory setting. New-generation diagnostic methods will most likely come from a better understanding of molecular virology and the use of cDNA probes.

Progress in dengue vaccine development has been very disappointing.

Major problems have been with safety and immunogenicity of the attenuated vaccine viruses. It is likely that production and use of a safe, effective, and economical vaccine and/or a polyvalent vaccine will have to await development of the technology to produce a genetically engineered vaccine.

In the area of epidemiology, one of the most important questions deals with risk factors for severe and fatal disease. Long-term prospective field studies are required to determine the relative importance of primary versus secondary dengue as a cause of DHF/DSS, and the role that strain variation among dengue viruses plays in the production of severe and fatal disease. Answers to these questions will be necessary before vaccines can be used in endemic areas for prevention and control of epidemic DHF. Until that time, emphasis should be placed on development of more effective surveillance systems that will provide a predictive capability for epidemic dengue and on more effective mosquito control measures.

Field studies will also be required to determine whether rural or forest maintenance cycles for dengue viruses actually occur in most endemic areas, and the significance of such cycles in the distribution and spread of epidemic dengue and DHF. Also, the role of endemic versus epidemic strains of dengue viruses and the role of mosquito vector competence as a selective mechanism for epidemic virus strains must be studied. Answers to these question have obvious important implications for long-term prevention and control of epidemic dengue and DHF.

Acknowledgments. The author acknowledges with gratitude Dr. Dennis W. Trent's contribution of the section on molecular epidemiology.

References

1. Rosen, L., 1984, The global importance and epidemiology of dengue infection and disease, in: t. Pang and R. Pathmanathan (eds.), *Proceedings International Conference on DHF, Kuala Lumpur, Malaysia, 1983,* pp. 1–6.
2. Technical Advisory Committee on DHF for the South East Asian and Western Pacific Regions, 1980, Guide for diagnosis, treatment and control of Dengue Hemorrhagic Fever, 2nd ed., World Health Organization.
3. Rodhain, F., 1983, Maladies transmises par les culicinés et urbanisation: Un exemple de coévolution, *Bull. Inst. Pasteur* **81**:33–54.
4. Siler, J.F., Hall, M.W., and Hitchens, A.P., 1926, Dengue: Its history, epidemiology, mechanism of transmission, etiology, clinical manifestations, immunity and prevention, *Phillip. J. Sci.* **29**:1–302.
5. Rudnick, A., 1965, Studies of the ecology of dengue in Malaysia: A preliminary report, *J. Med. Entomol.* **2**:203–208.
6. Rudnick, A., 1978, Ecology of dengue virus, *Aisan J. Infect. Dis.* **2**:156–160.
7. Rudnick, A., 1984, The ecology of the dengue virus complex in Peninsular Malaysia, in: T. Pang and R. Pathmanathan (eds.), *Proceedings International*

Conference on Dengue Hemorrhagic Fever, Kuala Lumpur, Malaysia, 1983,
pp. 7–15.
8. Anon, 1976, Dengue hemorrhagic fever in the Democratic Republic of Viet
Nam, *Dengue Newsl. SE Asian W. Pac. Regions WHO* **2**(1):1–6.
9. Cordellier, R., Bouchité, B., Roche, J.C., Monteny, N., Diaco, B., and Ako-
liba, P., 1983, Circulation selvatique du virus Dengue 2 en 1980, dans les
savanes sub-soudaniennes de Côte d'Ivoire, *Cah. O.R.S.T.O.M. Entomol.
Med. Parasitol.* **21**:165–179.
10. Roche, J.C., Cordellier, R., Hervy, J.P., Digoutte, J.P., and Monteny, N.,
1983, Isolement de 96 souches de virus dengue 2 a partir de moustiques cap-
turés en Cote-D'ivoire et Haute-Volta, *Ann. Virol.* **134E**:233–244.
11. Roberts, D.R., Peyton, E.L., Pinheiro, F.P., Balderrama, F., and Vargas,
R., 1984, Associations of arbovirus vectors with gallery forests and domestic
environments in southeastern Bolivia, *Bull. PAHO* **18**:337–350.
12. Gubler, D.J., Novak, R.J., Vergne, E., Colón, N.A., Vélez, M., and Fowler,
J., 1985, *Aedes (Gymnometopa) mediovittatus* (Diptera: Culicidae), a potential
maintenance vector of dengue viruses in Puerto Rico, *J. Med. Entomol.* **22**:469–
475.
13. Coz, J., Valade, M., Cornet, M., and Robin, Y., 1976, Transmission tran-
sovarieene d'un Flavivirus, le virus Koutango chez *Aedes aegypti* L., *C. R.
Acad. Sci. Paris D* **283**:109–110.
14. Rosen, L., Tesh, R.B., Lien, J.C., and Cross, J.H., 1978, Transovarial transmission of Japanese encephalitis virus by mosquitoes, *Science* **199**:909–911.
15. Aitken, T.H.G., Tesh, R.B., Beaty, B.J., and Rosen, L., 1979, Transovarial
transmission of yellow fever virus by mosquitoes (*Aedes aegypti*), *Am. J.
Trop. Med. Hyg.* **28**:119–121.
16. Hardy, J.L., Rosen, L., Kramer, L.D., Presser, S.B., Shroyer, D.A., and
Turell, M.J., 1980, Effect of rearing temperature on transovarial transmission
of St. Louis encephalitis virus in mosquitoes, *Am. J. Trop. Med. Hyg.* **29**:963–
968.
17. Kay, B.H., and Carley, J.G., 1980, Transovarial transmission of Murray Valley
encephalitis virus by *Aedes aegypti* (L.), *Aust. J. Exp. Biol. Med. Sci.* **58**:501–
504.
18. Tesh, R.B., 1980, Experimental studies on the transovarial transmission of
Kunjin and San Angelo viruses in mosquitoes, *Am. J. Trop. Med. Hyg.* **29**:657–
66.
19. Rosen, L., Shroyer, D.A., Tesh, R.B., Freier, J.E., and Lien, J.C., 1983,
Transovarial transmission of dengue viruses by mosquitoes: *Aedes albopictus*
and *Aedes aegypti, Am. J. Trop. Med. Hyg.* **32**:1108–1119.
20. Khin, M.M., and Than, K.A., 1983, Transovarial transmission of dengue 2 virus by *Aedes aegypti* in nature, *Am. J. Trop. Med. Hyg.* **32**:590–
594.
21. Hull, B., Tikasingh, E., de Souza, M., and Martínez, R., 1984, Natural transovarial transmission of dengue 4 virus in *Aedes aegypti* in Trinidad, *Am. J.
Trop. Med. Hyg.* **33**:1248–1250.
22. Watts, D.M., Harrison, B.A., Pantuwatana, S., Klein, T.A., and Burke, D.S.,
1985, Failure to detect natural transovarial transmission of dengue viruses by
Aedes aegypti and *Aedes albopictus* (Diptera: Culicidae), *J. Med. Entomol.*
22:261–265.

23. Gubler, D.J., and Rosen, L., 1976, Variation among geographic strains of *Aedes albopictus* in susceptibility to infection with dengue viruses, *Am. J. Trop. Med. Hyg.* **25**:318–325.

24. Gubler, D.J., Nalim, S., Tan, R., Saipan, H., and Sulianti Saroso, J., 1979, Variation in susceptibility to oral infection with dengue viruses among geographic strains of *Aedes aegypti, Am. J. Trop. Med. Hyg.* **28**:1045–1052.

25. Jumali, Sunarto, Gubler, D.J., Nalim, S., Eram, S., and Sulianti Saroso, J., 1979, Epidemic dengue hemorrhagic fever in rural Indonesia. III. Entomological studies, *Am. J. Trop. Med. Hyg.* **28**:717–724.

26. Rosen, L., Rozeboom, L.E., Gubler, D.J., Lien, J.C., and Chaniotis, B.N., 1985, Comparative susceptibility of various species and strains of mosquitoes to oral and parenteral infection with dengue and Japanese encephalitis viruses, *Am. J. Trop. Med. Hyg.* **34**:603–615.

27. Gubler, D.J., Suharyono, W., Lubis, I., Eram, S., and Sulianti Saroso, J., 1979, Epidemic dengue hemorrhagic fever in rural Indonesia. I. Virological and epidemiological studies, *Am. J. Trop. Med. Hyg.* **28**:701–710.

28. Gubler, D.J., Suharyono, W., Tan, R., Abidin, M., and Sie, A., 1981, Viraemia in patients with naturally acquired dengue infection, *Bull. WHO* **59**:623–630.

29. Gubler, D.J., Reed, D., Rosen, L., and Hitchcock Jr., J.C., 1978, Epidemiologic, clinical, and virologic observations on dengue in the Kingdom of Tonga, *Am. J. Trop. Med. Hyg.* **27**:581–589.

30. Gubler, D.J., Suharyono, W., Lubis, I., Eram, S., and Gunarso, S., 1981, Epidemic dengue 3 in Central Java, associated with low viremia in man, *Am. J. Trop. Med. Hyg.* **30**:1094–1099.

31. Sangkawibha, N., Rojanasuphot, S., Ahandrik, S., Viriyapongse, S., Jatanasen, S., Salitul, V., Phanthumachinda, B., and Halstead, S.B., 1984, Risk factors in dengue shock syndrome: A prospective epidemiologic study in Rayong, Thailand, 1. The 1980 outbreak, *Am. J. Epidemiol.* **120**:653–669.

32. Halstead, S.B., 1980, Dengue haemorrhagic fever—A public health problem and a field for research, *Bull. WHO* **58**:1–21.

33. Halstead, S.B., 1981, The Alexander D. Langmuir Lecture, the pathogenesis of dengue, molecular epidemiology in infectious disease, *Am. J. Epidemiol.* **114**:632–648.

34. Barnes, W.J.S., and Rosen, L., 1974, Fatal hemorrhagic disease and shock associated with primary dengue infection on a Pacific island, *Am. J. Trop. Med. Hyg.* **23**:495–506.

35. Scott, R.M., Nimmannitya, S., Bancroft, W.H., and Mansuwan, P., 1976, Shock syndrome in primary dengue infections, *Am. J. Trop. Med. Hyg.* **25**:866–874.

36. Rosen, L., 1977, The emperor's new clothes revisited, or reflections on the pathogenesis of dengue hemorrhagic fever, *Am. J. Trop. Med. Hyg.* **26**:337–343.

37. Vitarana, T., and Jayasekera, N., 1984, A study of dengue in a low DHF area—Sri Lanka, in: T. Pang and R. Pathmanathan (eds.), *Proceedings International Conference DHF, Kuala Lumpur, Malaysia, 1983*, pp. 103–109.

38. Gubler, D.J., Suharyono, W., Sumarmo, Wulur, H., Jahja, E., and Sulianti Saroso, J., 1979, Virological surveillance for dengue haemorrhagic fever in Indonesia using the mosquito inoculation technique, *Bull. WHO* **57**:931–936.

39. Suharyono, W., Gubler, D.J., Lubis, I., Tan, R., Abidin, M., Sie, A., and Sulianti Saroso, J., 1979, Dengue virus isolation in Indonesia, 1975–1978, *Asian J. Infect. Dis.* **3**:27–32.
40. Sumarmo, Wulur, H., Jahja, E., Gubler, D.J., Suharyono, W., and Sorensen, K., 1983, Clinical observations on virologically confirmed fatal dengue infections in Jakarta, Indonesia, *Bull. WHO* **61**:693–701.
41. Vezza, A.C., Rosen, L., Repik, P., Dalrymple, J., and Bishop, D.H.L., 1980, Characterization of the viral RNA species of prototype dengue viruses, *Am. J. Trop. Med. Hyg.* **29**:643–652.
42. Trent, D.W., Grant, J.A., Rosen, L., and Monath, T.P., 1983, Genetic variation among dengue 2 viruses of different geographic origin. *Virology* **128**:271–284.
43. Repik, P.M., Dalrymple, J.M., Brandt, W.E., McCown, J.M., Russell, and P.K., 1983, RNA fingerprinting g as a method for distinguishing dengue 1 virus strains, *Am. J. Trop. Med. Hyg.* **32**:577–589.
44. Webster, R.G., Laver, W.G., Air, G.M., and Schild, G.C., 1982, Molecular mechanisms in variation of influenza viruses, *Nature* **296**:115–121.
45. Palese, P., and Young, J.F., 1982, Variation of influenza A, B, and C viruses, *Science* **215**:1468–1474.
46. Nottay, B.K., Kew, O.M., Hatch, M.H., Heyward, J.T., and Obijeski, J.F., 1981, Molecular variation of type 1 vaccine-related and wild polioviruses during replication in humans, *Virology* **108**:405–423.
47. Kew, O.M., Nottay, B.K., Hatch, H.M., Nakano, J.H., and Obijeski, J.F., 1981, Multiple genetic changes can occur in oral poliovaccines upon replication in humans, *J. Gen. Virol.* **56**:337–347.
48. King, A.M.Q., Underwood, B., McCahon, D., Newman, J.W.I., and Brown, F., 1984, Biochemical identifications of viruses causing the 1981 outbreaks of foot and mouth disease in the United Kingdom, *Nature* **293**:479–480.
49. Trent, D.W., and Grant, J.A., 1980, A comparison of New World alphaviruses in the western equine encephalitis virus complex by immunochemical and oligonucleotide fingerprint techniques, *J. Gen. Virol.* **47**:261–282.
50. Trent, D.W., Monath, T.P., Bowen, G.S., Vorndam, A.V., Cropp, C.B., and Kemp, G.E., 1980, Variation among St. Louis encephalitis virus: Basis for genetic pathogenetic and epidemiologic classification, *Ann. N. Y. Acad. Sci.* **354**:219–237.
51. Yewdell, J.W., and Gerhard, W., 1981, Antigenic characterization of viruses by monoclonal antibodies, *Annu. Rev. Microbiol.* **35**:185–206.
52. Gubler, D.J., 1986, Application of serotype specific monoclonal antibodies for identification of dengue viruses, in: C. Yunker (ed.), *Arbovirus Cultivation in Arthropod Cells in Culture,* CRC Press, Boca Raton, Florida (in press).
53. Granoff, H., Frischauf, A.M., Simons, K., Leharach, H., and Delius, H., 1980, Nucleotide sequence of cDNA coding for Semliki Forest membrane glycoproteins, *Nature* **288**:236–241.
54. Dalgarno, L., Rice, C.M., and Strauss, J.H., 1983, Ross River virus 26S RNA: Complete nucleotide sequence and deduced sequence of the encoded structural proteins, *Virol.* **129**:179–187.
55. Strauss, E.G., Rice, C.M., and Strauss, J.H., 1984, Complete nucleotide sequence of the genomic RNA of Sindbis virus, *Virology* **133**:92–110.
56. Gubler, D.J., 1978, Factors influencing the distribution and spread of epidemic dengue haemorrhagic fever, *Asian J. Infect. Dis.* **2**:128–131.

57. Tabachnick, W.J., and Powell, J.R., 1979, A worldwide survey of genetic variation in the yellow fever mosquito, *Aedes aegypti, Genet. Res.* **34:**215–229.

58. Powell, J.R., Tabachnick, W.J., and Arnold, J., 1980, Genetics and the origin of a vector population: *Aedes aegypti*, a case study, *Science* **208:**1385–1387.

59. Tabachnick, W.J., Aitken, T.H.G., Beaty, B.J., Miller, B.R., Powell, J.R., and Wallis, G.P., 1982, Genetic approaches to the study of vector competency in *Aedes aegypti*, in: W. Steiner, W.J. Tabachnick, K.S. Rai, and S. Narang (eds.), *Recent Developments in the Genetics of Insect Disease Vectors*, Stipes, Champaign, Illinois, pp. 413–432.

60. Gubler, D.J., Novak, R., and Mitchell, C.J., 1982, Arthropod vector competence—Epidemiological, genetic, and biological considerations, in: W.W.M. Steiner, W.J. Tabachnick, K.S. Rai, and S. Narang (eds.), *Recent Developments in the Genetics of Insect Disease Vectors*, Stipes, Champaign, Illinois, pp. 343–378.

61. Wallis, G.P., Tabachnick, W.J., and Powell, J.R., 1984, Genetic heterogeneity among Caribbean populations of *Aedes aegypti, Am. J. Trop. Med. Hyg.* **33:**492–498.

62. Halstead, S.B., Nimmannitya, S., and Cohen, S.N., 1970, Observations related to pathogenesis of dengue hemorrhagic fever. IV. Relation of disease severity to antibody response and virus recovered, *Yale J. Biol. Med.* **42:**311–328.

63. Guzmán, M.G., Kouri, G.P., Bravo, J., Soler, M., Vázquez, S., Santos, M., Villaescusa, R., Basanta, P., Indan, G., and Ballester, J.M., 1984, Dengue haemorrhagic fever in Cuba. II. Clinical investigations, *Trans. R. Soc. Trop. Med. Hyg.* **78:**239–241.

64. Rosen, L., and Gubler, D.J., 1974, The use of mosquitoes to detect and propagate dengue viruses, *Am. J. Trop. Med. Hyg.* **23:**1153–1160.

65. Igarashi, A., 1978, Isolation of Singh's *Aedes albopictus* cell clone sensitive to dengue and chikungunya viruses, *J. Gen. Virol.* **40:**530–544.

66. Varma, M.G.R., Pudney, M., and Leake, C.J., 1974, Cell lines from larvae of *Aedes (Stegomyia) malayensis* Colless, and *Aedes (S.) pseudoscutellaris* (Theobald) and their infection with some arboviruses, *Trans. R. Soc. Trop. Med. Hyg.* **68:**374–382.

67. Kuno, G., 1982, Dengue virus replication in a polyploid mosquito cell culture grown in serum-free medium, *J. Clin. Microbiol.* **16:**851–855.

68. Tesh, R.B., 1979, A method for the isolation and identification of dengue viruses using mosquito cell cultures, *Am. J. Trop. Med. Hyg.* **28:**1053–1059.

69. Gubler, D.J., Kuno, G., Sather, G.E., Vélez, M., and Oliver, A., 1984, Use of mosquito cell cultures and specific monoclonal antibodies for routine surveillance of dengue viruses, *Am. J. Trop. Med. Hyg.* **33:**158–165.

70. Race, M.W., Williams, M.C., and Agostini, C.F.M., 1979, Isolation of dengue virus in the *Aedes pseudocutellaris* cell line (LSTM-AP-61), in: *PAHO Scientific Publication No. 375*, Pan American Health Organization, pp. 165–172.

71. Kuno, G., Gubler, D.J., Vélez, M., and Oliver, A., 1985, Comparative sensitivity of three mosquito cell lines for isolation of dengue viruses. *Bull. WHO* **63:**279–286.

72. Leake, C.J., Misalak, A., and Burke, D.S., 1984, Comparative isolation of dengue viruses from DHF patients by mosquito inoculation and on three mosquito cell lines, in: T. Pang and R. Pathmanathan (eds.), *Proceedings International Conference on DHF, Kuala Lumpur, Malaysia, 1983*, pp. 437–445.

73. Zhu, Guan-fu, Liu, Zi-hui, and Wang, Jin, 1984, Improved technic for dengue virus micro cell culture, *Chin. Med. J.* **97**:73–74.
74. Gentry, M.K., Henchal, E.A., McCown, J.M., Brandt, W.E., and Dalrymple, J.M., 1982, Identification of distinct determinants on dengue-2 virus using monoclonal antibodies, *Am. J. Trop. Med. Hyg.* **31**:548–555.
75. Henchal, E.A., McCown, J.M., Seguin, M.C., Gentry, M.K., and Brandt, W.E., 1983, Rapid identification of dengue virus isolates by using monoclonal antibodies in an indirect immunofluorescence assay, *Am. J. Trop. Med. Hyg.* **32**:164–169.
76. Kuno, G., Gubler, D.J., and Santiago de Weil, N., 1985, Antigen capture ELISA for the identification of dengue viruses, *J. Virol. Meth.* **12**:93–103.
77. Burke, D.S., Nisalak, A., and Ussery, M.A., 1982, Antibody capture immunoassay detection of Japanese encephalitis virus immunoglobulin M and G antibodies in cerebrospinal fluid, *J. Clin. Microbiol.* **15**:1034–1042.
78. Bundo, K., and Igarashi, A., 1984, Enzyme-linked immunosorbent assay (ELISA) on sera from dengue hemorrhagic fever (DHF) patients in Thailand, in T. Pang g and R. Pathmanathan (eds.), *Proceedings International Conference on DHF, Kuala Lumpur, Malaysia, 1983,* pp. 478–484.
79. Gadkari, D.A., and Shaikh, B.H., 1984, IgM antibody capture ELISA in the diagnosis of Japanese encephalitis, West Mile and dengue virus infections, *Ind. J. Med. Res.* **80**:613–619.
80. Marchette, N.J., Halstead, S.B., and Chow, J.S., 1976, Replication of dengue viruses in culture of peripheral blood leukocytes from dengue-immune rhesus monkeys, *J. Infect. Dis.* **133**:274–282.
81. Halstead, S.B., O'Rourke, E.J., and Allison, A.C., 1977, Dengue viruses and mononuclear phogocytes. II. Identity of blood and tissue leukocytes supporting *in vitro* infection, *J. Exp. Med.* **46**:218–229.
82. Scott, R.M., Nisalak, A., Cheamudon, M., Seridhoranakul, S., and Nimmannitya, S., 1980, Isolation of dengue viruses from peripheral blood leukocytes of patients with hemorrhagic fever, *J. Infect. Dis.* **141**:1–6.
83. Boonpucknavig, S., Bhamarapravati, N., Nimmannitya, S., Phalavadhtana, A., and Siripont, J., 1976, Immunofluorescent staining of the surfaces of lymphocytes in suspension from patients with dengue hemorrhagic fever, *Am. J. Pathol.* **85**:37–48.
84. Waterman, S.H., Kuno, G., Gubler, D.J., and Sather, G.E., 1985, Low rates of antigen detection and virus isolation from the peripheral blood leucocytes of dengue fever patients, *Am. J. Trop. Med. Hyg.* **34**:625–632.
85. Eckels, K.H., Brandt, W.E., Harrison, V.R., McCown, J.M., and Russell, P.K., 1976, Isolation of a temperature-sensitive dengue-2 virus under conditions suitable for vaccine development, *Infect. Immunol.* **14**:1221–1227.
86. Harrison, V.R., Eckels, K.H., Sagartz, J.W., and Russell, P.K., 1977, Virulence and immunogenicity of a temperature-sensitive dengue-2 virus in lower primates, *Infect. Immunol.* **18**:151–156.
87. Eckels, K.H., Harrision, V.R., Summers, P.L., and Russell, P.K., 1980, Dengue-2 vaccine: Preparation from a small-plaque virus clone, *Infect. Immunol.* **27**:175–180.
88. Bancroft, W.H., Top, F.H., Jr., Eckels, K.H., Anderson, J.H., Jr., McCown, J.M., and Russell, P.K., 1981, Dengue-2 vaccine: Virological, immunological, and clinical responses of six yellow fever-immune recipients, *Infect. Immunol.* **31**:698–703.

89. Bancroft, W.H., Scott, R. McN., Eckels, K.H., Hoke, C.H., Jr., Simms, T.E., Jesrani, K.D.T., Summers, P.L., Dubois, D.R., Tsoulos, D., and Russell, P.K., 1984, Dengue virus type 2 vaccine: Reactogenicity and immunogenicity in soldiers, *J. Infect. Dis.* **149:**1005–1010.

90. Bancroft, W.H., Scott, R. McN., Brandt, W.E., McCown, J.M., Eckels, K.H., Hayes, D.E., Gould, D.J., and Russell, P.K., 1982, Dengue-2 vaccine: Infection of *Aedes aegypti* mosquitoes by feeding on viremic recipients, *Am. J. Trop. Med. Hyg.* **31:**1229–1231.

91. Miller, B.R., Beaty, B.J., Aitken, T.H.G., Eckels, K.H., Russell, P.K., 1982, Dengue-2 vaccine: Oral infection, transmission, and lack of evidence for reversion in the mosquito, *Aedes aegypti, Am. J. Trop. Med. Hyg.* **31:**1232–1237.

92. Halstead, S.B., Diwan, A.R., Marchette, N.J., Palumbo, N.E., and Srisukonth, L., 1984, Selection of attentuated dengue 4 viruses by serial passage in primary kidney cells. I. Attributes of uncloned virus at different passage levels, *Am. J. Trop. Med. Hyg.* **33:**654–665.

93. Halstead, S.B., Marchette, N.J., Diwan, A.R., Palumbo, N.E., and Putvatana, R., 1984, Selection of attenuated dengue 4 viruses by serial passage in primary kidney cells. II. Attributes of virus cloned at different dog kidney passage levels, *Am. J. Trop. Med. Hyg.* **33:**666–671.

94. Halstead, S.B., Marchette, N.J., Diwan, A.R., Palumbo, N.E., Putvatana, R., and Larsen, L.K., 1984, Selection of attenuated dengue 4 viruses by serial passage in primary kidney cells. III. Reversion to virulence by passage of cloned virus in fetal rhesus lung cells, *Am. J. Trop. Med. Hyg.* **33:**672–678.

95. Halstead, S.B., Eckels, K.H., Putvatana, R., Larsen, L.K., and Marchette, N.J., 1984, Selection of attentuated dengue 4 viruses by serial passage in primary kidney cells. IV. Characterization of a vaccine candidate in fetal rhesus lung cells, *Am. J. Trop. Med. Hyg.* **33:**679–683.

96. Eckels, K.H., Scott, R.McN., Bancroft, W.H., Brown, J., Dubois, D.R., Summers, P.L., Russell, P.K., and Halstead, S.B., 1984, Selection of attentuated dengue 4 viruses by serial passage in primary kidney cells. V. Human response to immunization with a candidate vaccine prepared in fetal rhesus lung cells, *Am. J. Trop. Med. Hyg.* **33:**684–689.

3
Systems Approaches for Management of Insect-Borne Rice Diseases

Keizi Kiritani, Fusao Nakasuji, and Shun'ichi Miyai

Introduction

Many pathogens of insect-borne plant diseases are spread by homopterous insects, e.g., aphids, leaf- and planthoppers, and whiteflies. At least 180 species of aphids transmit 164 viruses, and 151 species of leaf- and plant-hoppers transmit 55 viruses and 40 other microorganisms (15). Of these, nine viruses and two mycoplasmalike organisms (MLO) that cause rice diseases are transmitted by leaf- and planthoppers (see Table 3.1). No aphid-borne viruses of rice diseases have been discovered. These rice diseases are distributed mostly in Asia, with hoja blanca in the New World. Two of nine rice virus diseases mentioned above have recently been discovered, ragged stunt in 1976 in Indonesia (18) and gall dwarf in 1979 in Thailand (54).

Rice dwarf virus (RDV) is transmitted by some leafhoppers. The causal relationship between the disease and a vector, *Nephotettix cincticeps*, was discovered as early as in 1900 in Japan (9). This was the first discovery of an insect-borne plant disease. Since then, pathological and physiological properties of the virus–vector relationship have been intensively studied as well as ecological traits of the vector population. Systems models have also been developed for predicting epidemics of RDV on the basis of the sufficient works of the pathosystem (45).

In the present chapter, we review mathematical theories in epidemiology of plant diseases, with special reference to insect-borne virus diseases. Recent contributions of systems approaches in RDV epidemiology are also discussed.

Keizi Kiritani, Division of Entomology, National Institute of Agro-Environmental Sciences, Yatabe, Tsukuba, Ibaraki 305, Japan.
Fusao Nakasuji, Entomological Laboratory, Faculty of Agriculture, Okayama University, Okayama 700, Japan.
Shun'ichi Miyai, Division of Information Analysis, National Institute of Agro-Environmental Sciences, Yatabe, Tsukuba, Ibaraki 305, Japan.

Ecology and Epidemiology

According to van der Plank (65), epidemiology is the science of disease in a population. Kranz (35) further mentions that epidemiology is the science of a population of pathogens in a population of hosts, and the disease resulting therefrom under the influence of the environment and human interference. In any such definition, epidemiology is considered as a science operating at the population level of the pathogen (69). However, it is very difficult to deal with pathogens at the population level, because individual microorganisms can be scarcely observed directly except on the surface of agar culture or under the microscope. Usually, we merely observe the theater on which pathogens play, namely infective hosts or organs of hosts. Therefore, epidemiology is rather considered as the science of the dynamics of infection, which is reflective of the host–parasite relationship or pathosystem. For example, the basic reproductive rate, one of the fundamental parameters in the epidemiology of human diseases, is never expressed in terms of the number of microorganisms, but rather as the number of infectious hosts (42). Direct population study is limited to the larger organisms, such as protozoa, helminths, hookworms, and nematodes (1).

Epidemiology of animal diseases has been markedly advanced partly due to their importance concerning human welfare and partly due to the accumulation of the sufficient data on epidemics of human diseases. In particular, mathematical epidemiology of animals is highly developed compared to that of plant diseases (e.g., 3). The delay in the study of plant diseases may be caused by the fact that few pathologists have been interested in ecological phenomena of disease in the field. Though Kranz (36, 37) emphasized the role of mathematical models and systems analysis in epidemiological research, the present status of the epidemiological study of plant diseases remains in the infant stage.

There are various images of epidemiology among phytopathologists (38). Some believe that the main subject of epidemiology is the variation among pathogens in their ability to attack hosts, in other words, race problems (e.g., 66). Harrison (16), in his lecture on plant virus ecology as the presidential address to the Association of Applied Biologists, described ecology as being almost synonymous to the pathology or physiology of hosts and pathogens. He mentioned only qualitative but not quantitative aspects of the virus–vector relationship.

Some pathologists have discussed r- and K-selection theory in relation to the traits of pathogens (e.g., 16, 64, 69). As is well known, r- and K-selection theory was proposed by MacArthur and Wilson (41) as the process of natural selection that operates on organisms invading and colonizing islands. Both r and K are parameters of the logistic equation, i.e., the intrinsic rate of increase and the carrying capacity, respectively. Many authors, however, emphasize only r or the transmission rate of each pathogen, but they scarcely mention the carrying capacity K or competitive ability of the pathogen. Thresh (64) considered that many virus diseases of annual crops are *r*-strategists, but virus diseases of perennial crops are *K*-strategists, because the former and latter have quick and slow rates of spread, respectively. As he mentioned, however, there is no evidence that spread is influenced by the presence of other viruses. Furthermore, Thresh (64) insisted that pathogens that are adapted for long-term survival in the host cause endemic disease and this type of pathogen is a *K*-strategist because stability and low transmission rate are associated with the exploitation of long-lived habitats. However, Anderson (1) predicted that the most effective infection, in other words, the highest infection density, is not brought about at the weakest toxicity, but at a moderate toxicity of the pathogen. Of course, too strong a virulence causes an extinction of the pathogen.

As mentioned above, ecological concepts have scarcely matured in the epidemiology of plant diseases. If we consider epidemiology as being the ecology of disease dissemination, concepts and techniques developed in ecology should be applicable to the study of epidemiology. For example, some authors utilize life table analysis in estimating the parameters of transmission of insect-borne viruses (7, 62). In particular, systems approaches to the epidemiology of insect-borne plant diseases should be one of the approaches toward their ecology-oriented management.

In the following sections, we reviewed advances in mathematical theory and systems analysis of epidemiology of plant diseases, with special reference to insect-borne virus diseases. The later half of the chapter is devoted to research on RDV we have conducted in Japan. It is quite surprising that this theoretical and experimental research on RDV epidemiology has scarcely been cited even in recent books and reviews on virus epidemiology (e.g., 36, 39, 43, 44, 55).

Mathematical Models in Epidemiology

Multiplication of Infection

Kranz (38) considered that mathematical epidemiology is composed of models on the modes of disease progress and gradients. This is, in ecological terms, equivalent to the spatiotemporal dynamics of disease. The pattern of increase in the number x of pathogens or infective organs (plants) is shown by an exponential equation on the assumption that there is no limitation of resources,

$$dx/dt = rx \tag{1}$$

which is solved as

$$x = x_0 e^{rt} \tag{2}$$

where r is the rate of increase and x_0 is the initial number (65). Equation (2) could be fitted to the early phase of the growth pattern of lesions on leaves.

More generally, the logistic equation is used under limited resources,

$$dx/dt = rx(1-x/K) \qquad (3)$$

which is solved as

$$x = K/(1+ke^{-rt}) \qquad (4)$$

where K is the maximum number of infections and $k = (K-x_0)/x_0$ (33). The following equation was proposed by van der Plank (65):

$$dx/dt = rx(1-x) \qquad (5)$$

where x is expressed in terms of the proportion of infections, and the following equation was proposed by Kiyosawa (33) on the basis of the study on epidemics of wheat rust by Burleigh et al. (5):

$$dx/dt = rx(X-x) \qquad (5)$$

where x and X are the number and the maximum number of spores, respectively. These equations are both essentially the same as the logistic equation (4).

Further, Kiyosawa (33) considered that susceptibility of host plants to disease decreases at a rate of $(1-t/T)$ as time t passes until T. On such an assumption,

$$dx/dt = rx(1-t/T) \qquad (t \leq T) \qquad (6)$$

which is solved as

$$x = x_0 \exp\{r[t-t^2/(2T)]\} \qquad (t \leq T) \qquad (7)$$

Kiyosawa (33) claimed the general usefulness of Eq. (7) in forecasting epidemics of rice blast.

Many other equations have been proposed for disease progress, estimating infection, incubation, and latent periods, and predicting effects of environmental factors (36). However, they are mostly empirical and specific models applicable only to a particular set of disease and environment.

According to Rose (56), infection of maize streak disease increases exponentially when density of the vector, *Cicadulina* spp., is high, while it is arithmetic when the density is low. Generally speaking, however, there is little research in which a theoretical model was applied to real data on the growth pattern of infection of insect-borne diseases.

Dissemination of Diseases

Pathologists who study insect-borne diseases have been much interested in the process of spread of infection.

Watson (67) presented an equation giving the proportion P of infective plants as follows:

$$P = 1-(1-p)^X \tag{8}$$

where p and X are the probability of transmission by a vector and the number of vectors, respectively. Frampton et al. (8) proposed a diffusion model to describe theoretically the spatial distribution of potato yellow dwarf. The solution under some limiting assumptions gives the following equation for the number of plants I that have become infective in time t:

$$I = Kt \exp(st - s^{1/2}x) \tag{9}$$

where x is the coordinate in the direction of insect flow, K is the average distance that an insect will move in the x direction, and s is a constant of integration. The data were well fitted by Eq. (9) when the number I in natural logarithms was plotted against x or $(t + \ln t)$ (8). A similar equation was empirically proposed by Gregory and Read (14). Earlier, Gregory (13) proposed a model to predict the number of infectious plants y from an analogy of Thompson's (63) parasitism model as follows:

$$y = N(1-e^{-x/N}) \tag{10}$$

where N and x are the numbers of host available and of spores, respectively. In Eq. (10), spores are assumed to be distributed randomly on each plant. The equation can be rewritten as follows:

$$x = N(\ln 10)[\log N - \log(N-y)] \tag{11}$$

When $N = 100$, y is expressed in terms of a percentage, and then x can readily be calculated for any value of $y\%$. Gregory (13) compiled a numerical table in which x values are shown for y varying from 1 to 99.9%. This table is called the multiple infection transformation.

Watson and Healy (68) investigated the natural spread of beet yellows and beet mosaic viruses, both transmitted by aphids. They gave the following model to predict the infection at intervals of 3–4 weeks:

$$z_{t+1} = z_t+(1-z_t)(1-e^{-Nk}) \tag{12}$$

where

$$k = p[(1-z_t)^a + z_t a - 1]/z_t$$

In Eq. (12), z_t is the proportion of infective plants in the field at time t, and N, p, and a are the number of vectors per plant, the probability of virus transmission after a visit by an infectious vector, and the number of plants visited by a vector, respectively. The factor Nk in Eq. (12) gives the expected number of potential infections caused by N aphids. Watson and Healy (68) roughly estimated the values of the parameters N, p, and a as one-tenth of the trap count, 0.5, and 5, respectively. Then, the expected percentage of infective plants was plotted against the observed values. A significant correlation was obtained between the two.

Recently, Kiritani and Nakasuji (29) demonstrated that the number of plants x infected by vectors can be predicted by the equation

$$x = N(1 - e^{-aXLP})$$ (13)

where N is the number of plants in the field, a is the efficiency of virus transmission by an infectious vector, X is the number of vectors, L is the mean duration of stay in the field, and P is the proportion of infectious vectors in the population (for the parameter P, see the next section). The model is essentially the same as Eqs. (10) and (12). Kiritani and Sasaba (30) divided the parameter a into two components, the number of plants visited by a vector and the susceptibility of the plant to the virus disease, and they estimated values of the parameters by experiment for RDV.

Proportion of Infective Individuals in Vector Population

No one had tried to deal mathematically with the proportion of infective insects in the context of epidemiology until Kono (34) proposed a simple equation to describe the generation by generation changes in the proportion of insects infected with persistent virus in the vector population. The proportion P_{n+1} in the $(n+1)$th generation of the vector is

$$P_{n+1} = P_n r(1 - W) + W$$ (14)

where r is the rate of transovarial transmission of virus from mother to progeny and W is the rate of acquisition feeding of virus by the vector. The equilibrium value of the proportion P_e is obtained by setting $P_n = P_{n+1} = P_e$ in Eq. (14):

$$P_e = W/[1 - r(1 - W)]$$ (15)

The proportion P_n eventually reaches this equilibrium value P_e as long as the values of W and r remain constant for the vector population. Kiritani (27) presents a diagram showing a series of isoclines of P_e values at different combinations of r and W values.

Kono's (34) model was originally developed to analyze the virus–vector relationship between rice stripe virus (RSV) and the small brown planthopper, *Laodelphax striatellus*. As seen in Table 3.1, the RSV scarcely affects its vector physiologically. On the other hand, RDV affects to some extent the longevity and fecundity of its vector, *Nephotettix cincticeps*. For this reason, a slight modification of the model is necessary in the latter case as follows:

$$P_{n+1} = P_n \alpha r(1 - W) + W$$ (16)

where α is the rate of adverse effect of the virus on the vector (49). It is possible to measure P_n for a vector that transmits a persistent or semi-persistent virus either by inoculating vectors to the test plants or by serological test.

TABLE 3.1. Epidemiological characteristics of rice diseases vectored by leaf- and planthoppers.

Disease	Pathogen	Distribution	Vector(s)[a]	Traits of pathogen in vector			Incubation period (days)		Percent of acquisition feeding	Percent of infected vectors in epidemic fields
				Persistent	Transovarial passage (%)	Deleterious effect to vector	In rice plant	In vector		
Stripe	Virus	East Asia (temperate)	*Laodelphax striatellus, Kakuna sapporonis, Ribautodelphax albifascia, Terthron albovittatus	Yes	Yes (42–96)	No or slight	14–54	5–21	20–30	5–52
Hoja blanca	Virus	Latin America (tropical)	*Sogatodes orizicola, S. cubanus	Yes	Yes (80–95)	Yes, severe	3–29	6–9	—	5–15
Dwarf	Virus	East Asia, Nepal (temperate)	*Nephotettix cincticeps, *N. nigropictus, Recilia dorsalis, N. virescens	Yes	Yes (58–95)	Yes, severe	12–20	10–28	8–36	4–14
Transitory yellowing	Virus	Southeast Asia (subtropical)	*N. nigropictus, *N. cincticeps, N. virescens	Yes	No	No	10–11	9–34	41–65	—
Grassy stunt	Virus	Southeast Asia (tropical)	*Nilaparvata lugens, Numata muiri, Nilaparvata bakeri	Yes	No	No	13–20	12–18	10–70	—
Gall dwarf	Virus	Thailand (subtropical)	*N. nigropictus, *R. dorsalis,	Yes	Yes (67–93)	No	14	10–18	12–96	19

TABLE 3.1. *Continued*

Disease	Pathogen	Distribution	Vector(s)[a]	Traits of pathogen in vector			Incubation period (days)		Percent of acquisition feeding	Percent of infected vectors in epidemic fields
				Persistent	Transovarial passage (%)	Deleterious effect to vector	In rice plant	In vector		
			N. virescens, N. malayanus, N. cinciceps							
Ragged stunt	Virus	Southeast Asia (tropical–subtropical)	*N. lugens, N. bakeri	Yes	No	No	14	3–18	14–76	—
Black-streaked dwarf	Virus	East Asia (temperate)	L. striatellus, R. albifascia, K. sapporonis	Yes	No	No	14–24	7–21	67–93	3–32
Tungro (waika, yellow-orange leaf, penyakit merah)	Virus	Southeast Asia (temperate–tropical)	*N. virescens, *N. nigropictus, *N. cinciceps, R. dorsalis	No (semi-persistent)	No	No	6–15	0	40–88	—
Yellow dwarf	MLO[b]	East to Southeast Asia (temperate–tropical)	*N. cinciceps, *N. virescens, N. nigropictus	Yes	No	No	30–58	25–39	88–96	13–80
Orange leaf	MLO[b]	Southeast Asia (subtropical–tropical)	*R. dorsalis	Yes	No	No	13–15	2–6	27–50	—

[a] Asterisk denotes principal vector(s).
[b] Mycoplasmalike organism.

Systems Approaches

Historical View

The Australian ecologist L.R. Clark and his co-workers proposed the new concept of a "life system" (6), composed of a "subject population and its effective environment which includes the totality of external agencies influencing the population." They attempted to single out a subject population and its effective environment from a complicated community in order to prevent researches on such ecological systems from sinking into a bog.

Broadbent (4) first calculated the correlation between the immigrating alate aphid population *Myzus persicae* and the spread of leafroll disease of potato. As mentioned previously, Watson and Healy (68) carried out similar research. These were indeed primitive, but they were precursors of current systems modeling and analysis approaches.

With regard to the vector population, a systems model of the population dynamics of the cabbage aphid, *Brevicoryne brassicae,* was first developed by Hughes and Gilbert (12, 21). This model, aiming at integrated pest management of the aphid, described the total life system of the aphid. Though no processes of virus spread were included in Hughes and Gilbert's model, their approach would be very promising if we could combine their model with an epidemiological model such as Watson and Healy's equation.

Concurrently, we have developed systems models to describe the total system of natural spread of infection of RDV, including the population dynamics of the vector and the virus transmission by the vector (47, 49, 58, 59). Most equations involved in the systems models were still empirical and parameter values used were specific to the epidemic area, i.e., Kochi Prefecture in southern Japan. In spite of such limitations, these models were useful for research in predicting the causes of epidemics in the area and in assessing the effectiveness of control measures.

Modeling of RDV Epidemiology

The epidemiological cycle or pathosystem of RDV is shown in Figure 3.1. The numbers of infective rice plants and of infective insects are determined as functions of the complex set of factors shown in this figure. Fundamentally, the pathosystem of RDV infection is composed of three elements, host plant, vector, and virus. The virus is transmitted to rice plants by vectors. On the other hand, the vector acquires the virus from infective plants as well as from its mother through eggs (9, 11). The virus, once acquired, is retained by the vector throughout its life span. It multiplies in the vector body (10, 24) and the multiplication affects adversely physiological traits of the vector (48, 53). All these factors are responsible for the determination of the rate of infective insects in the vector population.

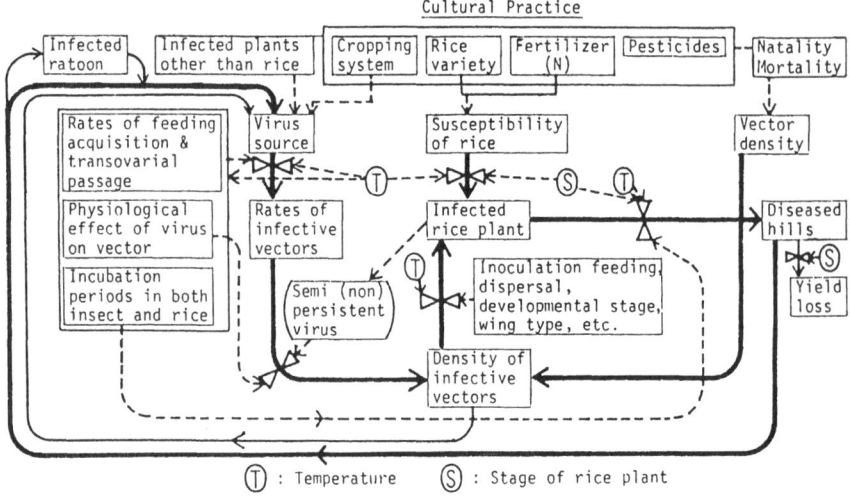

FIGURE 3.1. Epidemiological cycle of RDV [Kiritani (27)].

The population dynamics of the vector *Nephotettix cincticeps* has been intensively studied by Japanese entomologists (19, 20, 31, 40). The density is regulated through density-dependent processes, which mainly operate in the adult stage, i.e., dispersal and fecundity. Predators, e.g., lycosids, also play an important role in determination of the population density of the leafhopper (25). Sasaba *et al.* (57–59) developed systems models for forecasting the population changes in overwintering seasons and breeding seasons in Kochi Prefecture (Fig. 3.2). Numerical relationships between

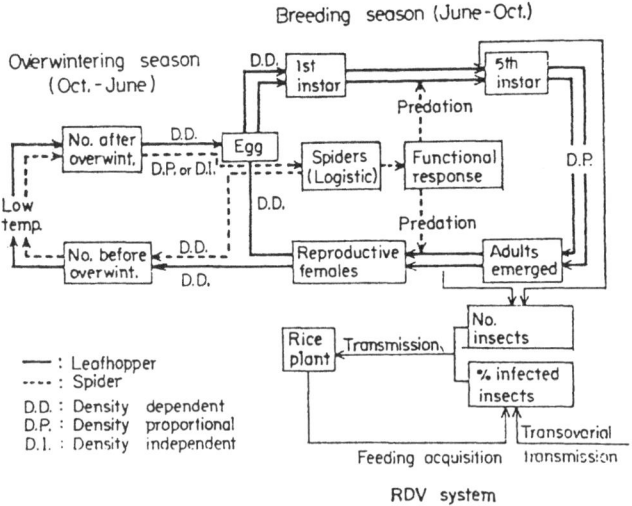

FIGURE 3.2. A flow chart of the systems model of *Nephotettix cincticeps* population dynamics [Sasaba (57)].

successive stages or time intervals were developed on the basis of the life table data for 7 years (Table 3.2). Density-dependent mortality, predation, and low temperatures in the winter determine the pattern of population change of the vector (Fig. 3.2). The seasonal trend of the spider population is given by a sigmoid curve. Extreme low temperatures kill both leafhoppers and spiders.

The model describing the processes of RDV transmission had been developed earlier than the population model of the vector (Table 3.3). The former is composed of two paths, transmission of the virus to the plant by infective vectors and acquisition feeding of the virus from infective plant by vectors. The intergenerational change in the proportion of infective insects is described by Eq. (16), in which the adverse effects of virus on the vector are incorporated by α. The combined parameter (αr), where r is the rate of transovarial transmission, can be estimated by the equation (Table 3.3) $\alpha r = -0.029T + 0.938$, using mean temperatures T in the field. The relationship came from measurements on experimental populations maintained for generations at different temperatures without any virus sources for acquisition feeding. The proportion of infective insects P_n in each generation was assessed by serological test. The higher the temperature, the greater is the rate of decrease of P_n between successive generations. The decreasing rate depends on the relative rate of the population increase of infective insects compared to healthy ones, i.e., the combined parameter αr, under the given conditions.

The rate W of acquisition feeding by the vector is related to the amount I_v of infective rice plants in the paddy field. The amount I_v is expressed in terms of the relative weight of the biomass of infective rice plants to healthy ones. Coefficients used in calculating the parameter I_v are given in Nakasuji and Kiritani (49) for three developmental stages of rice, the transplanting, tillering, and flowering stages.

The virus transmission in the paddy field actually occurs only twice: immediately after transplanting, by immigrating adults, and in the late tillering stage, by the ensuing progeny. No more infection occurs after about 70 days of planting, because rice plants become tolerant to the infection (22, 60). Initially, the process of transmission was described by two regression equations between the percentage of rice hills infected and the density of the infective vector population. In a later paper, however, the regression equations were replaced by Eq. (13) (29).

Nakasuji et al. (51) conducted some simulation tests by varying the values of the three parameters in the model, i.e., vector density, the coefficient for the efficiency of acquisition feeding, and that of virus transmission (Fig. 3.3). Changes in these parameter values correspond to killing of the vector, replacing the original population with another vector strain having different efficiency in the virus acquisition, and cultivating rice varieties resistant to the virus disease, respectively. The effects of varying the parameters are shown in Figure 3.3 in terms of the changes in percentage of infective insects and of infective rice hills. Both percentages increase rapidly with an increase of the vector density within the range

TABLE 3.2. Numerical relationships between successive time intervals.[a]

Time interval (symbol)		Numerical relationship and/or mortality factor	r^2
From	To		
Green rice leafhopper (breeding season)			
Egg (E)	First instar nymph (N_1)	$N_1 = 0.36E^{0.89}$	0.94
First instar nymph (N_1)	Fifth instar nymph (N_5)	Predation[b]	
		Constant survival rate (S_n)[b]	
Fifth instar nymph (N_5)	Adult emerged (A_e)	$A_e = 0.77 N_5$	0.99
Adult emerged (A_e)	Adult matured (A_m)	Predation[b]	
		Constant survival rate (S_a)[b]	
Adult matured (A_m)	Female matured ($A_♀$)	$A_♀ = 0.50 A_m$	0.99
Female matured ($A_♀$)	Egg (E)	$E = 192.27 (A_♀)^{0.41}$	0.56
Green rice leafhopper (overwintering season)			
Adult emerged in the third generation (A_3)	Nymph before wintering (N_b)	$N_b = 93.68 - 0.50A_3 - 5.80 L_m$	0.81
Nymph before wintering (N_b)	Nymph after wintering (N_a)	$N_a /N_b = 0.80 - 0.0045 N_b - 0.0076T$	0.94
Nymph after wintering (N_a)	Egg in the second generation (E_2)	$E_2 = -139.71 \log N_a + 229.32 \ (N_a > 3)$	
		$E_2 = 54.23 N_a \ (0 \le N_a \le 3)$	

		r
Lycosid spiders (breeding season)		
Lycosa invaded paddy field (L_0)	$L_m = 7.15 L_0 + 1.67$	0.98
Maximum number of Lycosa (L_m)	$L_t = \dfrac{L_m}{1 + (L_m/L_0 - 1)e^{-0.1t}}$	
Lycosid spiders (Overwintering season)		
Maximum number of Lycosa (L_m)	$S_{12} = -1.04 L_m + 10.49$	0.97
Lycosa before wintering (S_{12})	$S_2/S_{12} = 4.17 + 0.082 S_{12} - 0.076T$	0.97
Lycosa after wintering (S_2)	$L_0 = 0.062 S_2 - 0.001$	0.89
Microphantid spiders		
Oedothorax invaded paddy field (O_0)	$O_0 = 0.80$	
Maximum number of Oedothorax (O_m)	$O_m = -3.15 \log Z + 2.29$	0.81
	$O_t = \dfrac{O_m}{1 + (O_m/O_0 - 1)\,e^{-0.1t}}$	

[a] From Sasaba (57). T, Number of days below 0°C during December–February; L_0, mean number of Lycosa in July per hill; L_t, number of Lycosa per hill on tth day; L_m, mean number of Lycosa per hill on tth day; O_t, number of Oedothorax in July per hill; O_0, mean number of Oedothorax per hill on tth day; Z, mean number of Lycosa during July–August per hill; t, number of days after transplantation minus 30 days (hence, $t = 0$ means 30 days after transplantation when appreciable invasion of spiders was observed); r, Correlation coefficient.

[b] Sasaba et al. (59).

TABLE 3.3. The systems model of rice dwarf virus (RDV) epidemiology.[a]

Parameter	Symbol	Generation of *Nephotettix cincticeps* (variety of rice)[b]	Submodel
Proportion of newly infected rice hills			
By immigrating adults	$A_{T(I)}$	OW (E), third (L)	$A_{T(I)} = 0.113 \log N_A P + 0.491$
By ensuing progeny	$A_{T(II)}$	First (E)	$A_{T(II)} = 0.098 \log N_{L-1} P + 0.255$
Proportion of infected rice hills	$\left\{\begin{array}{l} A_I \\ A_{II} \end{array}\right.$	First (E), fourth (L) — Second third (E), fifth (L)	$A_I = A_{T(I)}$; $A_{II} = A_{T(I)} + A_{T(II)} - A_{T(I)} A_{T(II)}$
Loss in yield of unpolished rice by RDV infection	$\left\{\begin{array}{l} L_I \\ L_{II} \end{array}\right.$		$L_I = 0.800\, A_{T(I)}$; $L_{II} = 0.333\, A_{T(II)}$
Index of infected rice plants	I_v	First–fifth (E, L)	$I_v = \dfrac{bA}{(1 - A) + aA + bA}$, $A = A_I$ or A_{II}
Proportion of infected individuals	P_n	First–fifth (E, L)	$P_{n+1} = P_{n\,n}(\alpha r)_{n+1}(1 - W_{n+1}) + W_{n+1}$
Proportion of individuals acquiring RDV orally	W_n	First–fifth (E, L)	$W_n = 1 - e^{-2.355\, I_{v(n)}}$
Relative rate of increase of infected insects to noninfected ones	$\left.\begin{array}{l} {}_n(\alpha r)_{n+1} \\[4pt] T_{n+1} \end{array}\right\}$	First–fifth (E, L)	${}_n(\alpha r)_{n+1} = -0.029\, T_{n+1} + 0.938$
Mean temperature			

[a]From Nakasyji and Kiritani (49).
[b]E, Early-planted rice; L, late-planted rice; OW, overwintering generation.
N_A is the number of newly emerged adults per rice hill.
N_{L-1} is the number of first instar nymphs per rice hill.
P is the proportion of infected insects.
a = (fresh weight of healthy tillers in infected hills)/(fresh weight of healthy hills).
b = (fresh weight of infected tillers in infected hills)/(fresh weight of healthy hills).

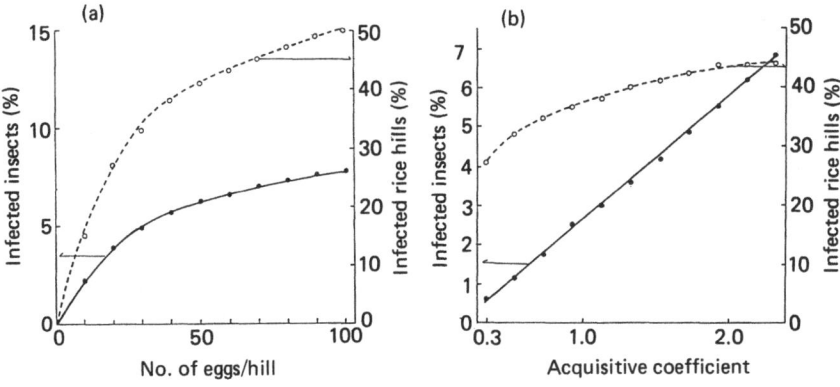

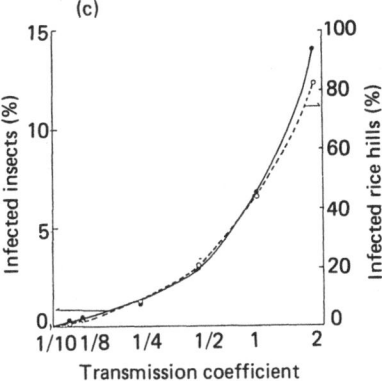

FIGURE 3.3. Simulation tests on the effects of varying control measures. (a) Relationship between the percentage of infected insects or that of infected rice hills and the egg density of the second generation. (b) Relationship between the percentage of infected insects or that of infected rice hills and the acquisition coefficient. (c) Relationship between the percentage of infected insects or that of infected rice hills and the transmission coefficient. [Nakasuji et al. (51).]

of low vector population densities (Fig. 3.3a). But they change little around the density normally observed in the epidemic areas. The simulation suggests that maintenance of vector density below a certain level is essential for the effective control of RDV disease.

In the second simulation (Fig. 3.3b), the percentage of infective insects increases linearly with increasing efficiency of acquisition feeding, while the percentage of infective rice hills is affected to a lesser extent. The efficiency of acquisition feeding is expressed by the acquisition coefficient a in the equation $W = 1 - \exp(-aI_v)$, whose standard value is 2.355 (Table 3.3). In the third simulation (Fig. 3.3c), both percentages increase exponentially with an increase in the efficiency of virus transmission. This efficiency is expressed by the transmission coefficients in the first two equations in Table 3.3, whose standard values are 0.113, 0.491, 0.098, and 0.255. The number on the abscissa is the multiplication factor of the standard values. Though the abscissa is plotted on log scale, the effect of the change in the coefficient of the virus transmission is more obvious than that of the vector density or the coefficient for the efficiency of acquisition feeding.

The simulation tests suggest as a result that the most effective control measure to suppress RDV epidemics is the cultivation of a variety of rice plant resistant to virus infection or the adoption of a cropping system that would reduce the coefficient of virus transmission. The course of the change in the prevalence of waika disease in Kyushu, Japan, strongly suggests that the result of the simulation test is much to the point. The newly discovered disease waika, which is transmitted semipersistently by *N. cincticeps,* extended as wide as 25,000 ha in 1973 in northern and mid-Kyushu, having started in 1967. Although chemical control of *N. cincticeps* and the cold winter of 1973–1974 both seem to have contributed to the decline of the disease, the disease was perfectly combatted by 1976 mainly due to the replacement of the susceptible rice variety by more tolerant ones (28, 61).

Insecticide spraying would not be so effective in suppressing epidemics of RDV unless simultaneous and area wide treatment, e.g., aerial application, were carried out (17, 46). Plowing the fallow paddy fields during the winter on a large scale, e.g., 300 ha in Kochi Prefecture, resulted in a complete suppression of RDV, probably through the decimation of the vector population (50). As Kiritani (26) suggested, it may be conceptually helpful to divide the control strategy of insect-borne rice virus diseases into long-term and short-term ones. Long-term control, such as winter plowing of fallow fields, does not necessarily aim at a decrease of the vector density below the control threshold within one cropping season. Even if the effect of control is small in a short period, it is sufficient to achieve a gradual decrease in population density where this type of control is continually applied on a large scale for a long enough period. As a result, virus diseases may be completely suppressed in a large area, as seen in the case of Kochi Prefecture. In contrast, short-term control, such as insecticide applications, must be effective in a given cropping season. As Kiritani (26) pointed out, a single control measure of RDV would not be sufficient to show a satisfactory effect, since the control threshold density of the vector is considerably low even under its endemic occurrence. By using Eq. (13), Kiritani (26) claimed the necessity of integrated use of as many compatible control methods as possible to control RDV.

Recent Advances in Mathematical Epidemiology of Rice Dwarf Virus

The system model developed by Nakasuji and Kiritani (49) was useful in predicting the occurrence of the disease and in comparing the effectiveness of various control measures (26, 51). However, the model is not suitable for use in theoretical studies of RDV epidemiology because it was developed empirically in a particular area.

Recently, infectious diseases of animals and humans have been studied from the viewpoint of population ecology by using differential equation models which describe the temporal changes of pathogen, vector, and host animal in number (1, 2). Adopting a similar approach, we developed a new mathematical model of RDV epidemiology (52).

A population of rice hills consists of healthy (noninfected), latent infected, and infectious ones, numbering, respectively, $X(t)$, $Y(t)$, and $Z(t)$ at time t. The total number of rice hills is denoted by $R(t)$, that is, $R(t) = X(t) + Y(t) + Z(t)$. Similarly, the vector population is composed of healthy (noninfected), latent infected, and infectious ones, whose number per rice hill are denoted by $U(t)$, $V(t)$, and $W(t)$, respectively. The total number of vectors is represented by $N(t)$, that is, $N(t) = U(t) + V(t) + W(t)$.

We assume that the net rate of the virus transmission to healthy rice hills is proportional to the density of infectious vectors W times the number of healthy rice hills X. Therefore, the rate can be expressed by αWX, where α is the coefficient of transmission efficiency. We further assume that the rice hills enter the infectious state from the latent infected one at the rate σ, so that the average latent period in rice plant is taken to be $1/\sigma$. These assumptions lead to the following simultaneous differential equations to describe the temporal change of the number of rice hills in each state:

$$dX/dt = -\alpha WX$$
$$dY/dt = \alpha WX - \sigma Y$$
$$dZ/dt = \sigma Y$$

In the present case, the total number of rice hills is constant because no hills die of the RDV infection.

The population change of the vector insect is rather complicated. The density of $N.$ $cincticeps$ is regulated well through density-dependent processes (19, 20, 31, 40). Generation by generation growth pattern of the population is nearly sigmoid, so we fit the following logistic equation to the population growth in the absence of RDV:

$$dN/dt = aN(1 - bN) - cN \qquad (17)$$

where a, b, and c are birth rate, density-dependent reduction rate of natality, and natural death rate, respectively. In the presence of RDV, the birth rate of latent infected vectors is not different from that of healthy vectors, but that of infectious ones is reduced considerably as described earlier. This depressed birth rate is denoted as a' to distinguish it from a. The other two parameters, b and c, are assumed not to be affected by RDV.

We assume that the net rate at which healthy vectors acquire the virus is proportional to the density of the healthy vectors U times the number of infectious rice hills Z. The rate is expressed by βUZ, where β is the coefficient of acquisition efficiency. Then, the temporal change in the number of healthy vectors U can be described by the following equation:

$$dU/dt = (U + V)a(1 - bN) + (1 - \gamma)Wa'(1 - bN) - cU - \beta UZ$$

where γ is the rate of transovarial transmission. Assuming that latent infected vectors become infectious at the rate λ, which is the reciprocal of the mean incubation period of the virus in the vector, we get the following differential equations for latent infected and infectious vectors, respectively:

$$dV/dt = \beta UZ - (c+\lambda)V$$
$$dW/dt = \gamma Wa'(1-bN) + \lambda V - cW$$

For convenience in numerical calculations, we use days as a time unit in the model. The duration of one cropping season of rice is assumed to be 150 days from planting ($t = 0$) to harvesting ($t = 150$). The parameter values in the model are roughly estimated from the data obtained by us and others, and are summarized in Table 3.4. Susceptibility of rice plants to virus infection differs greatly depending on the growing stage. Since rice plants lose their susceptibility to RDV about 70 days after planting (22, 60), we assume that the value of α becomes zero on and after the day 71 ($t = 70$).

By using this model, we simulated the temporal changes of the percentage of infectious rice hills ($100Z/R$) and that of infectious insects ($100W/N$) within one cropping season (52). In this simulation, the total number R of rice hills is conveniently fixed at 10^5 and initial values of X, Y, and Z are chosen as 1.0×10^5, 0.0, and 0.0, respectively. In the same way, the initial total number N of vectors per hill is determined to be 1.0 and initial values of U, V, and W are set at 0.95, 0.0, and 0.05, respectively. The result of the simulation is shown in Figure 3.4. . The percentage of infectious rice hills increases rapidly from the start and reaches an asymptotic value of 83% on about the day 100. On the other hand, the percentage of infectious insects decreases during about 50 days from the beginning and

TABLE 3.4. Parameter values used in the model.[a]

Symbol	Meaning	Value
α	Virus transmission coefficient to plants by infectious vectors	0.15 when $0 \leqslant t \leqslant 70$, 0.00 when $70 < t$
β	Acquisition feeding coefficient of virus by vectors	3.1×10^{-8}
γ	Rate of transovarial transmission of virus by vectors	0.84
σ	Reciprocal of latent period of virus in plant	6.7×10^{-2}
λ	Reciprocal of latent period of virus in vector	7.9×10^{-2}
a	Birth rate of healthy vectors	0.1
a'	Birth rate of infectious vectors	7.0×10^{-2}
b	Density dependent reduction rate of natality of vectors	2.0×10^{-2}
c	Death rate of vectors	3.3×10^{-2}

[a]From Nakasuji et al. (52).

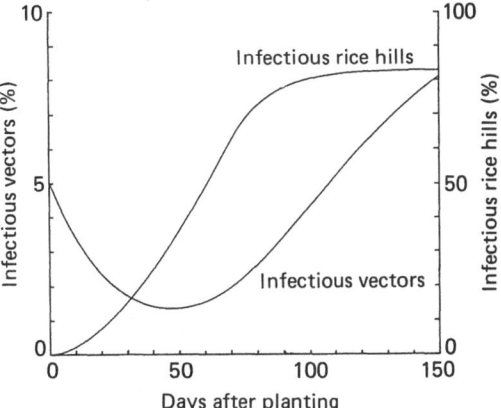

FIGURE 3.4. Calculated percentages of infectious rice hills ($100Z/R$) and of infectious vectors ($100W/N$) [Nakasuji et al. (52)].

increases thereafter. It reaches 8% at the end of the cropping season ($t = 150$). The trajectories in Figure 3.4 are quantitatively somewhat different from the observed ones, but qualitatively they show a typical pattern of changes observed in the epidemic areas. Therefore, we conclude that this model and almost all the parameter values may be reasonable enough to describe the dynamics of RDV epidemiology.

This model makes it possible to solve some interesting problems about RDV epidemiology. In many cases observed in the fields, the percentage of infectious insects never goes up to very high values or becomes zero, i.e., extinction of RDV, but it remains at relatively low values. Simulation tests using this model indicate that the trajectory of within-season change in percentage of infectious vectors reaches an equilibrium. Furthermore, the equilibrium trajectory obtained with those particular parameter values is stable because the same trajectory appears irrespective of the initial value of the percentage of infectious vectors (S. Miyai *et al.*, in preparation). This may explain why the infectious insects remain at such a low rate in the vector population without rising extremely or decreasing to zero.

The discrepancy in the geographical distribution between *N. cincticeps* and RDV has been a subject of dispute among researchers (27, 60). *Nephotettix cincticeps* is distributed in a more northerly part of Japan than RDV. Shinkai (60) claimed that northern races of *N. cincticeps* may be considerably inferior to southern ones in the ability of acquisition as well as transmission of RDV, but a clear explanation has not been given. Kiritani (27) considered that a lower turnover rate of transmission cycle due to prevailing low temperatures in northern Japan was responsible for this discrepancy. By using the present model, this problem also can be tackled. The density of *N. cincticeps* becomes thinner with higher latitude. In particular, the density in spring is usually very low in northern districts, so that there are few vectors during the early cropping season of rice plants in the fields. By changing the initial density of vectors and the values of the parameters a, b, and c in Eq. (17), we can describe the growth pattern of the vector population in northern Japan. The result of the simulation using this modified population growth is that the percentage of infectious insects decreases to zero, that is, RDV is extinct in the simulation (S. Miyai *et al.*, in preparation). This is similar to the well-known concept of threshold host density for disease maintenance in human epidemiology, that is, the density below which the disease will not be able to spread if introduced into a noninfected host population (1, 23). Similarly, if the vector density during the early cropping season is low enough, RDV cannot be established in the vector–rice plant system regardless of the affinity between the vector and RDV. This explanation is nothing but a hypothesis, but, judging from the fact that RDV has spread rapidly in southern Japan together with the increase in the density of *N. cincticeps*, it may be a satisfactory explanation.

The RDV–green rice leafhopper relationship is one of the most complicated among the relationships between a plant virus and its vector. For example, we can neglect the difference between a and a' in the relationship between rice stripe virus (RSV) and the small brown planthopper *Laodelphax striatellus* because RSV scarcely affects the vector physiologically (32). Black-streaked dwarf virus is not passed to the progeny of the vector *L. striatellus* through the ovary and also it does not affect the vector. In this case, we can use our model with $a' = a$ and $\gamma = 0$. Thus, this model will be useful in exploring the dynamic nature of insect-borne rice virus epidemiology besides that of RDV.

Conclusion

The transmission cycle of insect-borne virus diseases is composed of three different components in the ecosystem: host plant, insect vector, and virus. The interaction systems among these components are affected by various external as well as internal factors. Among the insect-borne virus diseases, the RDV transmission cycle offers the most complicated system, and includes transovarial passage of the virus and the effect of virus on the physiology of vectors. Presently we have no direct measure to eliminate viruses, and the control threshold of vector density is surprisingly low compared to that of phytophagous insect pests. This situation demands the integrated use of many compatible control measures in controlling insect-borne virus diseases. In particular, to attain a successful control, it is necessary to manage the virus–vector–host system. This system is so complicated that modeling approaches are inevitable. Though such approaches are still in the infant stage for insect-borne virus diseases, systems analysis will teach us how to achieve optimal management of their epidemics.

References

1. Anderson, R.M., 1981, Population ecology of infectious disease agents, in R.M. May, ed., *Theoretical Ecology,* 2nd ed., Blackwell, Oxford, pp. 318–355.
2. Anderson, R.M., and May, R.M., 1979, Population biology of infectious diseases: Part I, *Nature* **280**:361–367.
3. Baily, N.T.J., 1975, *The Mathematical Theory of Infectious Diseases and its Applications,* Griffin, London: 413 pp.
4. Broadbent, L., 1950, The correlation of aphid numbers with the spread of leaf roll and rugose mosaic in potato crops, *Ann. Appl. Biol.* **37**:58–65.
5. Burleigh, J.R., Romig, R.W., and Roelfs, A.P., 1969, Characterization of wheat rust epidemics by numbers of Uredia and numbers of Urediospores, *Phytopathology* **59**:1229–1237.
6. Clark, L.R., Geier, P.W., Hughes, R.D., and Morris, R.F., 1967, *The Ecology of Insect Populations in Theory and Practice,* Methuen, London: 232 pp.
7. Fine, P.E., and Sylvester, E.S., 1978, Calculation of vertical transmission rates of infection, illustrated with data on an aphid-borne virus, *Am. Nat.* **112**:781–786.
8. Frampton, V.L., Linn, M.B., and Hansing, E.D., 1942, The spread of virus diseases of the yellows type under field conditions, *Phytopathology* **32**:799–808.
9. Fukushi, T., 1934, Studies on the dwarf disease of rice plant, *J. Fac. Agric. Hokkaido Imp. Univ.* **37**:41–164.
10. Fukushi, T., 1935, Multiplication of virus in its vector, *Proc. Imp. Acad. Japan* **11**:301–303.
11. Fukushi, T., 1940, Further studies on the dwarf disease of rice plant, *J. Fac. Agric. Hokkaido Imp. Univ.* **45**:83–154.
12. Gilbert, N., and Hughes, R.D., 1971, A model of an aphid population—Three adventures, *J. Anim. Ecol.* **40**:525–534.
13. Gregory, P.H., 1948, The multiple-infection transformation, *Ann. Appl. Biol.* **35**:412–417.
14. Gregory, P.H., and Read, D.R., 1949, The spatial distribution of insect-borne plant-virus diseases, *Ann. Appl. Biol.* **36**:475–482.
15. Harris, K.F., 1980, Aphids, leafhoppers and planthoppers, in K.F. Harris and K. Maramorosch (eds.), *Vectors of Plant Pathogens,* Academic Press, New York, pp. 1–13.
16. Harrison, B.D., 1981, Plant virus ecology: Ingredients, interactions and environmental influences, *Ann. Appl. Biol.* **99**:195–209.
17. Hashizume, B., 1964, Studies on forecasting and control of the green rice leafhopper, *Nephotettix cincticeps* Uhler with special reference to eradication of the rice dwarf disease, Memoirs of the Association for Plant Protection of Kyushu, No. 2 77 pp. (in Japanese with English summary).
18. Hibino, H., Roechan, M., Sudarisman, S., and Tantera, D.M., 1977, A virus disease of rice (kerdil hampa) transmitted by brown planthopper, *Nilaparvata lugens* Stal, in Indonesia, *Contrib. Central Res. Inst. Agric. Bogor* **35**:1–15.
19. Hokyo, N., 1972, Studies on the life history and the population dynamics of the green rice leafhopper, *Nephotettix cincticeps* Uhler, *Bull. Kyushu Agric. Exp. Stn.* **16**:283–382 (in Japanese with English summary).

20. Hokyo, N., and Kuno, E., 1977, Life table studies on the paddy field population of the green rice leafhopper, *Nephotettix cincticeps* Uhler (Hemiptera: Cicadellidae), with special reference to the mechanism of population regulation, *Res. Popul. Ecol.* **19**:107–124.
21. Hughes, R.D., and Gilbert, N., 1968, A model of an aphid population—A general statement, *J. Anim. Ecol.* **37**:553–563.
22. Ishii, M., Yasuo, S., and Yamaguchi, T., 1970, Epidemiological studies on rice dwarf disease in Kanto-Tosan district, Japan, *J. Central Agric. Exp. Stn.* **14**:1–115 (in Japanese with English summary).
23. Kermack, W.O., and McKendrick, A.G., 1927, A contribution to the mathematical theory of epidemics, *Proc. R. Soc. Lond. A* **115**:700–721.
24. Kimura, I., 1962, Further studies on the rice dwarf virus—I, *Ann. Phytopathol. Soc. Japan* **27**:197–203 (in Japanese).
25. Kiritani, K., 1977, Systems approach for management of rice pests, in *Proceedings International Congress Entomol., 15th, Washington D.C., 1976*, pp. 591–598.
26. Kiritani, K., 1979, Pest management in rice, *Annu. Rev. Entomol.* **24**:279–312.
27. Kiritani, K., 1981, Spacio-temporal aspects of epidemiology in insect borne rice virus diseases, *Japan Agricultural Research Quarterly* **15**:92–99.
28. Kiritani, K., 1983, Changes in cropping practices and the incidence of hopper-borne diseases of rice in Japan, in R.T. Plumb and J.M. Thresh (eds.), *Plant Virus Epidemiology*, Blackwell, Oxford, pp. 239–247.
29. Kiritani, K., and Nakasuji, F., 1977, A systems model for the prediction of rice dwarf virus infection of middle-season rice, *Appl. Entomol. Zool.* **12**:118–123.
30. Kiritani, K., and Sasaba, T., 1978, An experimental validation of the systems model for the prediction of rice dwarf virus infection, *Appl. Entomol. Zool.* **13**:209–214.
31. Kiritani, K., Hokyo, N., Sasaba, T., and Nakasuji, F., 1970, Studies on population dynamics of the green rice leafhopper, *Nephotettix cincticeps* Uhler: Regulatory mechanism of the population density, *Res. Popul. Ecol.* **12**:137–153.
32. Kisimoto, R., 1966, Genetic mechanism determining the ability of the smaller brown planthopper *Laodelphax striatellus* to acquire rice stripe virus, *Ann. Phytopathol. Soc. Japan* **32**:90 (in Japanese).
33. Kiyosawa, S., 1972, Mathematical studies on the curve of disease increase, *Ann. Phytopathol. Soc. Japan* **38**:30–40.
34. Kono, T., 1966, Changes in the proportion of infective insects in the vector population, *Shyokubutsu-boeki* **20**:131–136 (in Japanese).
35. Kranz, J., 1973, Epidemiology, concepts and scope, in S.P. Raychaudhuri and J.P. Verma (eds.), *Current Trends in Plant Pathology*, University of Lucknow, Lucknow, pp. 26–32.
36. Kranz, J., 1974a, The role and scope of mathematical analysis and modeling in epidemiology, in J. Kranz (ed.), *Epidemics of Plant Diseases*, Springer-Verlag, Berlin, pp. 7–54.
37. Kranz, J., 1974b, Comparison of epidemics, *Annu. Rev. Phytopathol.* **12**:355–374.
38. Kranz, J., 1980, Comparative epidemiology: An evaluation of scope, concepts and methods, in J. Palti and J. Kranz (eds.) *Comparative Epidemiology*, Centre

for Agricultural Publishing and Documentation, Wageningen, the Netherlands, pp. 18–28.

39. Kranz, J., and Hau, B., 1980, Systems analysis in epidemiology, *Annu. Rev. Phytopathol.* **18**:67–83.

40. Kuno, E., 1968, Studies on the population dynamics of rice leafhoppers in a paddy field, *Bull. Kyushu Agric. Exp. Stn.* **14**:131–246 (in Japanese with English summary).

41. MacArthur, R.H., and Wilson, E.O., 1967, *The Theory of Island Biogeography,* Princeton University Press, Princeton, New Jersey. 203 pp.

42. Macdonald, G., 1957, *The Epidemiology and Control of Malaria,* Oxford University Press, London, 201 pp.

43. Maramorosch, K., and Harris, K.F. (eds.), 1979, *Leafhopper Vectors and Plant Disease Agents,* Academic Press, New York. 654 pp.

44. Maramorosch, K., and Harris, K.F., 1981, *Plant Diseases and Vectors. Ecology and Epidemiology,* Academic Press, New York. 368 pp.

45. Miyai, S., Kiritani, K., and Nakasuji, F., 1986, Models of epidemics of rice dwarf, in G.D. McLean, R.G. Garrett, and W.G. Ruesink (eds.), *Plant Virus Epidemics,* Academic Press, Sydney, pp. 459–480.

46. Nagai, K., Iwahashi, T., and Goto, S., 1970, On the control of the green rice leafhopper, *Nephotettix cincticeps* Uhler, in early spring, *Proc. Assoc. Plant Prot. Kyushu* **16**:15–17 (in Japanese).

47. Nakasuji, F., 1970, Epidemiology of rice dwarf virus transmitted by the green rice leafhopper, *Jpn. J. Appl. Entomol. Zool.* **14**:157–162 (in Japanese).

48. Nakasuji, F., and Kiritani, K., 1970, Ill-effects of rice dwarf virus upon its vector, *Nephotettix cincticeps* Uhler (Hemiptera: Deltocephalidae), and its significance for changes in relative abundance of infected individuals among vector populations, *Appl. Entomol. Zool.* **5**:1–12.

49. Nakasuji, F., and Kiritani, K., 1972, Descriptive models for the system of the natural spread of infection of rice dwarf virus (RDV) by the green rice leafhopper, *Nephotettix cincticeps* Uhler (Hemiptera: Deltocephalidae), *Res. Popul. Ecol.* **14**:18–35.

50. Nakasuji, F., and Kiritani, K., 1976, Causes of the outbreaks of rice dwarf virus disease, *Shokubutsu-boeki* **30**:48–52 (in Japanese).

51. Nakasuji, F., Kiritani, K., and Tomida, E., 1975, A computer simulation of the epidemiology of the rice dwarf virus, *Res. Popul. Ecol.* **16**:245–251.

52. Nakasuji, F., Miyai, S., Kawamoto, H. and, Kiritani, K., 1985, Mathematical epidemiology of rice dwarf virus transmitted by green rice leafhoppers: A differential equation model, *J. Appl. Ecol.* **22**:839–847.

53. Nasu, S., 1963, Studies on some leafhoppers and planthoppers which transmit virus diseases of rice plant in Japan, *Bull. Kyushu Agric. Exp. Stn.* **8**:153–345 (in Japanese).

54. Omura, T., Inoue, H., Morinaka, T., Saito, Y., Chettanachit, D., Putta, M., Parejarearn, A., and Disthaporn, S., 1980, Rice gall dwarf, a new virus disease, *Plant Disease* **64**:795–797.

55. Palti, J., and Kranz, J. (eds.), 1980, *Comparative Epidemiology,* Centre for Agricultural Publishing and Documentation, Wageningen, the Netherlands. 122 pp.

56. Rose, D.J.W., 1974, The epidemiology of maize streak disease in relation to population densities of *Cicadulina* spp., *Ann. Appl. Biol.* **76**:199–207.

57. Sasaba, T., 1974, Computer simulation studies on the life system of the green rice leafhopper, *Nephotettix cincticeps* Uhler, *Rev. Plant Prot. Res.* **7**:81–98.
58. Sasaba, T., and Kiritani, K., 1975, A system model and computer simulation of the green rice leafhopper populations in control programmes, *Res. Popul. Ecol.* **16**:231–244.
59. Sasaba, T., Kiritani, K., and Urabe, T., 1973, A preliminary model to simulate the effect of insecticides on a spider–leafhopper system in the paddy field, *Res. Popul. Ecol.* **15**:9–22.
60. Shinkai, A., 1962, Studies on insect transmissions of rice virus diseases in Japan, *Bull. Natl. Inst. Agric. Sci. C* **14**:1–112 (in Japanese with English summary).
61. Shinkai, A., 1977, Rice waika, a new virus disease, found in Kyushu, Japan, in *Symposium Virus Diseases of Tropical Crops, Tropical Agriculture Research Center, Tokyo, 1976,* pp. 123–127.
62. Sylvester, E.S., and Richardson, J., 1966, Some effects of temperature on the transmission of pea enation mosaic virus and on the biology of the pea aphid vector, *J. Econ. Entomol.* **59**:255–261.
63. Thompson, W.R., 1924, Théorie mathématique de l'action des parasites entomophages et le facteur du hasard, *Ann. Fac. Sci. Marseille Ser. 2* **2**:69–89.
64. Thresh, J.M., 1980, An ecological approach to the epidemiology of plant virus diseases, in J. Palti and J. Kranz (eds.), *Comparative Epidemiology,* Centre for Agricultural Publishing and Documentation, Wageningen, the Netherlands, pp. 57–70.
65. Van der Plank, J.E., 1963, *Plant Diseases: Epidemics and Control,* Academic Press, New York. 349 pp.
66. Wallace, J.M., and Murphy, A.M., 1938, Studies on the epidemiology of curly top in southern Idaho, with special reference to sugar beet and weed hosts of the vector *Eutellix tenellus, Tech. Bull. Dept. Agric.* **624**:1–46.
67. Watson, M.A., 1936, Factors affecting the amount of infection obtained by aphid transmission of the virus Hy III, *Phil. Trans. R. Soc. Lond. B* **226**:457–489.
68. Watson, M.A., and Healy, M.J.R., 1953, The spread of beet yellows and beet mosaic viruses in sugar-beet root crop. II. The effects of aphid numbers on disease incidence, *Ann. Appl. Biol.* **40**:38–59.
69. Zadoks, J.C., and Schein, R.D., 1980, Epidemiology and plant-disease management, the known and the needed, in J. Palti and J. Kranz (eds.), *Comparative Epidemiology,* Centre for Agricultural Publishing and Documentation, Wageningen, the Netherlands, pp. 1–17.

4
Aphid Vector Monitoring in Europe

Yvon Robert

Introduction

Since World War II, there has been an increasing demand on agriculture for higher and higher yields and better food quality. Various schemes have been devised to take up this challenge. Among the choices adopted for an increasing intensification of agriculture were breeding of new, highly productive and edible cultivars, intensive use of nitrogen fertilizers, and frequent and widespread use of broad-spectrum pesticides as a matter of routine without taking account of the actual identity, presence, abundance, and real harmfulness of the pests, together with profound modifications in agricultural landscape and practices resulting from the use of more and more advanced farm machinery. In several regions of Europe, hedges were removed over considerable areas, as in eastern Britain and to a lesser extent in western France (Brittany), leading to an "opening up" of the bocage landscape with eventually a possible decrease in its potential as a reservoir for both pests and their enemies. In parallel with this, monoculture has tended to replace mixed agriculture and there has been a significant trend toward sharp reductions in the genetic diversity of available cultivars along with a shortening of crop rotations and modifications in the sowing dates and length of the growing period.

It is of course very difficult to ascertain the actual effect of all these changes on overall aphid potentialities, including their biology, their ecology, and their relationships with their host plants, i.e., the likelihood of their becoming more damaging to the crops both as "sapsuckers" and as virus vectors. While there is no doubt that chemical and cultural control of aphids as well as improved forecasting techniques have become more and more reliable and successful as a whole, this was achieved without taking full account of the fact that aphids are "*r*-selected" (80) and show

Yvon Robert, Institut National de la Recherche Agronomique, Centre de Recherches de Rennes, Laboratoire de Recherches de la Chaire de Zoologie, Domaine de la Motte-au-Vicomte, B.P. 29-35650 Le Rheu, France.

a remarkable aptitude to adapt and evolve according to new conditions: they can migrate and transport viruses over long distances (58), they are very efficient at colonizing new biotopes, and they may do so in a very erratic manner (118), building up enormous ephemeral populations, self-selecting new biotypes with adapted biological properties, and overcoming cultivar resistance; moreover, several species, such as *Myzus persicae* (Sulz.), have become resistant to all known aphicides to such an extent that in many areas of Britain it is now hardly possible if at all to find any susceptible aphid (109). Consequently, one can anticipate the recording of aphid and virus disease outbreaks at any time and at any place, probably because (1) they are better known and recorded since farmers and technicians have acquired a better technical knowledge of the problems, being more vigilant as a result of economic constraints and undoubtedly also due to pressure from chemical companies, (2) the media keep on the look-out for sensational events, and finally (3) more scientists are working on the subject.

In any event, some words and expressions in recent accounts do illustrate what the reality has been like: the "aphid plague" and the "green invasion" (43, 110) refer to the extraordinarily huge mass flight of the rosegrain aphid, *Metopolophium dirhodum* (Wlk), over western Europe in 1979, and the "yellows plagues" to the exceptionally intense spread of sugar-beet yellowing viruses in 1949, 1957, and 1974 in Britain, with 40–60% of the plants being infected on a national average (40, 144), or in 1959, 1967, and 1974 in France (67); the "potato disaster years" usually refer to (1) the apparently uncontrollable spread of potato virus Y (PVY) over the whole of Europe in 1976 (92, 138), with the exception of Scotland, where in the same year potato leaf roll virus (PLRV) predominated (50), and (2) to the heavy infection of seed potatoes with PLRV in 1970 in Brittany (104), while in Scotland at the same time a nil acreage rejection for PLRV was noted (50). Many other examples could be given regarding other crops, such as carrots infected with carrot motley dwarf virus (CMDV) (143) and winter cereals with barley yellow dwarf virus (BYDV), which apparently has increased in importance during the last 15 years. Now the obvious cause for concern to any decision-maker is to know if we are still willing to tolerate such a depressing situation, where, although higher technologies become available, "aphid-borne viruses can spread as widely in the 1970s as in the 1940s, in spite of our efforts" (41)? If the answer is clearly "No," there immediately arises the problem of knowing why we have failed and how we can expect to deal with this. Relying solely on pesticides has often proved a failure if they are used as routine insurance treatments done at improper times and if in addition they are intended to control virus spread, since just killing the vectors does not mean *ipso facto* avoiding their propagating the virus!

Alate aphids have long been known to be mainly responsible for disseminating virus diseases probably more as a result of their numbers and their activity than of their intrinsic vector efficiency (90): they may bring viruses early into the crop if these are not seed-borne and/or spread them within the crop if seed-borne, and this primary infection may serve in turn as a reservoir for apterous populations and alates to initiate secondary infection. The timing of these successive events must be clearly appreciated, recorded, and interpreted in order to improve any forecasting system, and this should be done at different scales, from field level up to a regional, national, even international one: obviously it is not very realistic to think it possible to get information from, and give advice to, individual fields, but a forecasting system based on aphid monitoring alone operated in well-situated and representative sites can prove to be suitable on a regional scale. So far, this has been and is still the situation in many parts of Europe, especially as regards potato and sugar-beet systems of prediction for virus disease spread. But it also has been progressively realized that a national or international monitoring network would probably help make earlier and more accurate warnings if it could actually anticipate long-range aphid migrations and transport of virus.

Some interesting circumstancial evidence was indeed recorded in Britain at Rothamsted Experimental Station (north of London) as early as 1947, when very large numbers of *M. persicae* invaded well-grown potato crops about mid-July (12), resulting in unusually large amounts of leaf roll virus infecting a great number of plants (13); this flight was thought first to have come from early potatoes in eastern England (17), but wind backtrack analysis suggested a possible continental origin (58), which was then put forward on many occasions as the main reason by English and Swedish researchers to explain, for instance, unusual spread of leaf roll (14) and yellows (41, 89, 144, 148) and by the English and their press at the time of the 1979 "green invasion," although in this latter case, contrary evidence was given (22, 27) based on both meteorological records and data given by the new European network of 12.2-m suction traps EURAPHID, which continuously monitors aerial aphid populations over certain European countries, mainly Britain, France, the Netherlands, Belgium, and to a lesser extent Switzerland and Italy (Fig. 4.1). Although virus spread forecast relying on aphid monitoring alone has proved for some decades to be fairly satisfactory on the whole as far as potato and sugar-beet aphids and viruses are concerned, such monitoring does not appear to be sufficiently accurate to predict the spread of other viruses, such as BYDV, where the proportion of infective alate aphids entering a crop may be as important a consideration as their total number for initiating primary infection. It follows that an aphid vector monitoring (AVM) scheme that takes into consideration all the epidemiological components is required and the general trend is to extend this idea in the future to as many cases as possible.

The purpose of this chapter is to give information on the present situation of AVM in Europe, concentrating on aphids as virus vectors and more precisely on alate individuals. I will define first the aims and requirements of such an AVM scheme and give some historical background to explain the situation achieved so far, then take some examples actually in progress, and finally discuss some prospects, hopes, and limitations in order to try and make the most of the systems now in operation.

Aims and Requirements of Aphid Vector Monitoring (AVM) in Europe

Aims

DEFINITION

Recently Dunnet (30), referring to environmental monitoring, stated that "monitoring and surveillance were interchangeable nouns and that they included detection, measurement and interpretation of changes in environmental variables together with the capacity for predicting likely future changes." I will rely on this definition, although it may be not so evident to consider monitoring as including prediction, but rather as being only a prerequisite to it: the interface between monitoring and forecasting is not always sharply defined, but in fact, those who are in charge of monitoring usually also release forecasts or at least intend to do so in the near future (114). To adapt Dunnet's statement to the present concern requires only adding "aphid and virus" before "environmental variables," as AVM is one of the main components of most relevant plant virus disease epidemiological approaches. Moreover, it must be made quite clear that AVM is likely to differ from aphid monitoring alone, which is only part of it, and the difficulty is indeed to know how to incorporate the virus component into the system, as not only has the virus, as such, to be taken into account, but also the whole range of its interrelations with vectors (e.g., specificity, mode of transmission), hosts (e.g., cultivars, phenology, aphid and virus reservoirs, cultural practices), and environment (physical and biological). Therefore, there is nothing surprising about the fact that the relationships between the numerical importance of aphid vectors and the ultimate spread of viruses are much less simple than, say, that between aphids in their own right and yield decrease.

It must be emphasized that some effort has been made for some time now to try and define more sharply these relationships and fill in gaps in the models, but until recently, forecasting at a practical level was still mainly based on the overall relationship, if any, between aphid numbers caught in traps and/or counted on plants and the final subsequent number of infected plants as a consequence of aphid monitoring only, some improvement being shown when weather data were entered. This holds for potato and sugar-beet virus diseases, but does not for BYDV on cereals nor for cucumber mosaic virus (CMV) on muskmelon, where the proportion of infective aphids in the field had to be taken into consideration, but it should be stressed also that the areas involved ranged from thousands of hectares down to less than 1 ha. Consequently, all that can be said, with humility, is that the general purpose of AVM until now has been nothing more than try to assist in determining the spatial and temporal distribution of viruses spread by aphids so that decision-makers can better choose more appropriate and up-to-date measures to control this spread, and that it has been done differently for different countries and crops. The soundness and accuracy of ensuing forecasts should be greatly improved, but paraphrasing Taylor (120), who wrote that "pest monitoring is still in its infancy," we must remind ourselves that AVM has only just been born, especially on a European scale.

CONSTRAINTS

One of the first constraints we have to face is undoubtedly the absolute necessity *to monitor continuously* throughout the year whatever crop is involved and to do it regularly for many years: the obvious reasons for this are that (1) aphids may fly at any time of the year, even in winter on some occasions, for anholocyclic species, (2) different crops are always available at different physiological stages, and (3) several species of aphids may usually infect them with virus at the same time or at different periods with variable success. Such a continuous monitoring is the only way to assess any departure from the average during a season and this average in turn cannot be anticipated after only a few years because of the great resilience of biological parameters and changing weather and environment. Most of the time, this constraint comes up against human, economic, and even political problems. For obvious short-term practical reasons, the farmer is impatient to use the monitoring data, as he dreads the risk to his crop, and he may condemn this "unproductive" and costly scheme before data interpretation becomes available. Moreover, administrators and politicians and, generally speaking, financiers can hardly understand that such an ecological problem needs to be studied for years "routinely." To ensure that monitoring will continue is a permanent challenge because people's motivations depend too often upon the phytosanitary situation of the previous year: if the latter has been "bad," they are prone to give help; in the opposite case, they forget it. This means that AVM should be entirely professionally operated, as was stressed by Taylor (119) referring to the Rothamsted Insect Survey (RIS) suction trap network for aphid sampling and monitoring, and it must be supported by long-term funding.

A second constraint is *a space-scale one*. Aphids in flight ignore regional, national, and continental boundaries and it is clearly impossible to know how far alate invaders have come or how long they are going to fly when leaving a crop. It is well established, too, that transmission efficiency of semipersistent and persistent viruses is retained for most if not all the life of an aphid (73, 86) and that nonpersistent viruses can still be transmitted to susceptible plants after some hours of uninterrupted flight (23, 24), a time which is largely sufficient to explain that part of virus influx into a crop may arise from distant origins. There is need for a warning system to be set up at least nationwide, but in Europe, where agricultural landscape, cultivated crop, and wild plant distribution as well as agricultural practices are less than homogeneous within and between countries, the crucial choice is to decide how dense a monitoring network should be ideally to give pertinent advice at both lowest acreage scale and cost. Experience has shown that no general rule can apply to different crops suffering from aphid-borne virus diseases, i.e., no minimum standard area can be assessed from the outset as being "covered" by one specific site of monitoring as has been shown possible when suction trap samples have remained remarkably similar when cumulated over several days for distances up to 20–50 km between traps (122). Among the possible reasons for the difficulty in estimating this area is probably the fact that virus reservoir distribution and incidence may differ in space and time from aphid distribution, and this, together with differences in the aphid biological cycle and phenology on the one hand, and crop distribution and phenology on the other hand, makes the damage thresholds likely to be very different kilometers apart. This can be easily understood from an example: in Switzerland, where mountains introduce a large heterogeneity, a site for monitoring potato aphid-borne viruses liable to be spread in seed potato crops is not likely to "cover" an area similar to that "covered" by one set up in Dutch polders. It follows that the choice of sites to monitor virus spread should primarily rely upon previous experience on past and current virus status, but for practical reasons the solution has often been to utilize sites where aphid monitoring is already done as part of a large-scale national or international network such as EURAPHID and to focus on aphids and viruses at a more regional scale with some specialization on a crop basis, in, e.g., sugar-beet areas in eastern England and southern Sweden, seed-potato areas in Scotland, Dutch polders, Brittany, and French-speaking Switzerland, and winter cereals in relation to BYDV in the London basin, southwest of England, Wales, and western France.

An integrated monitoring system is intended to enable warnings to be released at the same time for all virus-susceptible crops and an ensuing third constraint is *to monitor continuously as many species as possible, if not all*. This is especially relevant to nonpersistent virus transmission and spread: indeed, if we are almost sure that only resident aphids, i.e., those able to colonize a crop and then feed and breed on it, have sufficient time to transmit persistent and semipersistent viruses such as PLRV, BYDV, beet mild yellowing virus (BMYV), and beet yellows virus (BYV), it has become no less obvious now for some years that nonpersistent viruses, which require seconds to minutes to be both acquired and transmitted, may be disseminated by a great many species of transient aphids alighting to probe for a short while before taking off again very soon, provided their number and activity are sufficient to compensate for their low intrinsic transmission efficiency. This is somewhat documented as regards CMV (65) and PVY, since *M. persicae* was shown not to be always the main agent involved in spreading these viruses in the field, although it was the best at transmitting them under laboratory conditions: the example of PVY having worsened the sanitary status of seed potatoes in Europe between 1974 and 1976 is very relevant in this respect. The peach potato aphid, which is quite easy to breed indoors and, unlike other species, such as *Aulacorthum solani* (Kltb), is not toxic to indicator plants (98), is therefore very suitable for virus transmission experiments, and it has indeed been tested most frequently in the laboratory by virologists for its virus transmission ability (63). Moreover, it also proved to transmit about 120 viruses (31) with an overall high rate of infectivity. Consequently, it has normally been inferred on this basis that this species should be considered as the best disseminator under field conditions and for this reason was taken into account almost alone in forecasting natural virus spread. No doubt this probably holds for a good many cases, but circumstantial evidence has now accumulated which shows that enormous numbers of "poor" vectors can disseminate as much virus as less numerous but more efficient ones.

If we had not monitored all species of aphids in 1974–1976, we would not have understood how PVY° spread with such an intensity in the near absence—at least in Brittany—of potato aphids and particularly *M. persicae,* as compared to other years (92, 103), and we would not have been able either to ascribe most of it to a cereal aphid, *Rhopalosiphum padi* (L.), if we had not been trapping aphids since 1967 (97), thus enabling us to compare from one year to another the average numbers of alate aphids caught in comparable situations (92). But there was only one reference in the literature, and with poor evidence, of the ability of *R. padi* to transmit PVY (48) and measures were not taken in time to try and control the disaster we suspected might occur, first because, with the exception of mineral oils, no appropriate measures other than cultural ones ("roguing" infected plants as soon as they show symptoms and haulm burning) were available during the crop season, and second because we had never met such a situation before. Last but not least, there was a similar pattern in the whole of mainland Europe, which was "flooded" with PVY° and/or PVYⁿ according to country. It seems relevant that a number of papers were subsequently published making a reappraisal of many aphid species other than *M. persicae* as vectors of this virus both in the laboratory and in the field (64, 107, 139), and giving the infection pressure concept a new role in helping with forecasts (108, 112, 137, 138).

A fourth constraint lies in the necessity *to check the actual significance of forecasts* based on such AVM. The economic thresholds which have been largely promoted by the integrated pest control concept and are getting better known for some crops (cereals) when aphids are considered as pests in their own right are very difficult to assess during crop season as regards aphids as virus vectors because (1) the presence of aphids on the plants does not imply *de facto* that virus is about to be transmitted and spread, (2) it is sometimes difficult to ascribe accurately the respective roles of aphids as pests in their own right and as vectors of virus in the ensuing yield decrease, (3) virus disease symptoms are usually displayed too late under our temperate climatic conditions, and (4) apart from seed potato certification schemes, where tolerance levels not to be exceeded in crops to be entered for certification refer merely to a percentage of plants infected by any one virus and not to a possible decrease in yield, decision-making in other crops, such as sugar beets and cereals, is based on thresholds supposed to be calculated as a given percent of infected plants above which control measures would be economically profitable to avoid excess of yield decrease (42, 61). These thresholds are eminently variable for a given crop according to time and space. First, since usually the younger the plant, the more susceptible it is to virus infection by aphids, it can be anticipated that (1) few efficient aphids probing or feeding on young crops may result in disseminating nonpersistent or persistent virus in it, respectively, as efficiently as numerous aphids occurring later, or (2) "poor" vectors are more liable to get a chance to play a role before mature plant resistance increases, hence the previously mentioned necessity to monitor all aphids continuously and to assess crop phenology. Second, different areas display different thresholds just because sources and abundance of inoculum are neither evenly distributed nor available at the same time. It follows that AVM has to be checked in as large a range of situations as possible: this requires more than a mere regional scale and also a multidisciplinary and international cooperation.

This extended cooperation depends largely on a final constraint, which is *the necessity for data and interpretations to be freely and confidently exchanged* between institutions and countries and this can only be achieved (1) if data bases and banks are started and become available and (2) if countries involved do not hinder access to them. For historical reasons, in some of European countries, aphid trapping has been initiated and operated under the aegis and with the aid of more or less private funds from agricultural extension or inspection services or organizations; this refers particularly to seed potato production. In this world of harsh export competition, the more seed is sold, the more money is put back into this technical tool liable to help improve seed quality. As a consequence, the merchant lobby has developed a protectionist behavior requiring that aphid data should not be released outside and leading to such a point that in 1976 some declared that their own country had no problem from PVY, an island of healthy seed being lost in an ocean of highly infected plants! Such short-sighted behavior conflicting with the scientific utility of not keeping aphid records secret was experienced once again recently (10). Regrettably, this may delay data interpretation and understanding and thus the release of improved forecasts, but solutions have been looked for and found, such as data sharing only in the following year, since AVM is not yet supposed to help finalize forecasting, but to be still only experimental.

It may appear that so many constraints would impede the feasibility and running of such a system: they probably do so partly, but the historical backgrounds to aphid trapping in Europe show that monitoring in different European countries is the result of a slow but steady step-by-step progress by trial and error which requires much patience. It may easily be imagined that it will take some more time to reach a true European integration of AVM, but it is in progress.

Requirements

It is anticipated that interpretation of AVM data and their use for forecasting will be rather tricky and this is one reason among others why it was considered right from the start as a research subject keeping in close touch with field realities. As was judiciously stressed by Taylor (122), who was referring to aphid aerial sampling by the Rothamsted Insect Survey (RIS), it is impossible if not dangerous to decide how to devise and develop such a system only by extrapolating laboratory biological results or by using sampling programs elaborated by theorizing statisticians: a good biological example can be that reported above of *M. persicae* and nonpersistent viruses. On the other hand, classical statistical analysis does not apply either to most data because of the nonlinearity of biological parameters. It follows that setting up such a monitoring network as was intended to assist in forecasting at different levels of space and time will have to meet some requirements and any improvement will be possible ònly if we can solve—eventually by laboratory research—new problems bound to arise from the field. This topic is a perfect illustration that specialization without basic general knowledge is meaningless.

First, to be able to select the more relevant aphid vectors, *a knowledge of the overall aphidofauna* of a given region is required and it must be as thorough as possible so as to point up similarities and differences between regions and countries. This census must be, as far as possible, both qualitative (list of species) and quantitative. In Europe so far we have been very fortunate to have skilled taxonomists to make this census available in many parts, and help given by D. Hille Ris Lambers in the Netherlands, V.F. Eastop and H.L.G. Stroyan in Britain, F. Leclant and G. Remaudière in France, and O.E. Heie in Denmark was invaluable to those in charge of alate aphid monitoring: for example, alates of some 320 species had already been trapped and identified over the British Isles in 1979, out of a total of about 590 species as previously recorded (129). Every year brings new records.

Second, there is also a need for *a better understanding of aphid bioecology and virus disease epidemiology.*

Aphids may play a different role as virus vectors according to whether they are holocyclic or anholocyclic. Holocyclic populations when leaving their primary hosts at the beginning of the spring are not likely to be loaded with any virus of agricultural importance to herbaceous crops, since usually woody primary hosts are not reservoirs for these viruses. In contrast, anholocyclic forms may have overwintered viviparously on virus reservoirs in mangold clamps or greenhouses or, outside, on winter crops, volunteers, regrowths, or susceptible weeds, and they may bring virus straight after departing from them for their spring contamination flight. In this case, winter conditions resulting in more or less successful overwintering play a great role. So any knowledge of these possible alternatives is of paramount importance, particularly to sugar-beet and seed-potato crops liable to become at risk more or less soon after emergence when they are the most susceptible to virus infection. In autumn, anholocyclic alate aphids may return to winter-sown crops and herbaceous weeds to feed and overwinter and they will remain on them a longer time than winged gynoparae looking for their primary host: this is especially relevant to autumn-sown cereals. It must be emphasized that anholocycly and holocycly usually exist together in a given region and it is the balance between the two possibilities according to year and place that may be most important.

In addition to differences in their method of overwintering, aphids display various intrinsic and adaptive possibilities for producing alatae at different periods of the year and in different places. As a consequence, they show a seasonal periodicity in flight numbers, size, and timing: different species have been shown to have different numbers of periods of flight activity (ranging from one to three) (45, 101) and these numbers, as well as the size and timing of the flight cycles, may fluctuate according to year, environment, host-plant availability, and latitude and longitude effect (96, 127). It is therefore essential to know all these biological traits in a region before even thinking of doing any interpretation or taking any decision concerning control measures. In order to follow these spatial and temporal variations, efficient systems of continuous sampling were needed, and in Europe the choice for such sampling devices has been very progressive and has been made in different ways by different nations, as I will show later. Anyway, whatever choice has been taken, such a census of aphid aerofauna and their host plants together with records of their distribution and phenology of flight are readily available for some countries, for a large amount of data have already been published (46, 94, 100, 101, 102, 115, 130, 131) or at least are stored in databases or banks waiting for processing. Data of this kind are very useful and relevant to a study of aphids as far as direct damage is concerned, provided they have been collected in such a way that they can be easily handled and quickly interpreted statistically so as to assist with providing consistent information to decision-makers on past, present, and if possible expected levels of infestation and timing, as an outcome of research (114).

But such knowledge is far from being sufficient when virus dissemination is involved. In this instance, it is essential that a monitoring system be at least able, in addition, first to assess (1) what proportion of which aphid species (or clone or biotype) is likely to alight on the crops, (2) when, (3) according to what spatial pattern of distribution on different scales, (4) for how long, and (5) what for (probing, feeding, breeding); and second, to estimate (1) what proportion of these individuals is likely to ultimately transmit some virus or other to the plants, and (2) to what extent. Consequently, a further specific virus monitoring should ideally be coupled with the former to provide relevant information about the overall virus situation on a fairly large scale with regard to virus strains and isolates, and virus influx at different periods of the year according to virus reservoirs, their distribution, their importance as donors, and their successive availability in time. This is also progressing.

As mentioned previously, it is impossible to rely entirely on laboratory experiments to anticipate what is to happen in the field. This is probably due to the inevitable bias introduced by researchers themselves in their own choices, which, for example, resulted in neglect of the appraisal of a great many aphid species as vectors of nonpersistent viruses as well as of plants as virus receptors and donors; this is also probably because genuine epidemiologists are still too few as compared to entomologists and virologists. Conversely, more and more is known on the kinetics of the viruses in the plants under controlled conditions and on their distribution in the vectors because the advent of new highly sensitive serological techniques, such as enzyme-linked immunosorbent assay (ELISA) and immunosorbent electron microscopy (ISEM), has increased the means of detection and suggested new ways for assessing the presence of virus-carrying aphids in catches (88).

Once more it is clear that all these requirements have been very unevenly statisfied so far, but it must be emphasized that alate aphid monitoring has also shown the need to fill many gaps in our overall knowledge to help with interpretation and to improve warnings: when possible relationships between climate, alate aphid trap samples, local apterous and winged populations, and virus dissemination were sought, it was evident that in different regions where aphid bioecology and virus status differ, the parameters involved did not have the same weight, and therefore any forecast must be adapted in time and space to local conditions. This means that two forecasts may be closer to each other for two regions of two different countries than for two regions of the same country, but this can be appreciated only if a true integration of our knowledge on an international scale is achieved.

Historical Background to Aphid Trapping

To understand how AVM has met with varying success in time and space in Europe, it is necessary to examine briefly some of the main historical events which have led to the present situation. For more than 40 years now, there have been many stages in the choices and improvements of appropriate techniques brought into operation to investigate the movements of flying aphids to, within, and from the crops where they have long been supposed to spread viruses (28). I will consider two main trends in methodology, one toward impaction traps (sticky and water yellow traps) (128), giving skewed measures (54), the other toward filter traps (suction traps), giving absolute measures.

Impaction Traps: Sticky and Water Yellow Traps

A development in the period between 1948 and 1951 that was crucial for the adoption of new techniques was the finding that aphids were strongly attracted by yellow color. In Britain, Broadbent (12) found that sticky traps (18) coated with primuline yellow caught about 1.5 times as many alates *(M. persicae)* as black-coated ones, and that they were more attractive than the white ones in operation since 1941. These traps, from 1946 onward, consisted of a cylinder 30 cm long and 12.7 cm in diameter, painted yellow from 1948 onward, and around which was wrapped a transparent plastic cover coated with a greasebanding preparation; they were fixed to a post and sited over crops with their base 1 m from the ground. At the same time, in Germany, Moericke (75) clearly demonstrated that the peach potato aphid was able to distinguish between different groups of colors, and for that reason he proposed a new type of trap, which has since been known as the Moericke yellow water trap (76). This trap consisted at first of a circular basin 22 cm in diameter and 6 cm high, with its bottom and inner wall, up to 1 cm from the upper rim, painted yellow; later he recommended that the trap should be square (34 cm × 34 cm) (77). This encouraged people to take up the challenge of virus disease control, especially in potatoes (leaf roll and Y viruses) in sugar beets (yellows viruses), and to a lesser extent in carrots (motley dwarf virus) (142), BYDV being taken into account later, since it was not until 1968 that trapping was initiated in Wales to study virus diseases of Gramineae (1).

It is now interesting to stress how different countries chose different orientations.

In the British Isles, yellow water traps were never used to monitor aphid vectors, with one exception in eastern Scotland, where rectangular trays 45.7 × 30.5 × 5.1 cm with their bottom 68.6 cm from the ground were operated for 3 years (1954–1956) in potato crops (32). In that country, sticky traps were retained as being the more useful tool (28), for several obvious reasons: Broadbent (13) indeed showed a significant correlation to exist between leaf roll spread and the numbers of alate *M. persicae* caught on sticky traps throughout the potato-growing season in ware-growing areas, and he was followed by Hollings (49), who extended the previous conclusions to virus Y both in ware and seed potato areas of England and Wales; moreover, Watson and Healy (141) established a good relationship between the numbers of *M. persicae* trapped on sticky traps at the beginning of the growing season in May–June and yellows incidence; lastly, Heathcote (37) concluded from a comparison of sticky and water yellow traps and of suction traps that cylindrical, "Hansa" yellow-painted sticky traps were the more convenient, although not the most efficient at catching aphids. As a result, sticky traps have been widely used in Britain for decades: for example, traps were operated temporarily in potato crops between 1950 and 1953 in northeast Scotland (111), between 1957 and 1961 at Harpenden (20), and during 6 years (1941–1947) in six different areas of the British Isles (15), and, in addition to some other but more occasional use for methodological purposes (12, 37, 38, 45), one trap was operated every year in potato crops at Rothamsted Experimental Station from 1942 up to 1961 (15, 17, 39), and 12 were in use in sugar-beet crops in eastern England from 1960 to 1973 (39, 40), plus one at Broom's Barn and one at Rothamsted until 1981 and 1980, respectively (G. Heathcote, personal communication). These 38-year-long records of aphid catches [mainly *M. persicae, Aphis fabae* Scop., *Cavariella aegopodii* (Scop.), *Brevicoryne brassicae* (L.)] have yielded invaluable information on flying aphid fauna, time of flight, and seasonal fluctuations in numbers, and have long been used to help predict virus disease incidence.

Conversely, *on the Continent,* sticky traps were not favored; they were replaced by yellow water trays and operated almost exclusively in potato crops for virus disease epidemiological studies and forecasting purposes. In the Netherlands, Hille Ris Lambers (47) decided soon after Moericke's paper that a network of yellow water traps $49.5 \times 32.5 \times 8$ cm with their bottom 60 cm above the soil, painted Hansa yellow, should be set up in Dutch seed potato crops, and starting with six traps in each of 22 fields in 1952, the number of trays was increased to 220 in 1955 at the rate of two per field, reached 250 in the late 1960s (48), and has now decreased to 64 (137). Unfortunately, no data over the period 1951–1969 were ever published, apparently because of trade concerns (47), and consequently all this information is lost to the scientific community.

In French-speaking Switzerland, as early as 1952, six sites were equipped with two circular traps each (25 cm in diameter and 7 cm high) and trapping operated until 1969 in relation to potato seed multiplication. It was then discontinued (25), but started again on a new basis in 1977 after it had become obvious that virus spread in 1976 was mainly due to transient noncolonizing aphids (78). These traps, $60 \times 60 \times 12$ cm, are raised to keep them at the same height as the potato haulms during the crop season (W. Gehriger, personal communication).

In France, yellow water traps were not used much until 1967, when I set up a network of trays in Brittany (western France) as part of an extensive project on potato virus disease epidemiology and aphid ecology (94). In some places, traps are operated almost throughout the year, from March to December ("permanent" traps), and this is supplemented by "temporary" traps operating for about 3 months (from April to the end of July) during the potato crop season (95). Right from the beginning, size $(60 \times 60 \times 10$ cm) and height (70 cm from the ground) as well as color ["buttercup yellow" characterized by a diffusion spectral range between 5000 and 6300 Å with a wavelength of 5800 Å at the maximum of diffusion and an absence of low-wavelength radiation, especially in the blue (19)] have been kept unchanged as far as possible. Most of the data have already been published (94, 97, 101–103, 105). As a consequence of continuous trapping over 17 years, we now have at our disposal probably the longest record of aphid catch data by yellow water traps, and although far from being completely processed, these data constitute the framework for integrated aphid vector warning systems.

Regrettably, and contrary to what might have been expected, no consistent data seem to be available for West Germany, although aphid trapping with 60 × 40 × 8 cm traps is apparently run for warning services to potato and sugar-beet virus control (79). This is not the case in eastern Europe, where yellow traps, 24 cm in diameter, have been widely used since the early 1950s in, for example, different parts of Poland (34, 150, 151) in close relation to epidemiological studies of potato virus diseases, and some collaborative and comparative work between different countries (Poland, East Germany, Czechoslovakia) was also done for some years (35). Furthermore, a forecasting network was set up in East Germany in 1970, which comprises 33 sites of aphid trapping in potato fields out of 86 localities contributing to it (29).

It appears that sticky traps, which were the first to be used extensively and standardized, are probably going to be replaced in Britain by suction traps. This is not the case with yellow water traps, although comparisons between regions and countries are seemingly very difficult, for two reasons: first, the trap size varies, those used most often being either circular and between 22 and 26 cm in diameter, or square with sides between 30 and 60 cm long, or rectangular; second, yellow paints also differ and can be divided into two categories, Hansa yellow (canary yellow or British standard 0-001) (128) and buttercup yellow, the former being used in the Netherlands, the latter in France, Belgium, and Switzerland and possibly in future in Britain. In spite of the many criticisms that have been leveled at this kind of trap, especially on account of their being attractive and giving skewed measures of the aerial aphid fauna, they seem to be suitable for some aspects of virus vector studies, since they are likely to catch those aphids in their landing phase and they can be a good complementary tool to suction traps.

Filter Traps: Suction Traps

From 1947, Johnson and Eastop (59) used a trap consisting ''of a powerful electric fan, 18 inches in diameter, sucking air through a muslin cone suspended in the top of a vertical duct, which was 5 ft above ground level'' to study aerial aphid density as part of a program on aphid movements. This suction trap, which sucked in air at a rate of 3285 m^3 hr^{-1}, was operated continuously from July to November and allowed for the first time the seasonability of different species to be seen. The design of the trap was then much improved (55, 116) and a full range of traps of different fan sizes (60) for different purposes were later made available (128). Until the early 1970s, 9-in. (~23-cm) traps were most often used, alone or combined with larger ones, to study specifically the aerial density of insects in general at different levels above the ground (119) as well as their patterns of distribution near artificial (69) or natural windbreaks (70, 71). Some work was done on aphids (57, 68, 106), sometimes as part of methodological experiments on trapping efficiency (1, 37, 56), but seldom, if ever, toward epidemiological studies of aphid vectors (36): for this purpose, different designs of suction traps have more recently been used to allow live aphids to be caught in order to assess the proportion of those infective (4, 66, 81, 87); these aphids are usually trapped at ground level or within 1.5 m above the soil, i.e., at crop level.

Very different in its initial purpose was the development in 1963 of a new type of suction trap, which began to sample aphids continuously in 1964 at Rothamsted at a height of 12.20 m (33): "The 40-ft trap was designed to monitor current aerial populations of small insects, such as aphids, and their movements, to provide information needed to warn of changes that might affect crops, and to show the pattern of movement into and about the country. The height of 40 ft was chosen to sample the generally dispersing populations rather than the local ones." So wrote Taylor and French (126) about this new tool, which was intended to be used together with light traps "in a national census of flying insects" as part of the Rothamsted Insect Survey (RIS) with the aim of providing "continuous information on abundance and distribution of pests and other insects in different regions throughout the year and year after year" and with the hope "that the census will eventually lead to warning system forecasting the likelihood of pest outbreaks." Starting with one trap in 1964, the number was increased to seven in 1967, 11 in Britain plus one in the Netherlands by the end of 1969, 16 by the end of 1970, and 18 in 1973; in 1977, 20 such traps were operated in Britain, one in Denmark, two in the Netherlands, and one in France. Their number has continued to increase (119, 120, 130). In 1978, after it had been shown that aphid control in France would probably benefit from such a scheme (96), some funds became available from the Délégation Générale à la Recherche Scientifique et Technique (DGRST) to set up a network of traps: this was done under the aegis of the Association de Coordination Technique Agricole (ACTA) as part of a closer cooperation in agriculture between research (Institut National de la Recherche Agronomique, INRA) plant protection (Service de la Protection des Végétaux, SPV), and extension per crop (Instituts Techniques) organizations. From 1978, seven traps were operated and the network took the name of ACTAPHID. It was extended to Belgium in 1980 and 1983. By the end of 1983, 24 traps were in service in the British Isles, two in Sweden, three in the Netherlands, two in Belgium, ten in France, one in Switzerland, two in northern Italy, and one in Sicily (Fig. 4.1). The Commission of the European Communities (CEC) supported some aspects of this venture and at a meeting held at Rothamsted

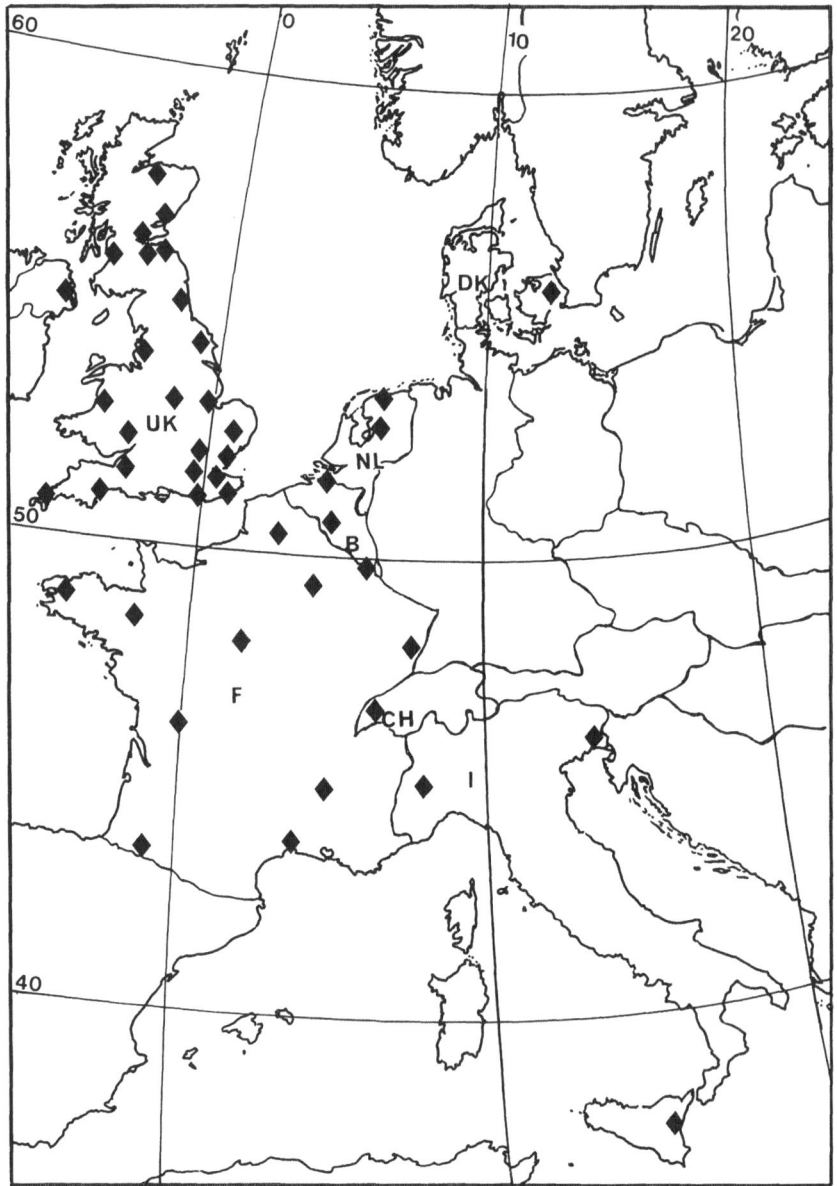

FIGURE 4.1. The European EURAPHID network of 12.2-m suction trap sites.

at the end of 1980, an informal group of people and organizations met to exchange views and information and achieve a better integration of the different existing systems on a European scale. This was proposed and called EURAPHID (123). This collaborative organization implies free exchange of data for scientific purposes and, at more or less long term, one hopes, may help international forecasting systems to be set up.

An *Aphid Bulletin* giving the weekly numbers of 31 or 32 species or species groups has been issued weekly from early spring until late autumn by the RIS since 1968 for aphids caught in Britain and by ACTAPHID since 1978 for those caught in France and in Belgium. Twenty-seven species are shared by both bulletins. Each weekly report is sent by the RIS and ACTAPHID to about 150 people or organizations, but it was shown that the information is finally passed to an average of 30 persons per recipient, at least for the *RIS Bulletin* (5). Any interpretation of these raw data has been considered so far as having to be made under the responsability of the *Bulletin* recipients themselves, but since 1980, the RIS has undertaken some more detailed interpretation released as an *Aphid Commentary* issued with the *Bulletin*. This *Commentary* was restricted, at the outset, to five species, *Aphis fabae, Metopolophium dirhodum, Phorodon humuli* (Schrk), *Rhopalosiphum padi,* and *Sitobion avenae* (Fab.), then extended in 1981 to potato aphids, *Aulacorthum solani, Macrosiphum euphorbiae* (Thomas), and *Myzus persicae* (114), and in 1982 to *Rhopalosiphum insertum* (Wlk). The interpretation involved refers mainly to a comparison of the current average sample for each of nine given regions of Britain with the corresponding sample in space and time of the previous year and with the long-term average sample: this, of course, has been made possible because many of the suction traps set up in that country have been continuously run for more than 10 years, enabling the RIS to make calculations and comparisons from a significant range of time (130). Moreover, attention was focused from October 1982 onward to the problem of BYDV being spread to winter cereals by *R. padi:* the *Commentary* gives the different values of the Infectivity Index elaborated by Plumb *et al.* (85) for four sites of the RIS trapping network. This new topic notably enlarges the range of possibilities offered by suction trapping from the initial purpose, which was to use suction traps to sample aerial populations (119).

Achievements

As mentioned previously, until recently, breeders have taken too little account of phytopathological problems when breeding for new cultivars, since they were instead concerned with improving yields and looking for better response and adaptability of varieties to local agrometeorological constraints and new farming methods (shorter rotations, increasing use of herbicides, minimum cultivation techniques, etc.). New problems were often created when new crops and different sowing dates became available. One of the best examples refers to autumn-sown cereals being drilled earlier and earlier (November versus October, even September), although it is well known from aphid monitoring that the earlier they are sown, the more they may become at risk from BYDV, since this period is the very one when the return (or remigration) flight of *R. padi,* one of the most efficient vectors, takes place, and the farmer is not liable to return to later sowing dates unless he realizes that his crop is becoming nothing more than a straw mat and that relying on insecticides as a last resort is not a panacea.

In this respect aphid monitoring has already done much to face such problems or to be ready to deal with new ones: seasonal flight activity in a given area as a species characteristic, adaptive alterations in flight phenology according to latitude and longitude, and the effect of some main components of the weather upon (1) the timing and magnitude of the different flights and (2) the aphids' ability to overwinter anholocyclically have been defined and knowledge of these bioecological features is consistently improving.

I will first give various examples of this and exemplify the role of AVM with three examples of virus disease forecasting, in sugar-beet, seed potato, and winter cereal crops.

Better Knowledge of Aphid Bioecology

SEASONAL FLIGHT ACTIVITY

Species Characteristics

Every species of aphid shows a cyclical production of alatae to achieve its biological cycle and although it is known from laboratory experience that different biotypes of a single species may differ in this respect, it appears that in a given region, a species can be defined by its overall seasonal flight periodicity, in which two components must be considered as important: first, the number of cycles of flight activity, and second, the time when this activity takes place. This can be established only from a long period of trapping with, if possible, different kinds of traps set at different heights (1, 45, 101, 119). Figure 4.2 gives some relevant examples of mean flight activity for some species caught in large yellow trays over 17 years in Rennes: in this area, all species except perhaps *P. humuli* are largely anholocyclic and, as will be shown later, the effect of winter and early spring weather, together with the fact that aphid generations overlap considerably, makes it rather difficult to separate the different cycles of flight, as it smooths the curves of catches. However, for example, it is obvious that *C. aegopodii* displays a single maximum peak of activity in the second half of May (Lewis and Taylor standard weeks 20–22) (72) and a very restricted one in October–November, whereas *P. humuli* flies mainly in June–July (weeks 24–27), but not in autumn: this latter case can be called "monomodal" (101) and has been shown to occur typically only in areas where cultivated hops do not exist (129). Conversely, *M. persicae* has three distinct cycles of flight activity: the first, around mid-May (week 20), is to be assimilated with the contamination flight (emigration flight for holocyclic individuals) (96), which makes the aphids leave their over-

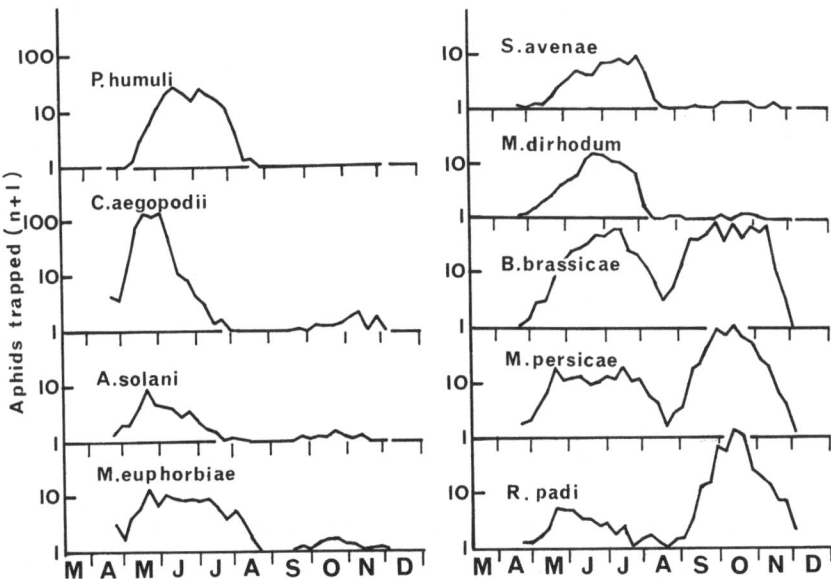

FIGURE 4.2. Seasonal mean flight activity of some aphid species (mean of weekly yellow trap catches in Rennes for 17 years).

wintering hosts to colonize crops; the second around mid-July (week 28), corresponds to the dissemination flight; and the third is the return flight (remigration flight for holocyclic individuals). Similar examples can be given for other species.

Any departure from this average both in time and in numbers gives a first indication as to whether given crops may be at risk or not, if enough well-documented instances have been recorded and interpreted in the past. This is illustrated in Figure 4.3, where weekly numbers of *M. persicae*

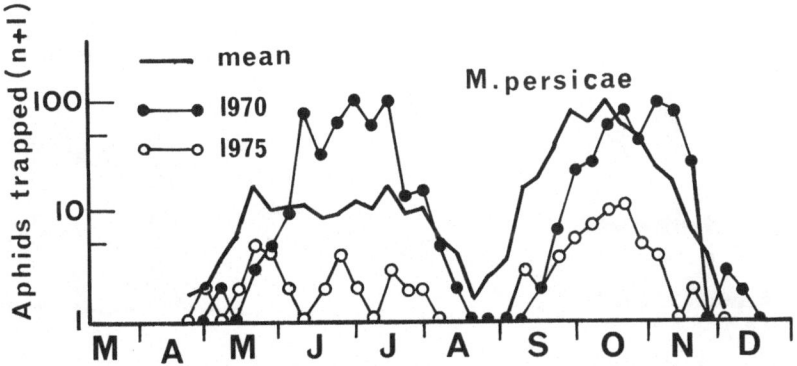

FIGURE 4.3. Weekly numbers of alate *Myzus persicae* caught in yellow traps in Rennes in 1970 and 1975 as compared with mean numbers for 17 years.

caught in the years 1970 and 1975 are compared with the long-term average: in 1970 ten times as many alatae were trapped weekly in June, resulting in large amounts of PLRV being spread in the potato areas further west in Brittany, whereas in 1975, numbers among the lowest ever recorded resulted in no problem to any one crop. This kind of reasoning is used to elaborate weekly interpretations for the RIS *Aphid Commentary* (see above).

Adaptive Alterations and Latitude–Longitude Effect

For aphids as for other insects, the warmer the climate, the shorter the mean generation time, within certain limits of temperature. As a consequence of the "effet de groupe" for aggregative species and/or of crucial physiophenological stages of the plants, the alatae production is liable to differ according to different meteorologically characterized areas in relation to aphid specific biological cycles: this may result in changes (1) either in the number of flight cycles or in the time between cycles, and (2) in the timing of different flights during the period of the year where temperatures exceed thresholds for flight. So, the latitude–longitude effect can probably be best delineated as the mean annual flight periodicity of

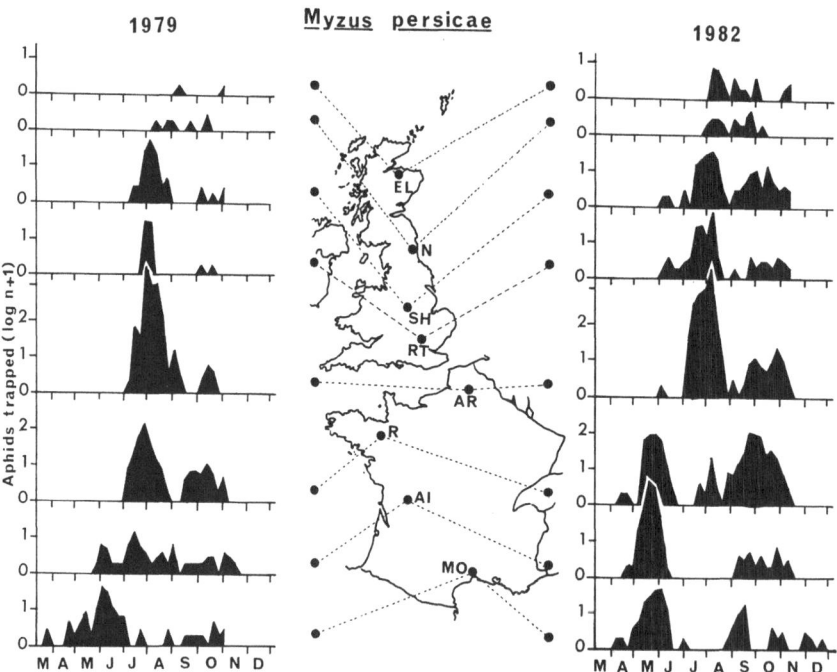

FIGURE 4.4. Periodicity and size of *Myzus persicae* flights according to latitude and year (weekly 12.2-m suction trap catches). MO, Montpellier; AI, Aigre; R, Rennes; AR, Arras; RT, Rothamsted; SH, Shardlow; N, Newcastle; EL, Elgin.

a given species of aphid in a given area specified by its latitude, longitude, and ensuing relevant meteorological variates. When this varying flight phenology is known, the timing and period when crops are grown can be related to it to assess maximum potential risk of infestation by alatae and infection by virus.

Figures 4.4 and 4.5 give two examples of such changes in timing, number, and amplitude of flights as expressed by the number of individuals of *M. persicae* and *R. padi* caught by EURAPHID suction traps from the south of France (Montpellier) to the north of Scotland (Elgin)—a range of about 15 deg. of latitude—for two different years: with the exception

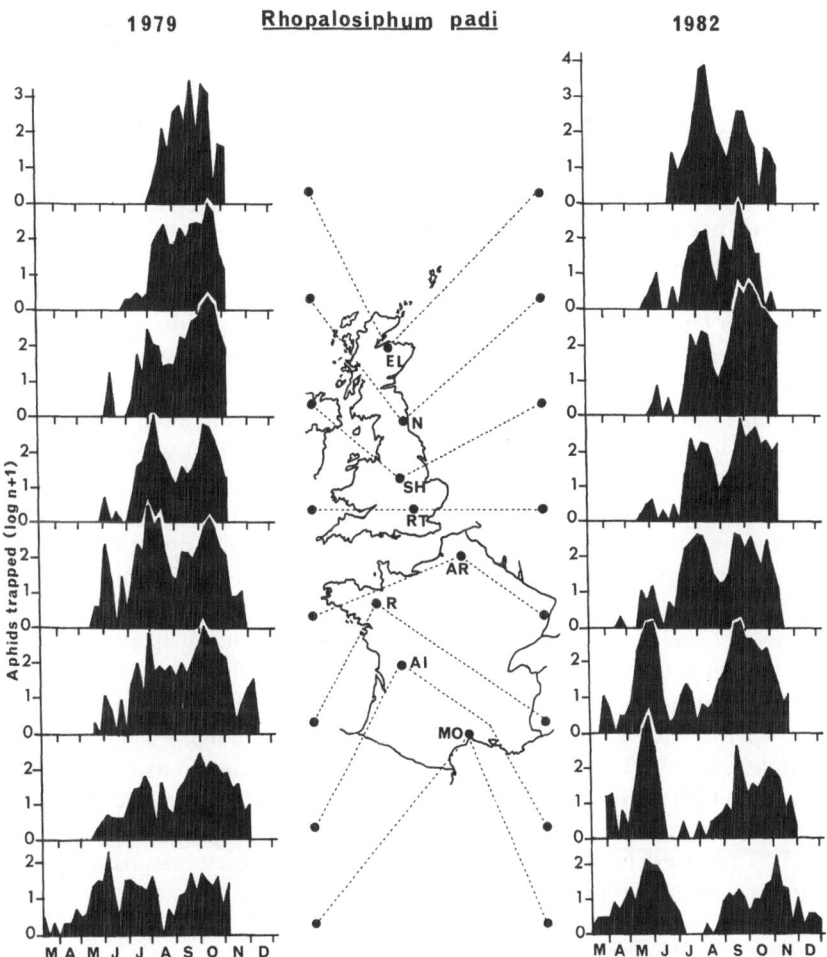

FIGURE 4.5. Periodicity and size of *Rhopalosiphum padi* flights according to latitude and year (weekly 12.2-m suction trap catches). MO, Montpellier; AI, Aigre; R, Rennes; AR, Arras; RT, Rothamsted; SH, Shardlow; N, Newcastle; EL, Elgin.

of the south of France, the 1978–1979 winter was largely colder than the long-term average, in contrast to the rather mild 1981–1982 one. Without giving details, it can be seen that the overall timing of the first flight may largely differ both between years and sites, being obviously earlier in 1982 than in 1979 and in the south than in the north. This has serious implications for virus disease epidemiology. For instance, most seed potatoes usually emerge in May, whatever the latitude, which means that a coincidence is generally to be observed between crop emergence and first contamination flight, say, in Brittany (103), but not in Scotland, where, on average, *M. persicae* colonizes crops relatively late in their development (133): the marked trend observed in Scotland to earlier migration of *M. persicae* toward mid-May from 1971 to 1976 (113, 134) was sufficient to make PLRV spread intensively in the whole area during this period. Once more, any departure from the average seems to be of paramount importance for any warning to be produced.

Since 1970, the RIS has used a mapping procedure to "display, analyse and interpret" the spatial nature of the data provided by the Survey (152): at first restricted to the British data, the procedure was extended to mainland data in 1978 as soon as ACTAPHID suction traps began to be operated and data became available. In nearly 15 years, a considerable number of maps have been computed and published (e.g., 27, 118, 120, 122, 125, 129, 131), using the SYMAP V program developed by the Laboratory for Computer Graphics, Harvard. Because aphid trapping is done daily, maps have usually been produced with a daily, weekly, monthly, or yearly frequency, depending on the objective. Figure 4.6 shows comprehensively how the latitude–longitude effect operates on flight phenology all through the year in the case of the black bean aphid *A. fabae* in western Europe and how the dynamic seasonal distribution occurs, resulting in a rather complex but consistent "wave" of aphids extending progressively from the southeast of France to the north of Britain until August (weeks 31–33) with a significant continuity from both sides of the Channel and the south of the North Sea, emphasizing the great potential of EURAPHID for short-term forecasts if a true European integration was ever possible in the future for aphid warning systems. Such a continuity can also be seen in Dewar *et al.* (27) in connection with the 1979 "green invasion" by *M. dirhodum*. It can be easily anticipated that such maps using programs such as SYMAP V or SURFACE II developed by the Kansas Geological Survey and based on the theory of regionalized variables may be useful not only for research objectives (152), but also as an elegant pedagogical tool to be used in television broadcasts for advisory purposes: such a study is already in progress by the RIS (114).

Effect of Weather

Aphid Overwintering

In temperate countries, one of the most important features of the aphid life cycle is the way they overwinter. Most species of agricultural signif-

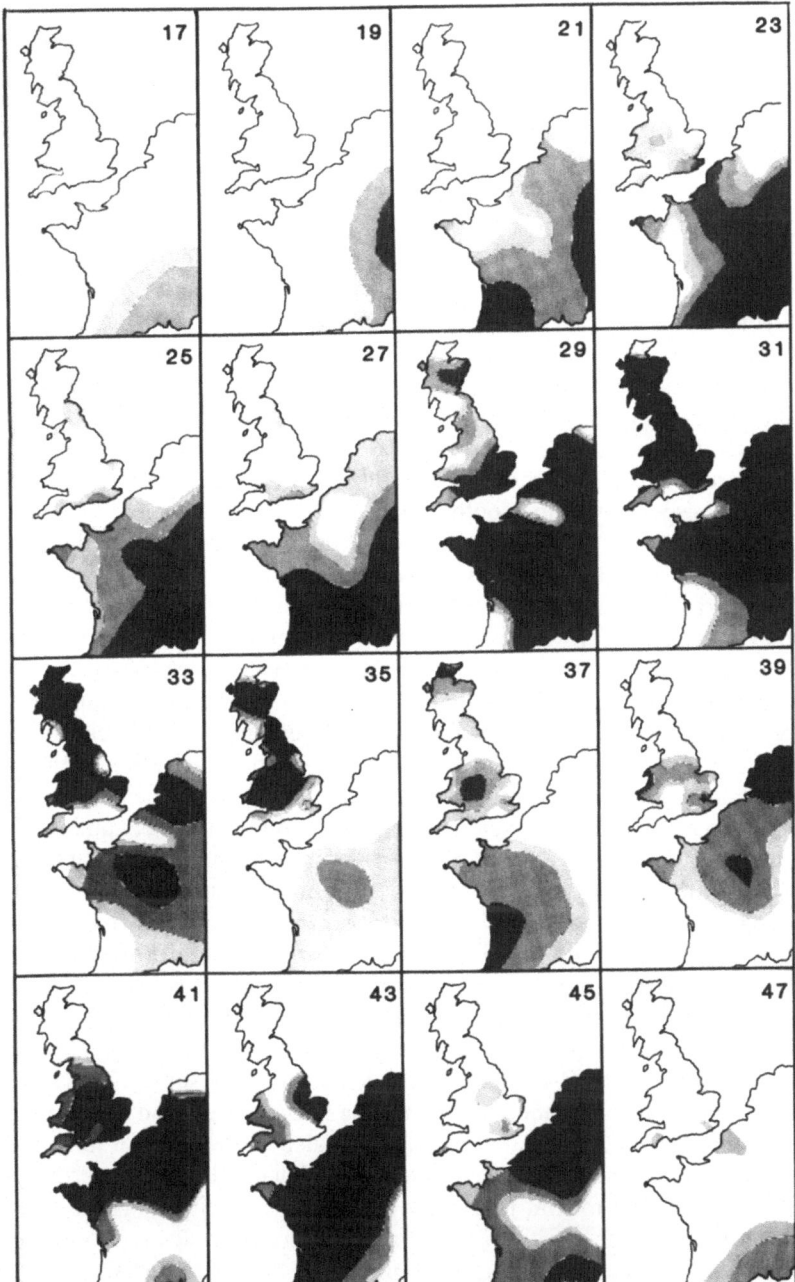

FIGURE 4.6. Dynamic seasonal migration system of *Aphis fabae* in western Europe in 1978 (alternating weeks 17–47) (after Taylor & French, 1981).

icance are heteroecious and holocyclic and/or anholocyclic. In the former case, overwintering takes place in the egg stage on primary hosts; in the latter case, overwintering is nothing less than continuing agamic viviparous reproduction, which occurs throughout the year. These two ways may occur in the same region for such important species as *A. fabae, M. dirhodum, M. persicae,* and *R. padi,* but in western Europe, the balance between the two can be different according to latitude and longitude, e.g., meteorological possibilities for survival outside, availability of primary hosts, possibilities for survival indoors, etc. Aphid monitoring has again greatly contributed to a better knowledge of local situations, often in an indirect way, especially when understanding was required, but, as was pointed out by Bouchery (11), every interpretation of aerial samples for forecasting should be done separately according to the holo- or anholocyclic status of a given aphid: this has not always been feasible or easy! However, it is possible to illustrate this with two examples.

Aphis fabae has been shown to be entirely holocyclic in Britain and to overwinter especially in the south, where its primary host, spindle, is found on calcareous soils (146). The analysis of suction trap catches and flight phenology shows that contrary to what is observed in Scotland, where only two rather late cycles are displayed, in southern England, three periods occur, of which the first is clearly interpreted as emigration from spindle to beans and sugar beet (120). Moreover, Taylor (120) has shown a significant correlation in the south between the size of the autumn re-migration and that of the succeeding spring emigration: this can be explained by the fact that in that part of England, overwintering in the egg stage is successful and mortality between autumn and spring, if any, is usually constant from one year to the other, both to the eggs and to subsequent apterous generations before emigrants appear. Y. Bouchery (unpublished) has found similar trends in Alsace (eastern France), where *A. fabae* is also holocyclic, as has Wiktelius (149) in Sweden for *R. padi* overwintering on *P. padus.* This good relationship between autumn and subsequent spring migration as monitored by suction traps in southern England is now largely utilized as an aid to an early forecast of the likelihood of the previously estimated economic threshold for chemical control on spring field beans (145) being exceeded or not next spring. This is then completed by an assessment of the timing of the emigration cycle related to meteorological conditions for flight and also of the level of infestation reached in the crop at the onset of population buildup (147). This procedure is complementary to that originally chosen of sampling eggs on spindle during early winter, assessing the time they hatch, and monitoring subsequent fundatrigenic populations building up before emigrants leave, all of which require extensive host sampling of both spindle and beans. It is also much less time-consuming and relies on standardized data. This success in the development of an efficient forecasting system is the outcome of extensive collaborative work between research and advisory services and it has been favored by the strict holocycly of *A. fabae* overwintering in a rather limited area on just one primary host and being a direct damaging pest on its own right. So far, it does not apply to sugar-beet infestation by this species as a virus vector and it remains a model.

It may be rather surprising to learn that *M. persicae*, which is considered to be anholocyclic all over Great Britain (16), also has its most successful overwintering area in the south of that country, but, of course, for quite different reasons from *A. fabae*. This has been shown from two indirect demonstrations: the first nearly ignored the aphid as it was deduced from multiple regressions between certain winter and spring weather data specifically recorded at Rothamsted and sugar-beet yellows infection in the whole eastern English sugar-beet area; the second came from the mapping of seasonal changes in the spatial pattern of *M. persicae*. It has long been known that in the sugar-beet area, the more "severe" the winter, the less the prevalence of *M. persicae* and yellows in the crops (53, 140). Progressively the meteorological variates that relate to the ensuing percentage of yellowing viruses by the end of August were determined, and in 1975, Watson *et al.* (144) published an extensive paper on the subject, based on a 21-year-long series of data and showing that (1) this percentage depends largely on the accumulated number of frost days (below $-0.3°C$) in January–March and on the mean temperature in April, (2) regression equations calculated separately for each of 17 geographically separate factory areas always gave a better fit when based on Rothamsted meteorological data than on those of local weather stations, and (3) mean yellows incidence for the different areas declined from south to north, as did the significance of frost days, while at the same time the April temperatures became more important.

They suggested that few aphids overwintered in eastern and northern beet areas, being killed by cold winters, and that infestations were initiated south of the river Thames and perhaps—from the Continent, Rothamsted lying in such a situation that it would represent weather conditions for aphid survival and monitor migrations northward. This hypothesis lacked supporting evidence until Taylor (121) demonstrated a slight increase in *M. persicae* populations in the extreme southeastern part of England during the winter: this was done by analyzing difference maps between autumn, when *M. persicae* has a random aerial distribution over Britain, and spring, when it shows a patchy distribution with more aphids in the south. These maps all showed a maximum density in an area south of Rothamsted. If the observed spring migration is assumed to be the single consequence of a local production of emigrants, it can clearly be inferred from these findings that maximum survival occurs in that part of England. From there, as had been anticipated by Watson *et al.* (144), migrant aphids could progressively fly northward every year to colonize crops and spread viruses. Taylor's analysis may support this evidence. A causative explanation was sought by this author, who found that this "pocket" with high survival was characterized by low winter rainfall (<400 mm) combined with moderate temperatures (>7.5°C). Conversely, the aphid limitations in the west would be mainly due to rainfall and in the north to temperature. These conclusions differ from those of previous authors, who suggested that mild climate would maximize survival of overwintering populations of *M. persicae*. Whatever the real reasons limiting this survival and which differ from what apparently takes place in Brittany, where parasites are involved (94, 99), this research has shown the potential displayed by the mapping system. More recently, Taylor (124), following the same procedure, scanned some cereal aphid overwintering main areas in Britain, showing large differences between *Sitobion avenae* and *M. dirhodum,* which display the same trend as *M. persicae* to overwinter in the south, and *R. insertum* and *R. padi,* which nearly "disappear" each winter.

These differences between species stress once more the need for a better knowledge of aphid bioecology, and this mapping technique has the great merit of giving some pointers for research direction to improve our interpretation. We really need, in the future, European maps of holo/anholocycly distribution, of relevant weather variates for different aphid species, and of corresponding virus distribution as a complement to aerial sampling maps, at least until we understand the actual causes of changes in aerial distribution.

Timing of First Flight

This timing is of paramount importance when alate aphids colonize young susceptible crops, since this is likely to favor population buildup. But it is important also for virus spread in regions where anholocyclic emigrants carrying non-seed-borne viruses from external winter reservoirs colonize young crops, which are most susceptible to virus infection, and so initiate primary foci of infection, and/or when anholo/holocyclic alatae,flying from one plant to another spread seed-borne viruses within the crop. The role of primary infection has sometimes been underestimated, but now it is well recognized as being important in determining the amount of virus at harvesting time. For most virus vectors overwintering anholocyclically, weather conditions during winter and early spring are likely to affect first flight timing by allowing winter aphid populations to maintain themselves or build up, and produce alatae that may fly as soon as their takeoff threshold is exceeded: so far, this has been expressed in terms of daily temperatures accumulated over periods that differ according to local conditions of latitude and longitude.

The respective roles of both components (accumulated mean daily temperatures and threshold for flight) are well shown in Figures 4.4, 4.5, and 4.7. As already described, in 1979 the winter was much colder than in 1982. For example, the sum of the mean daily temperatures over the three months of January–March in Rennes was 410°C in 1979 and 620°C in 1982. In these conditions, anholocyclic populations are not likely to produce winged individuals at the same time on winter hosts and indeed this is what was observed under natural conditions (Figs. 4.4 and 4.5). But it is not sufficient for individual alatae to be ready to take off, for under our

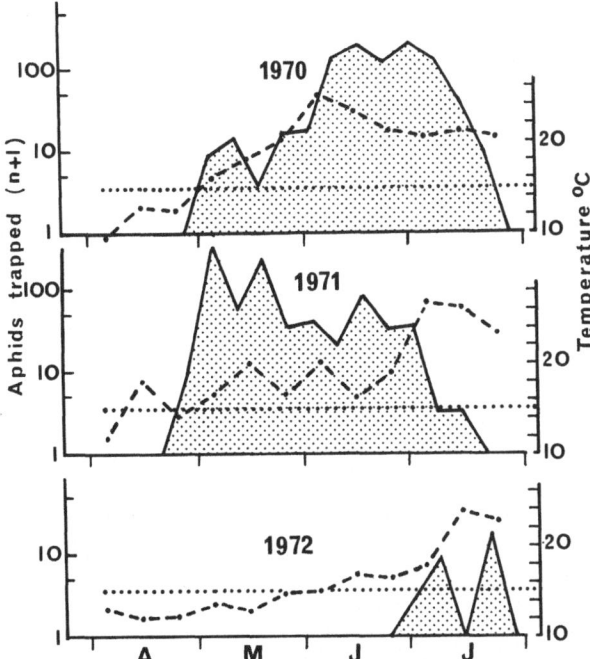

FIGURE 4.7. Role of the takeoff threshold temperature (15°C) early in the season as a limiting factor to *Myzus persicae* flight (mean maximum temperatures per 10 days; weekly yellow trap catches at Elliant in 1970–1972).

temperate climates, "unusual" weather in April and May can prevent aphids from flying away: in 1972, the winter was rather mild in western Brittany (579°C accumulated from January to March inclusive), but neither *M. persicae* nor many other aphids could take off for a while, partly because their takeoff threshold was never exceeded (Fig. 4.7), i.e., roughly 15°C, as previously determined for potato aphids from 60 × 60 cm trap catch data (103, 104); in contrast, in 1970 and 1971 (491 and 466°C accumulated, respectively), right from the beginning of May (1970) or mid-April (1971), maximum temperatures were above 15°C on average, resulting in an early flight.

Similar results were recorded in southeast Scotland (113, 134, 135): significant correlations were found between winter temperatures (expressed as daily sums above and below a threshold of 5°C and used as the difference between them or expressed as hours spent below temperature thresholds ranging from 5 to 0°C to improve the regression coefficient according to aphid species) and first catches of each of nine species of aphids [*M. persicae, M. euphorbiae, A. solani, S. avenae, Sitobion fragariae* (Wlk), *Metopolophium festucae* (Theob.), *R. insertum, M. dirhodum, R. padi*], with different periods showing greater significance according to species and suction trap site involved: the months most frequently shown to be significant were February for the first seven species and December for the last two, which coincided with a different mode of overwintering (anholocycly for the first group, holocycly for the second). Conversely, Wiktelius (149) could not find any significant correlation between the start of migration of *R. padi, M. dirhodum,* or *S. avenae* and temperature during winter, and, like most authors, he ascribed this lack of link to holocycly in Sweden, arguing that the ambient temperature would not influence an egg in diapause. However, Thomas *et al.* (132), looking for such correlations between weather factors and the spring migration of a purely holocyclic aphid *P. humuli,* which gives rise to a varying number of apterous virginoparous generations (six to eight) after egg hatching, actually found associations between the date of first catch of alate emigrants and maximum temperature in late March and early April together with periods of rainfall in mid-January and mid-April. A good knowledge of the aphid's biology enabled them to ascertain that temperatures in late March and early April were critical, as they could accelerate or retard the date of appearance of the first alatae in the second fundatrigenic generation in the latter half of May or early June, since they have an effect on the mean generation time.

These examples taken from Brittany, Scotland, Sweden, and southern England may indicate the formidable task that awaits those who want to make the most of any AVM intended for forecasting purposes, but they also show that such prospects are not beyond the scope of possibility as our knowledge of aphid biology improves and computing facilities increase.

Magnitude of Flight

We know from experience that in addition to the timing of first flight, the larger the number of aphid vectors, the more likely crops will be at risk from virus spread if other epidemiological requirements are satisfied: the greatest danger will generally come from any one synchronization between the appearance of an early flight and its magnitude to initiate foci of primary infection. This is one of the reasons why correlations between meteorological variates and subsequent numbers of aphids trapped by a given system may help forecast potential dangers more accurately. Looking for such correlations for 9 years in potato areas of western Brittany, we found (103) a significant correlation between the number of alate *A. solani* and *M. persicae* caught in 60 × 60 cm yellow traps and the sum of daily mean temperatures accumulated between 10 and 30 April. By the end of April, some warning and advice can be released to potato growers regarding the likelihood of their crop becoming heavily infested, because I have shown (94) a significant correlation between the number of alatae of *A. solani, M. euphorbiae,* and *M. persicae* caught between 20 and 30 May in 60 × 60 cm yellow traps, 70 cm from the ground, and the number of nymphs of each species laid by the end of May on potato leaves.

In 1975, outbreaks of *S. avenae* populations occurred on wheat in many parts of France, resulting in large yield losses. In Brittany this happened only in 1976. In 1974–1976, comparatively high numbers of alatae of *R. padi* were trapped in Brittany in yellow traps, resulting in potato virus Y spread (92), and we tried to understand the causes. The maize crop area has increased considerably in the 1970s in that region, allowing *R. padi* to maintain rather large populations until late autumn and so increasing its opportunities for overwintering on autumn-sown cereals and spreading BYDV. However, it was found that again, winter "mildness" was apparently largely responsible for the large numbers caught in spring. Significant correlations were found in Rennes and farther to the west both with *R. padi* (Figs. 4.8a, 4.8c) and *S. avenae* (Fig. 4.8b) between numbers of aphids caught in yellow traps from April to July inclusive and the sum of daily mean temperatures accumulated in January–March. This overall relation has been found to hold from 1967 to 1983 (17 years): of course, it is still a somewhat rough guide to the possible reasons for explaining fluctuations in numbers caught in spring, but it is open to improvement.

Probably more unusual and unexpected is the highly significant relation

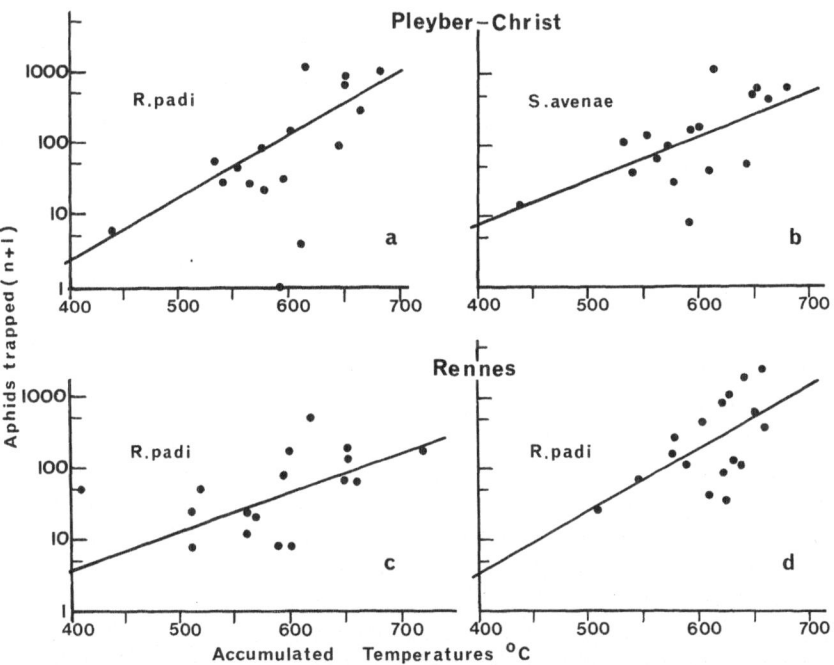

FIGURE 4.8. Relationships between numbers of alate *Rhopalosiphum padi* caught from April to July inclusive in (a) Pleyber-Christ and (c) Rennes, and (b) those of *Sitobion avenae* in Pleyber-Christ, and the sum of daily mean temperatures accumulated in January–March; (d) relationship between numbers of alate *R. padi* caught from 1 September to 15 November in Rennes and the sum of daily maximum temperatures accumulated in September (yellow trap catches over the years 1967–1983).

we found (Fig. 4.8d) between the sum of daily maximum temperatures accumulated in the single month of September and the number of *R. padi* caught in yellow trays from September to November! (94, 100, 102). Here again explanations are being sought, but it appears that climatic factors in September are largely responsible for allowing enough alatae to be produced for the subsequent return flight and we are attempting to improve our understanding of that. Anyhow, this prediction by the end of September of what may happen during the next month may be of value in epidemiological studies of BYDV. So far, we have not found such relationships with suction trap catches in spring or in autumn, although some similar trends can be shown in Rennes: the reason is probably that either we have been trapping for too short a time (only 6 years, since 1978) or the climatic variates involved are not the same. However, A'Brook (3) in western Wales, in an impressive work based on multiple linear regression analysis on suction trap data, found similar conclusions to ours, e.g., associations of number of *S. avenae* caught in June with high January–February temperatures, and he suggested relationships between rainfall in July and soil moisture in August in relation to grass growth and the number of *R. padi* caught in October.

It may appear that somewhat disparate analyses have been attempted so far without actual preliminary consultations, everyone using his or her own biological and meteorological local data, but the fact that many similar trends appear to result in different regions and countries has prompted those who work in this area to share their experience in the EURAPHID venture to progress toward a better integration on a European scale. This brings us back to one of the problems which might have provided a barrier and which I have stressed above in the section on historical background, i.e., differences in interpretation from yellow and suction trap data.

SUCTION TRAPS OR YELLOW TRAPS?

My intention is not to make a decision for or against one system or the other, nor to compare the kind of information they each give. This is already documented (96, 128). However, I will give some as yet unpublished results from AVM. One of the crucial problems already mentioned is to try and get the least dense network of trapping by whatever system that can still give efficient information for forecasting purposes. Because (1) we have been trapping aphids in Brittany for 17 years now with the aid of large yellow water trays, which gave us satisfactory data for our immediate purposes, (2) it is actually not possible to increase the number of ACTAPHID suction traps, and (3) it is basically always desirable to release forecasts on the smallest possible geographical scale, we have compared, since 1978, suction and yellow trap catches from different sites in Brittany for all species trapped. One can see in Figure 4.9 the kinds of differences and similarities shown for two years by *M. persicae* and *R. padi*. Generally speaking, it confirms our original hypothesis that yellow

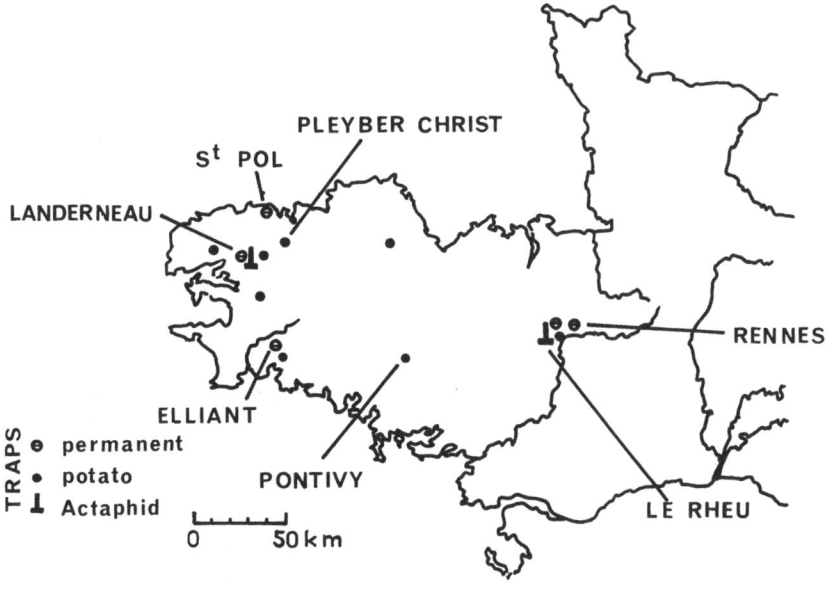

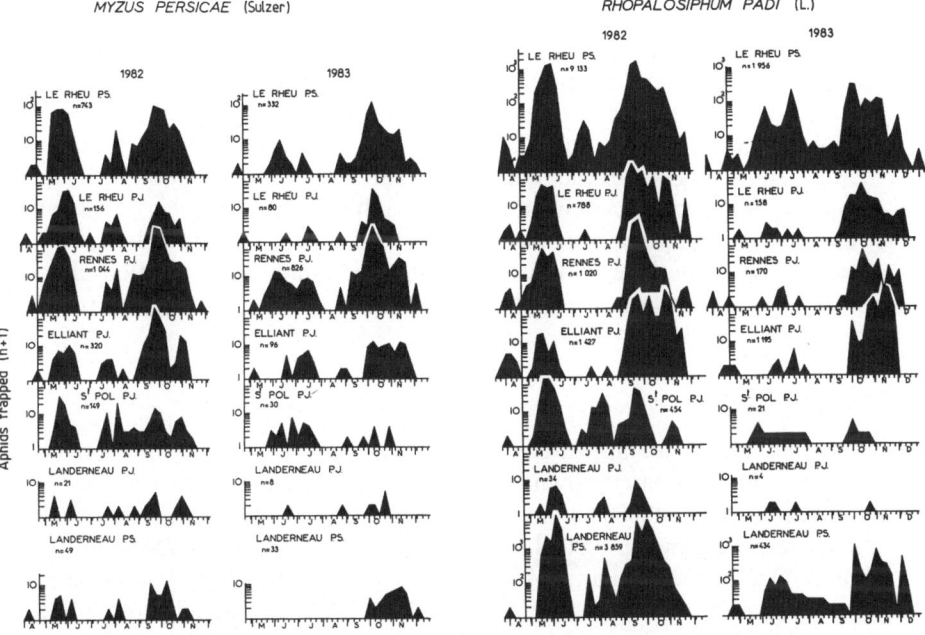

FIGURE 4.9. Network of yellow trap (permanent and potato) and 12.2-m suction trap sites in Brittany and comparison of weekly catches of *Myzus persicae* and *Rhopalosiphum padi* by both systems in 1982–1983 (PS, suction trap; PJ, yellow trap).

and suction trapping can be complementary systems in a given region, provided the yellow traps are large enough: under given conditions of size of trap, height of trapping, and choice of yellow color, water traps can match suction traps in locales where the latter cannot be operated. It may be possible to calculate transfer relationships between the two systems according to aphid species, environment, etc. Work is now in progress, but it needs to be checked in other geographical areas where aphid biological cycles and meteorological conditions differ: this has already been undertaken in the sugar-beet area of northern France in relation to beet yellows spread (around the Arras suction trap), in the south in relation to nonpersistent virus spread, such as virus Y and CMV (near the Montpellier suction trap), and in a slightly different way in French-speaking Switzerland; there are some similar prospects now for southern Britain. These possible good relationships of yellow catch data with suction catch data may be very useful in improving mapping and the efficacy of virus spread warnings in remote regions or regions with particular features (microclimate, altitude, localized aphid and virus reservoirs, etc.), as this may supplement, on a more limited scale, the information given on a regional scale by suction traps.

Some Achievements in Progress

In this section, I will only describe three contrasting examples of potential and actual AVM for improving forecasts. The first example deals with sugar beets; it refers to attempts to correlate numbers of aphids and virus incidence and is based on what is probably the longest record of aphid catches and yellows levels. The second one deals with seed potatoes and not only takes into account numbers of aphids, but also their estimated relative efficiency as virus vectors. The third one is intended to show another way to utilize both numbers of aphids and the proportion of infective ones to assess the risk of a crop to become infected, and refers to winter cereals.

Sugar Beets

In western Europe, large areas are occupied by sugar-beet crops: roughly 500,000–600,000 ha is grown in France, mainly in the northern part and the Paris basin, 400,000 ha in Germany, 200,000 ha in Great Britain, mainly in eastern England, 130,000 ha in the Netherlands, and 110,000 ha in Belgium. One of the regular problems facing growers has been yellowing viruses, namely the two non-seed-borne ones, BYV, semipersistently transmitted by *M. persicae,* and to a lesser degree by *A. fabae,* and BMYV, persistently transmitted by *M. persicae*. Both viruses are usually present in the crops, but their relative importance can fluctuate: BMYV used to predominate in Britain in the mid-1960s, but in the early 1970s, severe strains of BYV also spread widely (42). It is difficult to make an accurate assessment of yield losses due to each virus: for the whole of France, overall crop yield losses have been estimated at about 2.5 tons ha^{-1} per annuum on average over the last decade (21), rising to 4–5 tons in the north of the Paris basin. The earlier the infection, the greater the loss, and in Britain, Hull and Heathcote (52) showed that more than 20% infected plants at the end of August resulted in a significant decrease in crop yield, which was evaluated at 5% (44), a situation which, however, has been apparently rather rare since 1946: Heathcote (42) has shown that "during the past 30 years, a crop loss of more than 10 per cent has been caused in only three years and of more than 5 per cent in only six years," 1949, 1957, 1974 and 1952, 1961, 1975 (Fig. 4.10). It must not be forgotten that this assessment is based on national averages, which may conceal large heterogeneities between regions: for instance, from 1970 to 1975, 0.3–85.7%, with a mean of 30%, of the sugar-beet area had more than 20% of plants showing symptoms.

For these reasons, measures to control aphids, and possibly viruses, were introduced 30 years ago. From 1959, English growers were advised to spray with systemic insecticides (51) and from 1975 to use granular pesticides (44), and this may not be irrelevant to the widespread field cross-resistance of *M. persicae* to any insecticide now available (109). In Britain, a spray warning scheme for controlling sugar-beet yellows was devised by Hull (51) and is still in use: it is mostly based on daily aphid counts (mainly *M. persicae*) on plants by sugar-beet fieldmen and the threshold for spraying is about one green aphid per four plants, according to the area; in France it is from one to two aphids per ten plants. It is rather surprising that winged aphids are not much taken into account in issuing warnings although Watson and Healy (141) found a close relationship between the numbers of alate *M. persicae* caught on sticky traps and yellows incidence, and this was confirmed later by Heathcote (40). Figure 4.10 shows clearly the relation between *M. persicae* caught on a sticky trap at Rothamsted in May and June and the national yellows incidence at the end of August since 1946. The fit is rather good, with some remarkable exceptions in 1949, 1974, and 1975. G. Heathcote (personal communication) explained that the *M. persicae* catch was small in June 1975, but large in July, resulting in late spread of yellows. This was combined with insecticide resistance (41). He also ascribed the abnormal spread in 1974 to viruliferous aphids having overwintered in southern England, which "were carried into crops in an unusually strong southerly airflow in May"; another possibility is that they might have come from France (41). In France, too, as in the rest of Europe, yellows incidence was high in 1974 and 1975; unfortunately, no aphid trapping was being operated there, nor in 1966–1968, which again were "bad" years: it is interesting to observe what is probably simply a coincidence, that numbers of *M. persicae* caught in Britain were somewhat large at the same moment. Should we deduce that they crossed the Channel to infect Continental crops?

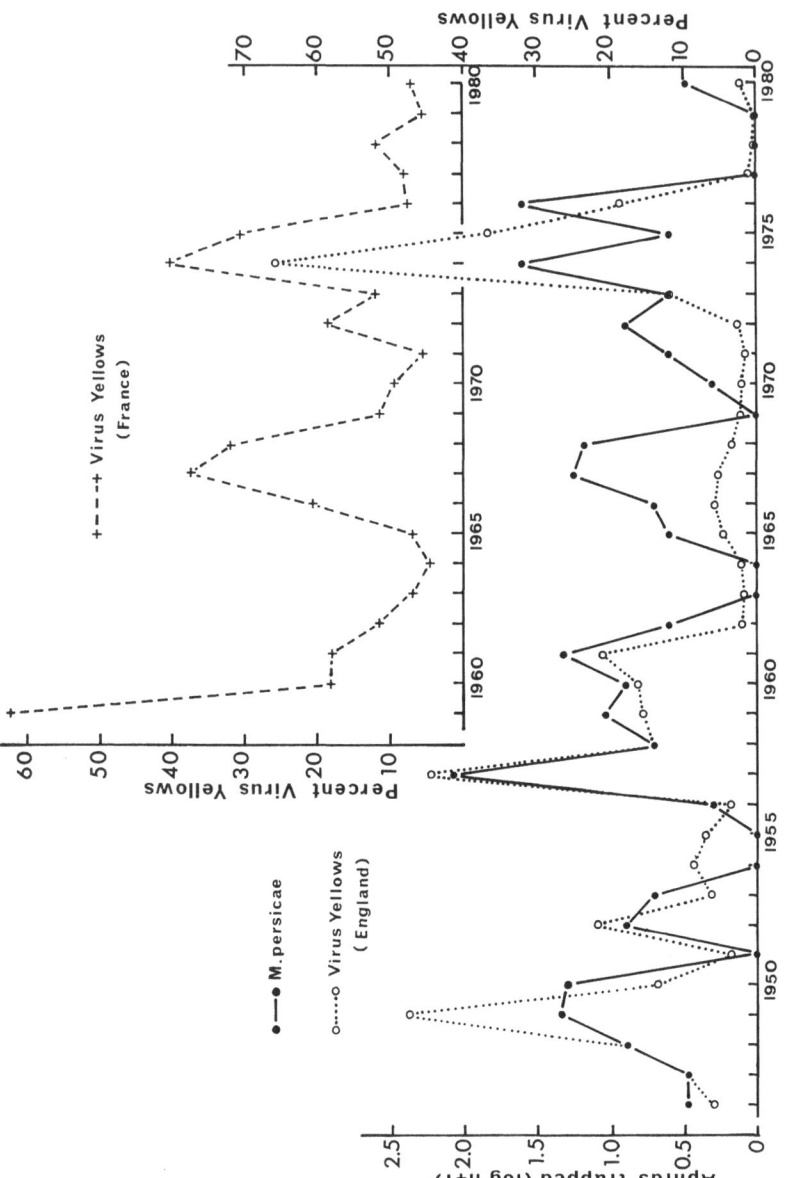

FIGURE 4.10. Mean percentage of sugar-beet plants infected with yellows at the end of August in France (1959–1980) and in England (1946–1980) and numbers of alate *Myzus persicae* caught by a sticky trap at Rothamsted in May and June (1946–1980). [Adapted from Le Cochec (67) and Heathcote (40) and unpublished data.]

Anyhow, growers need to make up their mind early in the season as to whether they should apply granules and later spray. AVM as it is now cannot provide answers quickly enough and therefore they continue to rely on Hurst–Watson equations, which are based on winter weather and on aphid counts. In northern France more attention is given to yellow and suction trap data at the beginning of the crop season, but this still needs to be improved.

SEED POTATOES

In Europe, the main seed potato areas are located in Scotland, the Netherlands, the north of France, and Brittany, and they cover roughly 24,000 ha in Great Britain, 30,000 ha in the Netherlands, and 13,000 ha in France. In contrast to sugar-beet viruses, potato viruses (PVY°, PVYⁿ, PLRV) are seed-tuber-borne and it must be emphasized that final virus incidence is related to (1) initial virus level at planting (hence the necessity to have healthy mother tubers and to do "roguing" practices to remove infected plants from the previous year as soon as they show symptoms), (2) virus influx into the crop during crop season (hence compulsory regulations to move ware crops away from seed crops), (3) earliness and magnitude of aphid flights, and (4) occasionally, subsequent rate of aphid multiplication (154). In that respect, AVM has long been used as an aid to decision-making to seed potato growers, and Hille Ris Lambers (48) could rightfully write that "the yellow trap . . . must have saved Dutch seed-potato growers many millions of guilders if one reckons that one day's difference in lifting of seed potatoes may rise the value of the harvest" by an average of 600 kg ha^{-1}: in that country, haulm destruction to prevent viruses inoculated to potato leaves from infecting tubers of the progeny relies on aphid trapping (see section on historical backgrounds) and for the highest grades of seed, it has to be done in a given region 8–10 days after yellow traps have caught two alates of *M. persicae* per day. This policy (1) is only practicable in regions where these aphids show nearly no flight activity until June and (2) is essentially relevant to years where PLRV is involved, since this persistently transmitted virus can only be spread by aphids such as *M. persicae* and to a decreasing extent *A. solani* and *M. euphorbiae* (90, 98) that can feed and breed on potatoes. Thus, in Brittany, haulm burning, which is compulsory for entering certification, occurs by the end of July at a date that takes into account the whole situation from planting time, since it has been shown that (1) in May, aphid colonization by *A. solani* and *M. persicae* may be fairly important (103) and may result in an early spread of PLRV (91), but (2) in these conditions, parasites [microhymenoptera *(Aphidiidae)* and fungi *(Entomophthoraceae)*] can soon control populations fairly well, resulting in hardly any dissemination flight (93, 99), and (3) dissemination flight may occur, if ever, only if contamination flight has been low because subsequent aphid population buildup is controlled too late.

AVM, based on suction and yellow trapping, is also useful to issue warnings for chemical control, and the potato aphid spray warning scheme in Scotland is an example of this (153, 154). However, warning schemes operating on the Continent have proved to be not as efficient as they used to be when from 1974 to 1976 inclusive, PVY incidence rose in huge proportions (92) for reasons given above. Figure 4.11 suggests possible relations between *M. persicae, R. padi,* and PLRV and PVY in western France. Since PVY can be nonpersistently transmitted by transient aphids, Van Harten (136, 137) made an attempt to improve the Dutch scheme for this virus by calculating daily vector pressures, taking into account both the relative efficiency factors of nine species known or supposed to be

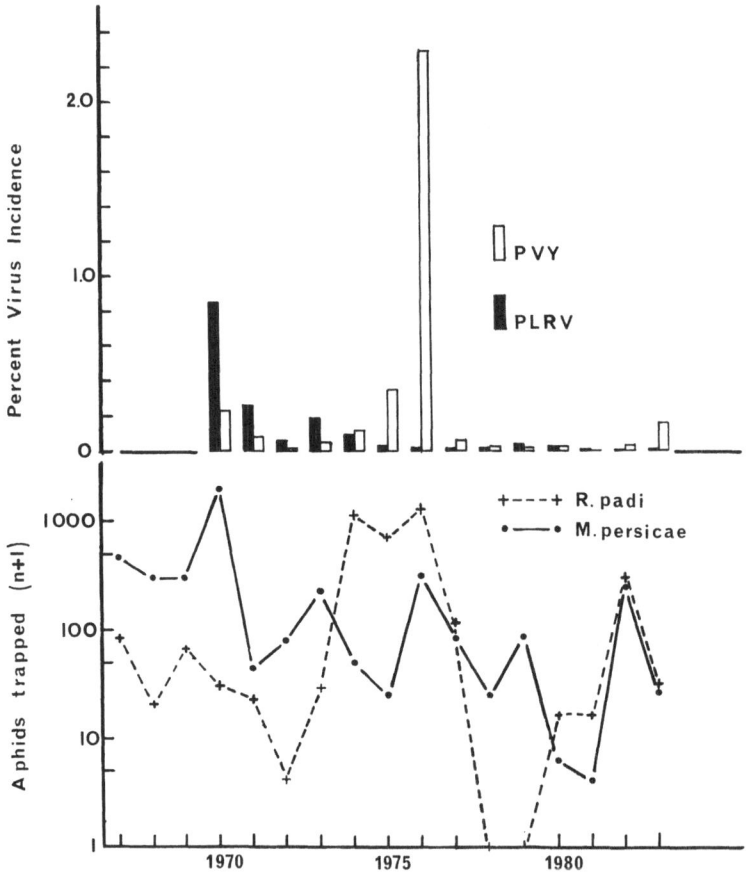

FIGURE 4.11. Relationships between the incidence of potato leaf roll virus (PLRV) and the sum of alate *Myzus persicae* yellow trap catches from April to July inclusive and between the incidence of potato virus Y (PVY) and the sum of alate *Rhopalosiphum padi* yellow trap catches from April to June inclusive in western Brittany (1967–1983). (Y. Robert and J. Quemener, unpublished data.)

PVY vectors and daily numbers of each species caught in Dutch 12.2-m suction traps. In comparing the accumulated vector pressures in each of ten years (1970–1979) with the average daily catches of *M. persicae* by yellow traps and virus incidence in relation to the dates of haulm destruction, he found that in six of these years the critical periods of both systems coincided, but in the "bad" years 1974–1976, his method was more efficient. In Sweden, Sigvald (112) devised a similar method based on the numbers of different species caught in yellow traps, their relative efficiency, and the time of their flight activity to take mature plant resistance into account, and he obtained a good correlation between this "total vector pressure" and PVY incidence for the years 1976–1980. It must be stressed that in both cases, weekly vector pressures were first checked against weekly infection percentages of bait plants [tobacco to PVY^n in the Netherlands (138); tobacco and potato to PVY^o in Sweden (108)], a procedure that was also used with some success to study CMV spread in southeastern France (74) despite serious limitations about the representativeness of such bait plants. While Van Harten's accumulated vector pressure has apparently not been put into practice in the Low Countries, AVM is being continued and improved in the area of seed potato production, if only because insecticide field resistance has become a permanent threat; in my opinion, this should be done in closer contact with sugar-beet schemes wherever possible, since very often potato and sugar-beet areas are intermingled and these crops share some of the same species, e.g., *M. persicae*. But potato and sugar-beet corporations do not seem to be acquainted with one another!

WINTER CEREALS

For more than a decade the attention of growers has been drawn to the increasing importance of BYDV infecting cereals, especially winter-sown ones. Many BYDV strains have been described as being persistently transmitted by different aphid species, especially *M. dirhodum, R. padi,* and *S. avenae,* and they differ in the damage they cause as expressed by yield decrease: *R. padi*-specific strains are among the most damaging. Yield losses are difficult to assess. For example, in France, at a similar rate of infection, they have been 0.9 ton ha^{-1} in winter wheat cv. Talent and 2.7–3 tons ha^{-1} in winter barley cv. Astrix (9). They also differ according to the percentage of tillers infected in different areas. In southwest England, in early-sown winter barley, they occurred when more than 10% tillers were infected and correlated well with disease incidence (61). In the French Atlantic southwestern area, 5–10% tillers infected resulted in yield losses of 0.6–0.9 ton ha^{-1}, 25% in yield losses of 0.9–1.1 tons ha^{-1}, and 50–60% in yield losses of 2–2.2 tons ha^{-1}, but in northern France, yield losses became significant only when more than 66% tillers were infected (7): these differences were ascribed to the origin of colonizing alate *R. padi,* those flying from cereal volunteers and regrowths being far more viruliferous than those from maize crops (8), hence the need to be able to ascertain the infectivity of incoming aphids (81) and their biology in relation to virus source distribution in time and space (26, 82, 84).

Since BYDV incidence cannot be predicted from aphid numbers alone, Plumb *et al.* (85) developed an Infectivity Index (II), which is best expressed as the product of alate *R. padi* numbers caught in 12.2-m suction traps by the proportion infective. This infectivity is assessed by allowing aphids caught alive in 1.5-m suction traps to feed on test plants (oat seedlings cv. Blenda) for 2–3 days. BYDV symptoms do not show until 2–4 weeks, which can introduce some limitations for predicting purposes, but in the future, ELISA and ISEM techniques applied to test plants could speed up the results. Like Van Harten's vector pressure index, the II can be calculated as a total for the whole of the autumn cereal aphid migration for year-to-year comparisons, but for prediction purposes, it is more valuable when it is expressed as a cumulative process calculated weekly from the sowing date onward (85). From a year-to-year comparison at Rothamsted (83) it was suggested that an autumn aphicide would become worthwhile if a threshold value of 50 was exceeded (85). This threshold value no doubt varies from one region to another, as shown from similar tests being done in western Wales since 1970 (2, 4), southwest England since 1976 (62), and the Midlands since 1982 (85). Kendall and Smith (62) have shown one of the possible reasons to be that during mild winters, such as prevailing in western maritime areas, secondary infection arises from wingless populations. This also occurs in western France. Once more, only continuous studies for many years will allow regional threshold values to be refined to improve virus control, and promising prospects are expected. A similar procedure has been used in southeastern France to assess CMV spread (65), but for BYDV, weekly exposition of bait plants to assess virus incidence, aphid infectivity, suction trap catches of *R. padi,* dates of plant emergence, and assessment of virus reservoir distribution are factors that were combined to produce a "decision-making tree" which has given rather satisfactory results (6, 7). Whatever the means chosen to predict and the need to control BYDV, AVM has become a basic requirement to improve forecasts.

Conclusion

In the context of integrated pest control, the use of insecticides must be kept to a minimum in quantity, time, and space because side effects and insecticide-resistant aphid strains are becoming more and more threatening. There is obviously a need to improve our forecasting systems, especially in the difficult area of virus disease epidemiology. In this review I have tried to show that AVM is basic, and has great potential, especially if it can become national or international. So far, though, we cannot say an integrated European AVM is actually operating, but in this part of the world, we have recorded for so long and accumulated such a large amount of relevant data that we have great hopes of achieving it. I have drawn attention first to some constraints we must face in order to succeed with such schemes. It is necessary to make sure that continuous monitoring throughout the year and for many years without a break is possible, although it might appear as a routine task without immediate outcome. Then aphid trapping and the study of virus disease distribution must be considered on the largest possible scale, so that a given region may benefit from data recorded in others and vice versa. Also, as many as possible, if not all, species of aphids should be monitored to make it possible at any time to take into account new problems and give a quick answer to growers' problems. From the start of any AVM, the problem of establishing specific economic thresholds cannot be ignored and the actual significance of forecasts must be assessed and checked in a multidisciplinary and international approach. Finally, and it is not the least important constraint, data and interpretations need to be freely and confidently exchanged and compared to make efficient forecasts rapidly available. I have drawn attention also to some requirements that have to be met before the interpretation of data can be of practical value. Knowledge of the overall aphidofauna is a requisite together with a better understanding of aphid bioecology and virus disease epidemiology: this refers mainly to (1) seasonal aphid flight activity and its adaptive evolution according to meteorological limitations as expressed by the latitude–longitude effect, (2) the type of biological cycle occurring in a given area (holocycly versus anholocycly) in relation to the effect of weather on aphid overwintering and both timing and magnitude of flight cycles, and (3) virus reservoirs and the dynamics of virus disease spread according to practices.

Different countries initially chose different techniques of trapping, such as sticky yellow traps in Great Britain and yellow water traps on the Continent, but recently they have also decided to cooperate within the EURAPHID venture, which now extends to many Western European countries, hence our expectation for more efficient and integrated monitoring in the future. Our hopes can probably be best placed on forecasts operated in sugar-beet, potato, and cereal crops and I have showed that so far, different techniques have been used both between crops and between countries. More and more attention is given to the role of different meteorological variates on aphid-transmitted virus diseases of crops in relation to aphid biology and to the infectivity of incoming alate individuals as expressed as an infection pressure index or an Infectivity Index. This is a step beyond simple aphid monitoring and it seems a promising way to develop models and improve the soundness of forecasts in the future. There remains some concern for the future with such forecasting systems resulting from AVM, relating to the need to make growers understand that a forecast should recommend whether or not chemicals should be used according to the situation. From experience, we know that this may be difficult unless we can rely with great confidence on warning schemes, which requires that they should be still further improved. This is a permanent challenge to our skill, and the consequences may be financially important: Bardner *et al.* (5) have calculated that "even the elimination of one spray round in a minor crop such as hops would pay for the entire annual cost of the suction trap scheme."

Acknowledgments. I wish to thank Dr. L.R. Taylor for permission to use the Rothamsted Insect Survey suction trap data and to reproduce Figure 4.6, and Dr. G.D. Heathcote (Broom's Barn Experimental Station) for permission to use some of his unpublished data. I am also very grateful to Dr. B.D. Smith (Long Ashton Research Station), who made my English more readable.

References

1. A'Brook, J., 1973, Observations on different methods of aphid trapping, *Ann. Appl. Biol.* **74**:263–277.
2. A'Brook, J., 1974, Barley yellow dwarf virus: What sort of a problem?, *Ann. Appl. Biol.* **77**:92–96.
3. A'Brook, J., 1981, Some observations in west Wales on the relationships between numbers of alate aphids and weather, *Ann. Appl. Biol.* **97**:11–15.
4. A'Brook, J., and Dewar, A.M., 1980, Barley yellow dwarf virus infectivity of alate aphid vectors in west Wales, *Ann. Appl. Biol.* **96**:51–58.
5. Bardner, R., French, R.A., and Dupuch, M., 1981, Agricultural benefits of the Rothamsted Aphid Bulletin, Rothamsted Experimental Station Report for 1980, Part 2, pp. 21–39.
6. Bayon, F., and Ayrault, J.P., 1983, La jaunisse nanisante de l'orge (J.N.O.): méthode actuelle de prévision des risques en automne, *Défense Végétaux* **223**:268–275.
7. Bayon, F., Ayrault, J.P., and Pichon, P., 1980, La jaunisse nanisante de l'orge. Symptômes, dégâts, cycles, moyens de lutte, améliorations, *Défense Végétaux* **204**:181–202.
8. Bayon, F., Ayrault, J.P., and Pichon, P., 1981, Utilisation pratique du test ELISA dans l'étude du BYDV en Poitou-Charentes, in: *Proceedings 3rd Conference on Virus Diseases of Gramineae in Europe,* Rothamsted Experimental Station, pp. 119–120.
9. Bayon, F., Ayrault, J.P., and Pichon, P., 1982, Epidémiologie de la jaunisse nanisante de l'orge (BYDV) en Poitou-Charentes, *Meded. Fac. Landbouwwet. Rijksuniv. Gent* **47**:1039–1052.
10. Bernard, J., 1982, Préface de l'Editeur, in: J. Bernard (ed.), *Euraphid Gembloux 1982,* Centre Recherche Agronomique de Gembloux, 91 pp.
11. Bouchery, Y., 1981, The biology of cereal aphids; overwintering and the spring migration, in: L.R. Taylor (ed.), *Euraphid Rothamsted 1980,* Rothamsted Experimental Station, Harpenden, p. 26.
12. Broadbent, L., 1948, Aphis migration and the efficiency of the trapping method, *Ann. Appl. Biol.* **35**:379–394.

13. Broadbent, L., 1950, The correlation of aphid numbers with the spread of leaf roll and rugose mosaic in potato crops, *Ann. Appl. Biol.* **37**:58–65.
14. Broadbent, L., 1969, Disease control through vector control, in: K. Maramorosch (ed.), *Viruses, Vectors and Vegetation,* Wiley, New York, pp. 593–630.
15. Broadbent, L., and Doncaster, J.P., 1949, Alate aphids trapped in the British Isles, 1942–1947, *Entomol. Mon. Mag.* **115**:174–182.
16. Broadbent, L., and Heathcote, G.D., 1955, Sources of overwintering *Myzus persicae* (Sulzer) in England, *Plant Pathol.* **4**:135–137.
17. Broadbent, L., and Heathcote, G.D., 1961, Winged aphids trapped in potato fields, 1942–1959, *Entomol. Exp. Appl.* **4**:226–237.
18. Broadbent, L., Doncaster, J.P., Hull, R., and Watson, M.A., 1948, Equipment used for trapping and identifying alate aphides, *Proc. R. Entomol. Soc. Lond. A* **23**:57–58.
19. Brunel, E., and Langouet, L., 1970, Influence des caractéristiques optiques du milieu sur les adultes de *Psila rosae* Fab. (Diptères Psilidés): Attractivité de surfaces colorées, rythme journalier d'activité, *C. R. Soc. Biol.* **164**:1638–1644.
20. Burt, P.E., Heathcote, G.D., and Broadbent, L., 1964, The use of insecticides to find when leaf roll and Y viruses spread within potato crops, *Ann. Appl. Biol.* **54**:13–22.
21. Cariolle, M., 1982, Les pucerons de la betterave, in: *Les pucerons des cultures,* ACTA, Paris, pp. 215–219.
22. Cochrane, J., 1980, Meteorological aspects of the numbers and distribution of the rose-grain aphid, *Metopolophium dirhodum* (Wlk.), over south east England in July 1979, *Plant Pathol.* **29**:1–8.
23. Cockbain, A.J., and Heathcote, G.D., 1965, Transmission of sugar beet viruses in relation to the feeding, probing and flight activity of alate aphids, in: *Proceedings 12th International Congress Ent. London, 1964,* pp. 521–523.
24. Cockbain, A.J., Gibbs, A.J., and Heathcote, G.D., 1963, Some factors affecting the transmission of sugar-beet mosaic and pea mosaic viruses by *Aphis fabae* and *Myzus persicae, Ann. Appl. Biol.* **52**:133–143.
25. Cornu, P., and Münster, J., 1975, Trente ans de contrôle de l'évolution des pucerons 1945–1975, Rapport pucerons Station féderale de Recherches Agronomiques de Changins, Nyon. 18 pp.
26. Dedryver, C.A., and Robert, Y., 1981, Ecological role of maize and cereal volunteers as reservoirs for gramineae virus transmitting aphids, in: *Proceedings 3rd Conference on Virus Diseases of Gramineae in Europe,* Rothamsted Experimental Station, pp. 61–66.
27. Dewar, A.M., Woiwod I., and Choppin de Janvry, E., 1980, Aerial migrations of the rose-grain aphid, *Metopolophium dirhodum* (Wlk.), over Europe in 1979, *Plant Pathol.* **29**:101–109.
28. Doncaster, J.P., and Gregory, P.H., 1948, The spread of virus diseases in the potato crop, ARC Report Series No. 7, HMSO, London. 189 pp.
29. Dubnik, H., 1978, Einschätzung der Befallsintensität bei Kartoffelblattlaüsen an Hand der Ergebnisse der Gelbschalenfänge 1970 bis 1977, *Nachrichtenbl. Pflanzenschutz DDR* **32**:79–82.
30. Dunnet, G.M., 1982, Ecology and everyman, *J. Anim. Ecol.* **51**:1–14.

31. Eastop, V.F., 1983, The biology of the principal aphid virus vectors, in: R.T. Plumb and J.M. Thresh (eds.), *Plant Virus Epidemiology*, Blackwell, Oxford, pp. 115–132.

32. Fisken, A.G., 1959, Factors affecting the spread of aphid-borne viruses in potato in eastern Scotland. II. Infestation of the potato crop by potato aphids, particularly *Myzus persicae* (Sulzer), *Ann. Appl. Biol.* **47**:274–286.

33. French, R.A., and Taylor, L.R., 1965, A continuous pest census for forecasting insect attack, *Report of the Rothamsted Experimental Station for 1964*, pp. 191–192.

34. Gabriel, W., Nuckowski, S., and Wislocka, M., 1965, Sept ans d'observations sur les pucerons de la pomme de terre en Pologne (1955–1961), *Parasitica* **21**:16–32.

35. Gabriel, W., Neitzel, K., Rasocha, V., Wójcik, A.R., Debus, R., Nuckowski, S., and Klinowsky, M., 1972, Die Beziehungen zwischen dem Auftreten der Vektoren und der Höhe des Virusbesatzes (Blattroll- und Y-Virus) bei Kartoffeln in Mitteleuropa, *Ziemniak*, pp. 27–63.

36. Haine, E., 1966, Beitrag zur Erforschung des Massenflugs virus-übertragender Blattläuse mit Hilfe neuer Fang- und Registrierverfahren, *Z. Angew. Entomol.* **58**:119–130.

37. Heathcote, G.D., 1957, The comparison of yellow cylindrical, flat and water traps, and of Johnson suction traps, for sampling aphids, *Ann. Appl. Biol.* **45**:133–139.

38. Heathcote, G.D., 1958, Effect of height on catches of aphids in water and sticky traps, *Plant. Pathol.* **7**:32–36.

39. Heathcote, G.D., 1966, The time of flight and the relative importance of *Myzus persicae* (Sulz.) and *Aphis fabae* Scop. in relation to the incidence of beet yellows as shown by trap catches at Rothamsted and Broom's Barn, *Bull. Entomol. Res.* **56**:473–480.

40. Heathcote, G.D., 1974, Aphids caught on sticky traps in eastern England in relation to the spread of yellowing viruses of sugar-beet, *Bull. Entomol. Res.* **64**:669–676.

41. Heathcote, G.D., 1977, Should we use aphicides? in: *Proceedings 1977 British Crop Protection Conference—Pests and Diseases,* Brighton, pp. 895–901.

42. Heathcote, G.D., 1978, Review of losses caused by virus yellows in English sugar beet crops and the cost of partial control with insecticides, *Plant Pathol.* **27**:12–17.

43. Heathcote, G.D., 1980, Were we invaded by greenfly?, *Trans. Suffolk Nat. Soc.* **18**:144–147.

44. Heathcote, G.D., 1983, Aphicidal persistence of aldicarb and possibilities for forecasting the need for its use before sowing sugar-beet crops, *Aspects Appl. Biol.* **2**:19–27.

45. Heathcote, G.D., Palmer, J.M.P., and Taylor, L.R., 1969, Sampling for aphids by traps and by crop inspection, *Ann. Appl. Biol.* **63**:155–166.

46. Heie, O.E., Philipsen, H., and Taylor, L.R., 1981, Synoptic monitoring for migrant insect pests in Great Britain and Western Europe. II. The species of alate aphids sampled at 12.2 m by Rothamsted Insect Survey suction trap at Tåstrup, Denmark, between 1971 and 1976, Rothamsted Experimental Station Report for 1980, Part 2, pp. 105–114.

47. Hille Ris Lambers, D., 1955, Potato aphids and virus diseases in the Netherlands, *Ann. Appl. Biol.* **42**:355–360.

48. Hille Ris Lambers, D., 1972, Aphids: Their life cycles and their role as virus vectors, in: J.A. de Bokx (ed.), *Viruses of Potatoes and Seed-potato Production*, Pudoc, Wageningen, 3, pp. 36–56.
49. Hollings, M., 1955, Aphid movement and virus spread in seed potato areas of England and Wales, 1950–53, *Plant Pathol.* **4**:73–82.
50. Howell, P.J., 1977, Recent trends in the incidence of aphid borne viruses in Scotland, in: *Proceedings Symposium on Problems of Pest and Disease Control in Northern Britain, Dundee*, pp. 26–28.
51. Hull, R., 1968, The spray warning scheme for control of sugar beet yellows in England. Summary of results between 1959–66, *Plant Pathol.* **17**:1–10.
52. Hull, R., and Heathcote, G.D., 1967, Experiments on the time of application of insecticide to decrease the spread of yellowing viruses of sugar beet, 1954–1966, *Ann. Appl. Biol.* **60**:469–478.
53. Hurst, G.W., 1965, Forecasting the severity of sugar beet yellows, *Plant Pathol.* **14**:47–53.
54. Irwin, M.E., 1980, Sampling aphids in soybean fields, in: M. Kogan and D.C. Herzog (eds.), *Sampling Methods in Soybean Entomology*, Springer-Verlag, New York, pp. 239–259.
55. Johnson, C.G., 1950a, A suction trap for small airborne insects which automatically segregates the catch into successive hourly samples, *Ann. Appl. Biol.* **37**:80–91.
56. Johnson, C.G., 1950b, The comparison of suction trap, sticky trap and townet for quantitative sampling of small airborne insects, *Ann. Appl. Biol.* **37**:268–285.
57. Johnson, C.G., 1952, The changing numbers of *Aphis fabae* Scop., flying at crop level, in relation to current weather and to the population on the crop, *Ann. Appl. Biol.* **39**:525–547.
58. Johnson, C.G., 1967, International dispersal of insects and insect-borne viruses, *Neth. J. Plant Pathol.* **73** (Suppl. 1):21–43.
59. Johnson, C.G., and Eastop, V.F., 1951, Aphids captured in a Rothamsted suction trap, 5 ft above ground level, from June to November, 1947, *Proc. R. Entomol. Soc. Lond. A* **26**:17–24.
60. Johnson, C.G., and Taylor, L.R., 1955, The development of large suction traps for airborne insects, *Ann. Appl. Biol.* **43**:51–61.
61. Kendall, D.A., and Smith, B.D., 1981a, Yield benefit from autumn control of aphids and barley yellow dwarf virus on winter barley, in: *Proceedings 1981 British Crop Protection Conference—Pests and Diseases*, Brighton, pp. 217–221.
62. Kendall, D.A., and Smith, B.D., 1981b, The significance of aphid monitoring in improving barley yellow dwarf virus control, in: *Proceedings 1981 British Crop Protection Conference—Pests and Diseases*, Brighton, pp. 399–403.
63. Kennedy, J.S., Day, M.F., and Eastop, V.F., 1962, *A Conspectus of Aphids As Vectors of Plant Viruses*, Commonwealth Institute of Entomology. 114 pp.
64. Kostiw, M., 1979, Transmission of potato virus Y by *Rhopalosiphum padi* L., *Potato Res.* **22**:237–238.
65. Labonne, G., Quiot, J.B., and Monestiez, P., 1982a, Rôle des diverses espèces de pucerons vecteurs dans la dissémination du virus de la mosaïque du concombre au niveau d'une parcelle de melon dans le Sud-Est de la France, *Agronomie* **2**:797–804.

66. Labonne, G., Fauvel, C., Leclant, F., and Quiot, J.B., 1982b, Description d'un piège à succion: son emploi dans la recherche des aphides vecteurs de virus transmis sur le mode non persistant, *Agronomie* 2:773–776.

67. Le Cochec, F., 1982, Sélection de la betterave sucrière pour la résistance à la jaunisse virale, *Sci. Agron. Rennes* 2:1–15.

68. Lewis, T., 1965, The effect of an artificial windbreak on the distribution of aphids in a lettuce crop, *Ann. Appl. Biol.* 55:513–518.

69. Lewis, T., 1967, The horizontal and vertical distribution of flying insects near artificial windbreaks, *Ann. Appl. Biol.* 60:23–31.

70. Lewis, T., 1969, The distribution of flying insects near a low hedgerow, *J. Appl. Ecol.* 6:443–452.

71. Lewis, T., 1970, Patterns of distribution of insects near a windbreak of tall trees, *Ann. Appl. Biol.* 65:213–220.

72. Lewis, T., and Taylor, L.R., 1967, *Introduction to Experimental Ecology,* Academic Press, London & New York, 401 pp.

73. MacCarthy, H.R., 1954, Aphid transmission of potato leafroll virus, *Phytopathology* 44:167–174.

74. Marrou, J., Quiot, J.B., Duteil, M., Labonne, G., Leclant, F., and Renoust, M., 1979, Ecologie et épidémiologie du virus de la mosaïque du Concombre dans le Sud-Est de la France. III. Intérêt de l'exposition de plantes-appâts pour l'étude de la dissémination du virus de la mosaïque du Concombre, *Ann. Phytopathol.* 11:291–306.

75. Moericke, V., 1949, Über den Farbensinn der Pfirsichblattlaus (*Myzodes persicae* Sulz.), *Anz. Schädlingsk.* 22:139.

76. Moericke, V., 1951, Eine Farbfalle zur Kontrolle des Fluges von Blattläusen, insbesondere der Pfirsichblattlaus, *Myzodes persicae* (Sulz.), *Nachrichtenbl. Dtsch. Pflanzenschutzdienst. (Braunschw.)* 3:23–24.

77. Moericke, V., 1955, Über die Lebensgewohnheiten der geflügelten Blattläuse (*Aphidina*) unter besonderer Berücksichtigung des Verhaltens beim Landen, *Z. Angew. Entomol.* 37:29–91.

78. Münster, J., and Cornu, P., 1977, Apparition anormalement précoce et massive de virus Y (mosaïque) dans les cultures de plants de pomme de terre en 1976. *Rev. Suisse Agric.* 9:71.

79. Naton, E., 1976, Die wichtigsten Blattläuse im Hackfruchtbau, *Bayer. Landesanst. Bodenkultur Pflanzenbau. Pflanzenschutzinformationen* 44:1–20.

80. Pianka, E.R., 1970, On r- and K-selection, *Am. Nat.* 104:592–597.

81. Plumb, R.T., 1976, Barley yellow dwarf virus in aphids caught in suction traps, 1969–1973, *Ann. Appl. Biol.* 83:53–59.

82. Plumb, R.T., 1977, Grass as a reservoir of cereal viruses, *Ann. Phytopathol.* 9:361–364.

83. Plumb, R.T., 1981, Aphid-borne virus diseases of cereals, in: L.R. Taylor (ed.), *Euraphid Rothamsted 1980,* Rothamsted Experimental Station, Harpenden, pp. 18–21.

84. Plumb, R.T., 1983, Barley yellow dwarf virus—A global problem, in: R.T. Plumb and J.M. Thresh (eds.), *Plant Virus Epidemiology,* Blackwell, Oxford, pp. 185–198.

85. Plumb, R.T., Lennon, E., and Gutteridge, R.A., 1982, Aphid infectivity and the Infectivity Index, Report of the Rothamsted Experimental Station for 1981, Part 1, pp. 195–197.

86. Ponsen, M.B., 1970, The biological transmission of potato leafroll virus by *Myzus persicae, Neth. J. Plant Pathol.* **76:**234–239.
87. Quiot, J.B., Marrou, J., Labonne, G., and Verbrugghe, M., 1979. Ecologie et épidémiologie du Virus de la Mosaïque du Concombre dans le Sud-Est de la France. Description du dispositif expérimental, *Ann. Phytopathol.* **11:**265–282.
88. Raccah, B., 1983, Monitoring insect vector populations and the detection of viruses in vectors, in: R.T. Plumb and J.M. Thresh (eds.), *Plant Virus Epidemiology,* Blackwell, Oxford, pp. 147–157.
89. Ribbands, C.R., 1964, The control of the sources of virus yellows of sugarbeet, *Bull. Entomol. Res.* **54:**661–674.
90. Robert, Y., 1971, Epidémiologie de l'Enroulement de la pomme de terre: Capacité vectrice de stades et de formes des pucerons *Aulacorthum solani* Kltb, *Macrosiphum euphorbiae* Thomas et *Myzus persicae* Sulz., *Potato Res.* **14:**130–139.
91. Robert, Y., 1976, Action de traitements insecticides endothérapiques sur la dissémination du virus de l'Enroulement de la Pomme de terre en Bretagne, *Phytiatrie-Phytopharmacie* **25:**187–200.
92. Robert, Y., 1978, Rôle épidémiologique probable d'espèces de pucerons autres que celles de la pomme de terre dans la dissémination intempestive du virus Y depuis 4 ans dans l'Ouest de la France, in: *Compte Rendu de la 7è Conférence trisannuelle de l'Association Européenne pour la Recherche sur la Pomme de terre (E.A.P.R.), Varsovie,* E. Kapsa (ed.), Instytut Ziemniaka, Bonin, pp. 242–243.
93. Robert, Y., 1979, Recherches écologiques sur les pucerons *Aulacorthum solani* Kltb., *Macrosiphum euphorbiae* Thomas et *Myzus persicae* Sulz. dans l'Ouest de la France. III—Importance du parasitisme par Hyménoptères *Aphidiidae* et par *Entomophthora* sur pomme de terre, *Ann. Zool. Ecol. Anim.* **11:**371–388.
94. Robert, Y., 1980, Recherches sur la Biologie et l'Ecologie des pucerons en Bretagne; application à l'étude épidémiologique des viroses de la pomme de terre, Thèse Doctorat d'Etat D.Sc., Université de Rennes 1. 243 pp.
95. Robert, Y., 1981, The operation of yellow water traps and suction traps and the interpretation of the data collected, in: L.R. Taylor (ed.), *Euraphid Rothamsted 1980,* Rothamsted Experimental Station, Harpenden, pp. 28–32.
96. Robert, Y., and Choppin de Janvry, E., 1977, Sur l'intérêt d'implanter en France un réseau de piégeage pour améliorer la lutte contre les pucerons, *Bull. Tech. Inf. Min. Agric.* **323:**559–568.
97. Robert, Y., and Dedryver, C.A., 1977, Remarques sur l'activité de vol des pucerons des céréales, en Bretagne, depuis 10 ans, *Ann. Phytopathol.* **9:**371–376.
98. Robert, Y., and Maury, Y., 1970, Capacités vectrices comparées de plusieurs souches de *Myzus persicae* Sulz., *Aulacorthum solani* Kltb. et *Macrosiphum euphorbiae* Thomas dans l'étude de la transmission de l'enroulement de la Pomme de terre, *Potato Res.* **13:**199–209.
99. Robert, Y., and Rabasse, J.M., 1977, Rôle écologique de *Digitalis purpurea* dans la limitation naturelle des populations du puceron strié de la pomme de terre *Aulacorthum solani* par *Aphidius urticae* dans l'Ouest de la France, *Entomophaga* **22:**373–382.

100. Robert, Y., and Rouzé-Jouan, J., 1975, Etude des populations ailées de pucerons des céréales, *Acyrthosiphon (Metopolophium) dirhodum* Wlk., *A. (M.) festucae* Wlk., *Macrosiphum (Sitobion) avenae* F., *M. (S.) fragariae* Wlk. et *Rhopalosiphum padi* L. en Bretagne de 1967 à 1975: Examen des possibilités de prévision des attaques, *C. R. Acad. Agric. Fr.* **16**:1006–1016.

101. Robert, Y., and Rouzé-Jouan, J., 1976a, Activité saisonnière de vol des pucerons *(Hom. Aphididae)* dans l'Ouest de la France. Résultats de neuf années de piégeage (1967–1975), *Ann. Soc. Entomol. Fr. (N.S.)* **12**:671–690.

102. Robert, Y., and Rouzé-Jouan, J., 1976b, Neuf ans de piégeage de pucerons des céréales, *Acyrthosiphon (Metopolophium) dirhodum* Wlk., *A. (M.) festucae* Wlk., *Macrosiphum (Sitobion) avenae* F., *M. (S.) fragariae* Wlk. et *Rhopalosiphum padi* L. en Bretagne, *Rev. Zool. Agric. Pathol. Veg.* **75**:67–80.

103. Robert, Y., and Rouzé-Jouan, J., 1978, Recherches écologiques sur les pucerons *Aulacorthum solani* Kltb, *Macrosiphum euphorbiae* Thomas et *Myzus persicae* Sulz. dans l'Ouest de la France. I. Etude de l'activité de vol de 1967 à 1976 en culture de pomme de terre, *Ann. Zool. Ecol. Anim.* **10**:171–185.

104. Robert, Y., Bonnemaison, L., and Quemener, J., 1974a, Facteurs épidémiologiques intervenant dans la transmission des viroses par les pucerons, in: *Maladies et parasites animaux de la pomme de terre,* Institut Technique de la Pomme de terre, Brochure 32, Paris, pp. 16–25.

105. Robert, Y., Rabasse, J.M., and Rouzé-Jouan, J., 1974b, Sur l'utilisation des pièges jaunes pour la capture de pucerons en culture de pomme de terre. I. Influence de la hauteur de piégeage, *Ann. Zool. Ecol. Anim.* **6**:349–372.

106. Robert, Y., Brunel, E., Malet, P., and Bautrais, P., 1976, Distribution spatiale de pucerons ailés et de diptères dans une parcelle de bocage, en fonction des modifications climatiques provoquées par les haies, in: J. Missonnier (ed.), *Les Bocages: Histoire, écologie économie,* EDIFAT-OPIDA, Echauffour, pp. 427–435.

107. Rydén, K., 1979, Havrebladlusen, *Rhopalosiphum padi,* kan sprida potatis virus Y, *Växtskyddsnotiser* **43**:51–53.

108. Rydén, K., Brishammar, S., and Sigvald, R., 1983, The infection pressure of potato virus Y° and the occurrence of winged aphids in potato fields in Sweden, *Potato Res.* **26**:229–235.

109. Sawicki, R.M., Devonshire, A.L., Rice, A.D., Moores, G.D., Petzing, S.M., and Cameron, A., 1978, The detection and distribution of organophosphorus and carbamate insecticide-resistant *Myzus persicae* (Sulz.) in Britain in 1976, *Pestic. Sci.* **9**:189–201.

110. Schaefers, G., Bent, G., and Cannon, R., 1979, The green invasion, *New Sci.* **1979**(9 August):440–441.

111. Shaw, M.W., 1955, Preliminary studies on potato aphids in North and Northeast Scotland, *Ann. Appl. Biol.* **43**:37–50.

112. Sigvald, R., 1981, A method of forecasting the spread of potato virus Y in Sweden, in: *Abstracts of Poster Meeting Plant Virus Disease Epidemics Oxford,* pp. 87–88.

113. Sparrow, L.A.D., 1976, Recent trends in the activity of aphids infesting potatoes in south-east Scotland in relation to virus incidence in the crop. *Scot. Hortic. Res. Inst. Assoc. Bull.* **11**:8–14.

114. Tatchell, G.M., 1982, Aphid monitoring and forecasting as an aid to decision

making, in: *Proceedings 1982 British Crop Protection Symposium,* pp. 99–112.

115. Tatchell, G.M., Parker, S.J., and Woiwod, J.P., 1983, Synoptic monitoring of migrant insect pests in Great Britain and Western Europe. IV. Host plants and their distribution for pest aphids in Great Britain, Rothamsted Experimental Station Report for 1982, Part 2, pp. 45–159.

116. Taylor, L.R., 1951, An improved suction trap for insects, *Ann. Appl. Biol.* **38:**582–591.

117. Taylor, L.R., 1955, The standardization of air-flow in insect suction traps, *Ann. Appl. Biol.* **43:**390–408.

118. Taylor, L.R., 1973, Monitor surveying for migrant insect pests, *Outlook Agric.* **7:**109–116.

119. Taylor, L.R., 1974, Monitoring change in the distribution and abundance of insects, Report of the Rothamsted Experimental Station for 1973, Part 2, pp. 202–239.

120. Taylor, L.R., 1977a, Aphid forecasting and the Rothamsted Insect Survey, *J. R. Agric. Soc. Engl.* **138:**75–97.

121. Taylor, L.R., 1977b, Migration and the spatial dynamics of an aphid *Myzus persicae, J. Anim. Ecol.* **46:**411–423.

122. Taylor, L.R., 1979, The Rothamsted Insect Survey—An approach to the theory and practice of synoptic pest forecasting in agriculture, in: R.L. Rabb and G.G. Kennedy (eds.), *Movements of Highly Mobile Insects: Concepts and Methodology in Research,* North Carolina State University, Raleigh, North Carolina, pp. 148–185.

123. Taylor, L.R., 1981, Editor's Preface, in: L.R. Taylor (ed.), *Euraphid Rothamsted 1980,* Rothamsted Experimental Station, Harpenden, p. 3.

124. Taylor, L.R., 1982, Changing aphid behaviour and distribution, in: J. Bernard (ed.), *Euraphid Gembloux 1982,* Centre de Recherche Agronomique de Gembloux, pp. 69–77.

125. Taylor, L.R., 1983, EURAPHID: Synoptic monitoring for migrant vector aphids, in: R.T. Plumb and J.M. Thresh (eds.), *Plant Virus Epidemiology,* Blackwell, Oxford, pp. 133–146.

126. Taylor, L.R., and French, R.A., 1968, A continuous census of flying insects, Report of the Rothamsted Experimental Station for 1967, pp. 194–198.

127. Taylor, L.R., and French, R.A., 1981, Synoptic aerial monitoring as a basis for an aphid forecasting system in Europe—EURAPHID, in: L.R. Taylor (ed.), *Euraphid Rothamsted 1980,* Rothamsted Experimental Station, Harpenden, pp. 9–12.

128. Taylor, L.R., and Palmer, J.M.P., 1972, Aerial sampling, in: H.F. Van Emden (ed.), *Aphid Technology,* Academic Press, London & New York, pp. 189–234.

129. Taylor, L.R., Woiwod, I.P., and Taylor, R.A.J., 1979, The migratory ambit of the hop aphid and its significance in aphid population dynamics, *J. Anim. Ecol.* **48:**955–972.

130. Taylor, L.R., French, R.A., Woiwod, I.P., Dupuch, M.J., and Nicklen, J., 1981, Synoptic monitoring for migrant insect pests in Great Britain and Western Europe. I. Establishing expected values for species content, population stability and phenology of aphids and moths, Rothamsted Experimental Station Report for 1980, Part 2, pp. 41–104.

131. Taylor, L.R., Woiwod, I.P., Tatchell, G.M., Dupuch, M.J., and Nicklen,

J., 1982, Synoptic monitoring for migrant insect pests in Great Britain and Western Europe. III. The seasonal distribution of pest aphids and the annual aphid aerofauna over Great Britain 1975–1980, Rothamsted Experimental Station Report for 1981 Part 2, pp. 23–121.

132. Thomas, G.G., Goldwin, G.K., and Tatchell, G.M., 1983, Associations between weather factors and the spring migration of the damson-hop aphid, *Phorodon humuli, Ann. Appl. Biol.* **102**:7–17.

133. Turl, L.A.D., 1978, Epidemiology of potato aphids in 1975–1977 with regard to the incidence of potato leaf roll virus in Scotland, in: P.R. Scott and A. Bainbridge (eds.), *Plant Disease Epidemiology* Blackwell, Oxford, pp. 235–242.

134. Turl, L.A.D., 1980, An approach to forecasting the incidence of potato and cereal aphids in Scotland, *European and Mediterranean Plant Protection Organization Bull.* **10**:135–141.

135. Turl, L.A.D., 1982, Forecasting colonisation of crops using aphid suction-trap data, in: J. Bernard (ed.), *Euraphid Gembloux 1982,* Centre Recherche Agronomique de Gembloux, pp. 79–82.

136. Van Harten, A., 1981, The use of suction trap data in the Netherlands, in: L.R. Taylor (ed.), *Euraphid Rothamsted 1980,* Rothamsted Experimental Station, Harpenden, pp. 22–24.

137. Van Harten, A., 1983, The relation between aphid flights and the spread of potato virus Y^n (PVY^n) in the Netherlands, *Potato Res.* **26**:1–15.

138. Van Hoof, H.A., 1977, Determination of the infection pressure of potato virus Y^n, *Neth. J. Plant Pathol.* **83**:123–127.

139. Van Hoof, H.A., 1980, Aphid vectors of potato virus Y^n, *Neth. J. Plant Pathol.* **86**:159–162.

140. Watson, M.A., 1966, The relation of annual incidence of beet yellowing viruses in sugar beet to variations in weather, *Plant Pathol.* **15**:145–152.

141. Watson, M.A., and Healy, M.J.R., 1953, The spread of beet yellows and beet mosaic viruses in the sugar-beet root crop. II. The effects of aphid numbers on disease incidence, *Ann. Appl. Biol.* **40**:38–59.

142. Watson, M.A., and Heathcote, G.D., 1966, The use of sticky traps and the relation of their catches of aphids to the spread of viruses in crops, Report of the Rothamsted Experimental Station for 1965, pp. 292–300.

143. Watson, M., and Serjeant, E.P., 1964, The effect of motley dwarf virus on yield of carrots and its transmission in the field by *Cavariella aegopodiae* Scop., *Ann. Appl. Biol.* **53**:77–93.

144. Watson, M.A., Heathcote, G.D., Lauckner, F.B., and Sowray, P.A., 1975, The use of weather data and counts of aphids in the field to predict the incidence of yellowing viruses of sugar-beet crops in England in relation to the use of insecticides, *Ann. Appl. Biol.* **81**:181–198.

145. Way, M.J., and Cammell, M.E., 1973, The problem of pest and disease forecasting-possibilities and limitations as exemplified by work on the bean aphid, *Aphis fabae,* in: *Proceedings 7th British Insecticide Fungicide Conference,*Vol. 3, Brighton, pp. 933–954.

146. Way, M.J., and Cammell, M.E., 1982, The distribution and abundance of the spindle tree, *Evonymus europaeus,* in southern England with particular reference to forecasting infestations of the black bean aphid, *Aphis fabae, J. Appl. Ecol.* **19**:929–940.

147. Way, M.J., Cammell, M.E., Taylor, L.R., and Woiwod, I.P., 1981, The use

of egg counts and suction trap samples to forecast the infestation of spring sown field beans, *Vicia faba,* by the black bean aphid, *Aphis fabae, Ann. Appl. Biol.* **98**:21–34.

148. Wiktelius, S., 1977, The importance of southerly winds and other weather data on the incidence of sugar-beet yellowing viruses in southern Sweden, *Swed. J. Agric. Res.* **7**:89–95.

149. Wiktelius, S., 1982, Flight phenology of cereal aphids and possibilities of using suction trap catches as an aid in forecasting outbreaks, *Swed. J. Agric. Res.* **12**:9–16.

150. Wislocka, M., 1970, The occurrence of aphids, potato virus vectors (1962–1967), *Ziemniak,* pp. 107–122 (in Polish).

151. Wislocka, M., 1972, The dynamics of aphids on potatoes at three localities in the years 1962–1971, *Biul. Instytut Ziem.* **10**:19–25 (in Polish).

152. Woiwod, I.P., 1982, Computer mapping of insect survey data, in: J. Bernard (ed.), *Euraphid Gembloux 1982,* Centre Recherche Agronomique de Gembloux, pp. 83–91.

153. Woodford, J.A.T., Shaw, M.W., McKinlay, R.G., and Foster, G.N., 1977, The potato aphid spray warning scheme in Scotland, 1975–1977, in: *Proceedings 1977 British Crop Protection Conference—Pests and Diseases,* Brighton, pp. 247–254.

154. Woodford, J.A.T., Harrison, B.D., Aveyard, C.S., and Gordon, S.C., 1983, Insecticidal control of aphids and the spread of potato leafroll virus in potato crops in eastern Scotland, *Ann. Appl. Biol.* **103**:117–130.

5
Nepoviruses of the Americas

Richard Stace-Smith and Donald C. Ramsdell

Introduction

The idea that plant viruses spread through the soil and infect healthy plants via the roots is nearly as old as the concept of viruses as causal agents of plant diseases. The first experimental proof that soil-inhabiting nematodes may serve as vectors of plant viruses was not obtained until 1958 when Hewitt *et al.* (51) published their significant findings linking soil transmission of grapevine fanleaf virus with the feeding of the ectoparasitic nematode *Xiphinema index*. This publication stimulated similar experiments on other soil-borne viruses, and by the early 1960s, several viruses with similar biological and physical properties were shown to have species of *Xiphinema* or *Longidorus* as vectors. In addition to grapevine fanleaf, the initial group of viruses with nematode vectors included tobacco ringspot, tomato ringspot, arabis mosaic, raspberry ringspot, strawberry latent ringspot, and tomato black ring. These viruses were indistinguishable morphologically and they were either serologically unrelated to each other or distantly related. Cadman (13) recognized that the viruses involved had close natural affinities and he proposed the group name NEPO viruses as a meaningful abbreviation of their two distinctive properties: nematode transmission and polyhedral-shaped particles. This group name has withstood the test of time, and although officially reduced to lower case letters, it now constitutes one of the 26 plant virus groups recognized by the International Committee on Taxonomy of Viruses (75). Tobacco ringspot virus (TobRSV) is identified as the type member of the nepovirus group and 25 other members and five possible members are included. Although most of the plant viruses vectored by nematodes have polyhedral particles, two have rod-shaped particles and are placed in the tobravirus group.

Richard Stace-Smith, Department of Plant Pathology, Agriculture Canada, Research Station, Vancouver, B.C. V6T 1X2, Canada.
Donald C. Ramsdell, Department of Botany and Plant Pathology, Michigan State University, East Lansing, Michigan 48824, USA.

This name is an abbreviation taken from tobacco rattle virus, the type member. The tobraviruses are transmitted by species of *Trichodorus* and *Paratrichodorus* and will not be discussed in this chapter. Readers wishing further information are referred to review articles by Martelli (74) and Harrison (48), where both nepoviruses and tobraviruses are discussed.

The nepoviruses have been the subject of a number of reviews over the past few years. Most articles are general and include discussions on the biology and epidemiology of both nepoviruses and tobraviruses. Two reviews deal specifically with nepoviruses: the first, by Harrison and Murant (49), is a description of the group; the second and more comprehensive, by Murant (89), examines properties, transmission, epidemiology, and control. These are both recent and current review articles and little purpose would be served in preparing another review of the advances that have been made over the past few years. One shortcoming of previous reviews is that, even though they summarize the world literature, they tend to stress the European nepoviruses. This is understandable, since the bulk of the research work has been done in Europe and the authors draw on their experiences with the European nepoviruses. We opted to restrict our coverage geographically to the Americas for two reasons: we are more familiar with the viruses that occur in this region, and we sense an increasing recognition of the importance of the nepoviruses in many of the crops grown in the Americas. It is the intention of this review to fill a niche in the literature and possibly to stimulate an awakening interest in this important group of plant viruses.

The Nepovirus Group

Characteristics of the Group

Purified preparations of nepoviruses subjected to rate zonal centrifugation in sucrose gradients usually reveal three classes of isometric particles. These particles are approximately 28 nm in diameter and often have hexagonal outlines. They sediment at approximately 55S, 90–120S, and 128S for the top (T), middle (M), and bottom (B) components, respectively. The T component is composed of empty protein shells, which are generally thought to consist of 60 identical subunits each with a molecular weight of approximately 55,000. The M and B components have the same protein shell as the T component, but differ in that they also contain strands of RNA. The RNA of the M component, designated RNA-2, varies in its molecular weight from 1.3×10^6 to 2.4×10^6 and the B-component RNA designated RNA-1, has a molecular weight of about 2.6×10^6. The percentage RNA in the three classes of particles is 0 (T), 27–40 (M), and 42 (B). Those viruses with low-molecular weight RNA may package two strands of RNA-2 in a protein shell, yielding a dense particle that cosediments with particles containing a single strand of RNA-1. Both RNA species are necessary for infection. In infected plants, virus particles reach a concentration of 10–50 mg/kg of tissue. Particles are reasonably stable and can be purified by a variety of techniques. They have thermal inactivation points of 55–65°C and remain infective for a few days to a few weeks. The viruses are readily sap-transmissible from woody or herbaceous hosts to herbaceous test plants, but are sap-transmitted with difficulty into woody hosts. Seed and pollen transmission are common. The viruses have a wide natural and experimental host range, causing ringspot or mottle symptoms on the inoculated leaves followed by symptomless infection.

American Nepoviruses

Since there are no universally accepted criteria for the separation of viral "species" from viral "strains," there is a divergence of opinion as to the number of distinct viruses that should be included in the nepovirus group. All those viruses that are recognized as members are sap-transmissible, are seed- and pollen-borne, cause diseases characterized by shock and recovery phases, and induce similar symptoms in herbaceous test plants. Morphologically, they all possess polyhedral particles that sediment as three components. Viruses that satisfy the above criteria but do not react serologically with other recognized members of the group are considered to be tentative members of the group. When a tentative member is found to be serologically related to a recognized member, should it be given a distinct name, or should it be considered a serotype or strain of the previously described virus? This problem has arisen a number of times and will undoubtedly continue to arise in attempts to classify the nepoviruses. Some authors have taken the conservative approach and have recognized as distinct only those viruses that are antigenically unrelated, with others being regarded as serotypes or strains. Using such a system of classification, Francki and Hatta (29) recognized tobacco ringspot virus as the type member and included only six additional viruses as distinct members. Grapevine fanleaf virus, for example, known to be distantly related serologically to arabis mosaic, was therefore considered a serotype or strain of arabis mosaic. While this grouping is justifiable, their listing of peach rosette mosaic virus as a strain of tomato ringspot is indefensible. These two viruses are serologically unrelated and with the information available, they should both be considered distinct members.

Given the current state of our knowledge as to which criteria are meaningful in plant virus classification, serological tests to demonstrate the degree of antigenic relatedness to known viruses are essential. Experience with the nepoviruses has shown that, if two viruses are related, the sedimentation values of their middle components are approximately equal, and thus the sedimentation pattern of a suspected nepovirus can provide clues as to the possible identity of the virus. Recognizing the inherent problems of excessive splitting in virus classification, we prefer to apply

distinctive names to those that show a sufficiently distant serological relationship to an accepted member. This applies specifically to Andean
potato calico and eucharis mottle, two viruses that are related serologically
to tobacco ringspot.

Another classification problem concerns the protein coat. Mayo *et al.*
(76) investigated the protein coats of three nepoviruses and found that
each consisted of a single species of polypeptide molecule with an estimated molecular weight of 55,000. The protein coat of several other suspected nepoviruses has since been shown to have a single species of polypeptide molecule also of molecular weight 55,000. This property is now
recognized as a meaningful criterion for classification. A problem has arisen in that a few viruses, typical of nepoviruses in all other aspects, have
two species of polypeptide coat proteins. The protein coat of cherry raspleaf, one of the North American viruses, has two species of polypeptides
with molecular weight 24,000 and 22,500. In this review, it is being considered a nepovirus, recognizing that it and others having more than one
polypeptide in the coat protein may not remain in the group.

A further consideration is whether those viruses that appear to have
originated elsewhere and to have been introduced to America more recently should be listed with American nepoviruses. Two such viruses,
grapevine fanleaf and cherry leaf roll, are known. Evidence suggests that
grapevine fanleaf virus has been indigenous in vineyards in the Mediterranean countries from the earliest ages of grape culture. From there, it
was introduced into Europe and subsequently to other parts of the world
in vegetatively propagated planting stock. The virus and its vector are
now common in grapevines grown in the Americas. Similarly, cherry leaf
roll virus is thought to have originated in Europe, but serological evidence
suggests that distinct strains may have arisen independently in Europe
and North America. These two viruses are both important plant pathogens
in America and are therefore included in our list of American nepoviruses.

When all the factors outlined above are taken into consideration, the
definitive and tentative American members of the nepovirus group stand
at 12. A list of the viruses and a few of their properties is summarized in
Table 5.1.

Exotic Nepoviruses in America

The nepoviruses are well adapted to dissemination within infected planting
stock to other parts of the world where the individual viruses are not
known to occur. First, a proportion of seed arising from infected plants
is infected and thus provides a means of spread in those crops that are
propagated by seed. Second, the wide herbaceous and woody host ranges
of many of the viruses in this group provides considerable opportunities
for nepoviruses to be present in hosts that are important in the international
plant trade. Third, most species that become infected with nepoviruses
tend to recover to the point where no obvious symptoms are expressed,

TABLE 5.1. Some properties of American members of the nepovirus group.

Virus	Synonym	Vector	Sedimentation coefficient[a]			Protein subunit (mol. wt. × 10³)
			T	M	B	
Tobacco ringspot	None	Xiphinema spp.[b]	53	91	126	57
Eucharis mottle	Eucharis strain of tobacco ringspot	Unknown	—	—	—	—
Andean potato calico	Potato black ringspot	Unknown	49	84	117	59
Grapevine fanleaf	None (in America)	Xiphinema index	50	86	120	54
Arracacha virus A	None	Unknown	50	92	125	53
Tomato top necrosis	None	Unknown	52	102	126	—
Cherry leaf roll	Elm mosaic, golden elderberry mosaic	Unknown	52	114	132	54
Peach rosette mosaic	Grape decline, grape degeneration	Longidorus diadecturus	52	115	134	57
Potato virus U	None	Unknown	55	117	135	58
Blueberry leaf mottle	American strain of grapevine Bulgarian latent virus	Unknown	52	120	128	54
Tomato ringspot	Peach yellow bud mosaic, grapevine yellow vein	Xiphinema spp.[b]	53	120	128	58
Cherry raspleaf	Flat apple virus	Xiphinema spp.[b]	56	96	128	22.5, 24

[a]T, top; M, middle; B, bottom components.
[b]Most vector studies have been done with nematode populations identified as Xiphinema americanum. However, X. americanum sensu lato is now considered to be a complex of many species (69) and clarification of the vector potential of the component species is required.

making it difficult for plant quarantine officials to suspect infected plant material. The combined effect of these characteristics would suggest that it is virtually impossible to prevent the international dissemination of ne- poviruses without placing an embargo on the exchange of planting stock. However, much to the credit of quarantine agencies, it appears that there have been remarkably few recorded instances of nepovirus introduction into America. Undoubtedly, some introductions have gone undetected, but in the major crops, dissemination by man appears to be minimal. Only three of the European nepoviruses have been isolated from planting ma- terial in America, and there is no evidence that these viruses have become established to any extent in planting stock in the Americas. There are two reports of the European nepovirus strawberry latent ringspot being re- covered from plants in America. The first was from a single sweet cherry tree located in the Niagara peninsula, which had been in the field for 16 years and was from a local budwood source grafted onto *Prunus mahaleb* understock (1). The virus source was thought to be the understock, which was believed to have originated in the United States. Since this virus is not known to occur in cherry in the United States, but does occur in cherry in Europe, it is more probable that this is an instance of chance importation in infected *P. mahaleb* seed originating from Europe. The other isolation was from parsley seedlings in California, where the source was seedlots imported from western Europe (46). Arabis mosaic virus, known to infect rhubarb in Europe, has been isolated from rhubarb plants in Ontario (W. Kemp, unpublished) and British Columbia (R. Stace-Smith, unpublished). In both instances, the infected planting stock is thought to have originated from Britain. The virus has also been isolated from a few grapevine clones imported to British Columbia under quarantine from Germany (R. Stace-Smith, unpublished). Arabis mosaic virus has been isolated from a dogwood tree in South Carolina, but no explanation was offered as to why a virus that is not endemic to North America was found in a native tree (7). More recently, tomato black ring virus has been isolated from grapevine clones in the Niagara peninsula (126). The virus has pre- viously been isolated from grapevines in Germany and the explanation for the occurrence of the virus in grapevines in Canada is that it was introduced by chance in infected stocks imported from Germany.

Nepoviruses as Disease Agents

Of the 12 nepoviruses listed in Table 5.1, only tobacco ringspot and tomato ringspot have wide geographical distribution in North America and are capable of infecting both wild and cultivated plants. The fact that these viruses are so widespread and are able to infect a variety of crops means that they are by far the most economically important nepoviruses in the Americas. The occurrence of devastating diseases attributed to these two viruses is correlated with the occurrence of high populations of nematode vectors belonging to the genus *Xiphinema*. If the viruses are disseminated in planting material to areas where *Xiphinema* spp. are either absent or present in low numbers, the disease problem is usually negligible. Five other nepoviruses, namely cherry leaf roll, cherry raspleaf, peach rosette mosaic, blueberry leaf mottle, and tomato top necrosis, may be of economic significance in restricted geographical areas and in one or two crops. Another virus, grapevine fanleaf, is restricted to grapevine and is of importance in some grapevine cultivars in both North and South America. Finally, eucharis mottle, Andean potato calico, arracacha virus A, and potato virus U are South American nepoviruses and their economic significance is unknown.

Tobacco Ringspot

Tobacco ringspot virus (TobRSV) fully justifies its designation as the type member of the nepovirus group, being described as early as 1927 (33) and, since that time, being used extensively in experimental studies. The virus has a wide host range, having been found naturally or transmitted experimentally to plants in at least 38 genera representing 17 families. It is associated with serious diseases in annual and perennial crops. Transmission by *Xiphinema americanum* was reported in 1962 (34). Several insect species may also serve as vectors (73). The virus is widespread in the eastern and southern United States. The major diseases associated with TobRSV will be discussed.

RINGSPOT DISEASE OF TOBACCO

TobRSV is widespread in the tobacco-growing areas of North America. In field-grown tobacco the virus causes concentric line patterns of chlorotic and necrotic tissue on the leaves. The line patterns are frequently circular or they may follow the veins as irregular wavy lines or streaks. When located in interveinal regions, spots are circular and the lines are necrotic; when centered near larger veins, the spots are irregular and symptoms follow the veins and their branches. The ringspots may be numerous or few and are often confined to one side of the plant. Symptoms may fail to develop on new leaves of systemically infected plants. Severely affected plants may be dwarfed, with small leaves of poor quality. The virus causes pollen sterility, which reduces the quantity of seed produced (73).

BUD BLIGHT OF SOYBEAN

This disease is common throughout the soybean-growing region of the eastern United States and extends into southern Ontario in Canada (118). Of the many diseases caused by TobRSV, bud blight of soybean is the most severe and causes the greatest losses; yield is reduced, depending on the number of plants infected and on the time of infection. Plants infected while less than 5 weeks old have shortened internodes and fewer nodes and as a result are severely stunted; plants infected when they are more mature show milder symptoms. The most striking symptom is curving of the terminal bud; other buds on the plant later become brown and brittle, hence the name bud blight. Brown streaks may develop in the stems and petioles of the larger leaves. Leaflets are dwarfed and tend to roll or cup. Pods are underdeveloped or aborted, and those that set before infection may develop dark blotches. Such pods generally do not produce viable seed and drop early. In addition, root nodule growth is reduced. The virus causes systemic infection in susceptible cultivars, moving from infected leaves to the tips of stems and into roots. Movement from roots to leaves is uncommon. Transmission to soybean by *X. americanum* is inefficient and even when it does occur, transmission from roots may be of no significance because the virus generally remains confined to this part of the plant. Leaf- and flower-feeding insects, such as thrips and beetles, may serve as vectors. Seed transmission is important for long-range dissemination and carryover of inoculum from season to season. However, even if seed transmission in soybean did not occur, carryover inoculum would be maintained in the seeds of many natural weed hosts.

BLUEBERRY NECROTIC RINGSPOT

This disease occurs in the major blueberry-growing regions of the United States (21,71). Cultivars of blueberry show considerable variability in their susceptibility. Those that are susceptible are stunted, unproductive, and show extensive twig dieback. New leaves produced in the spring have chlorotic or necrotic spots, rings, or line patterns. As the leaves mature, the chlorotic areas become necrotic and drop out, giving the leaf a tattered appearance. Severely affected leaves are reduced in size and deformed. Some cultivars that are less susceptible may show tip dieback followed by recovery and then produce new growth that is symptomless.

CUCURBIT RINGSPOT

TobRSV is prevalent in cucurbits in Texas (80) and Wisconsin (119). Naturally infected plants are stunted and show a leaf mottle accompanied by leaf malformation. Fruit set and size are greatly reduced. A diagnostic symptom of infected watermelon is the upright position of the distal ends of runners compared with the nearly prostrate position of vines of healthy plants. As the plant matures, leaves on infected plants become tattered and necrotic and internodes are shortened, producing a compact, bunchy plant with brittle leaves and stems. Many affected plants tend to recover and their symptoms become masked. Symptoms persist longer in squash than in other affected cucurbits. Failure of some infected plants to recover may indicate that they are infected with other cucurbit viruses in addition to TobRSV.

In addition to the diseases listed above, which represent the major ones caused by TobRSV, the virus occurs to a lesser extent in a wide range of wild and cultivated plants. The plants that have been found to be naturally infected with TobRSV include American spearmint (127), blackberry (112), cherry (124), apple (70), grapevine (37), geranium (64), ash (52), dogwood (111, 142), forsythia (142), autumn crocus (142), and elderberry (144).

Tomato Ringspot

The designation "tomato ringspot virus" was first proposed in 1937 for the causal agent of a disease of field tomatoes in Indiana (57). This disease was characterized by ringspot markings on the foliage and fruit. The same designation was applied to a virus that was originally described in 1936 as tobacco ringspot no. 2, a virus that occurred spontaneously in otherwise healthy Turkish tobacco seedlings growing in an experimental greenhouse in New Jersey (98). The two viruses were later found to be unrelated and the tomato ringspot (TomRSV) designation came to be used exclusively for tobacco ringspot no. 2. It was an unfortunate series of events that led to the designation of tomato ringspot as the vernacular name of a virus that does not occur naturally in tomato, but causes serious diseases in other crops.

TomRSV was originally distinguished from TobRSV on the basis of plant-protection tests. It was later found that TomRSV failed to react with antiserum against TobRSV, and, conversely, antiserum against TomRSV failed to react with TobRSV (130). Since that time, the use of plant virus serology has been the standard method for distinguishing these two viruses. In contrast to TobRSV, where the most severe problems are associated with annual crops, TomRSV is almost exclusively confined in nature to perennial crops. The virus constitutes a serious economic problem in those areas where the nematode vectors and the virus both occur.

Symptoms that are now known to be associated with TomRSV infection were recognized in raspberry for many years before it was demonstrated that TomRSV was the causal virus. The first description of a raspberry disease caused by TomRSV was from eastern Canada, where the disease was named yellow blotch-curl (16). The next was from the western United States, where the name raspberry decline was used for a disease that was characterized by the absence of leaf and cane symptoms but which caused a general decline in plant growth and productivity (145). More recent work leaves little doubt that the virus involved in both instances was TomRSV. The term "ringspot" was later used to describe a disease that was general throughout plantings in the western United States (140). The same disease was later found in British Columbia, where it was shown for the first time to be caused by TomRSV (122). The virus is now known to occur throughout the raspberry-growing areas of the United States and Canada. It is not known to occur naturally in South America, but it has been introduced to Chile in red raspberry planting stock from North America (4). Field spread of the disease is restricted to those areas where vectors of the genus *Xiphinema* occur.

Symptoms of TomRSV in red raspberry vary, depending on the cultivar, the duration of the infection, and the stage of growth of the plants being examined. Plants normally exhibit no symptoms in the year that they become infected, but, in the spring of the following year, some leaves on the primocanes show yellow rings, line patterns, or vein chlorosis. These symptoms may be considered a "shock reaction," as distinct foliar symptoms are rare in subsequent years. Chronic symptoms include delayed foliation in the spring, varying degrees of chlorosis on the leaves of the fruiting canes, yield reduction, a high proportion of misformed or crumbly fruit, and a general decline in vigor (123).

RINGSPOT DISEASE IN STRAWBERRY

TomRSV was first isolated from native *Fragaria chiloensis* and other wild plants along the coast of northern California (30). Although the infected native strawberry plants were symptomless, the disease was lethal in eight of nine commercial cultivars that were graft-inoculated (81). Natural infection of strawberry has only recently been reported (20), and surveys to determine the prevalence of the virus and its economic significance have yet to be conducted.

PRUNUS STEM PITTING AND DECLINE

A serious disease of peach, nectarine, sweet cherry, almond, and plum, called yellow bud mosaic in California (117), is caused by a strain of TomRSV (14). Typical yellow bud mosaic symptoms appear to be confined to California, but stem pitting, one of the symptoms associated with yellow bud mosaic in California (120), is commonly associated with TomRSV infection of peach and cherry in the eastern United States (121). The first symptom on newly infected peach or nectarine trees is the development of yellow blotches or spots on the leaf blades. They are irregular in outline, feather-edged, and usually follow the main veins. The blotches commonly occur toward the base of the leaf and predominantly on one side of the midrib. Blotching is accompanied by leaf distortion, which takes the form of pinching, puckering, corkscrewing, and lateral bending of the leaf blade toward the chlorotic areas. The chlorotic areas frequently die, resulting in holes and tattering. In chronically infected trees, the yellow bud mosaic symptoms are most clearly expressed in the spring as the new leaves expand. Growth of some buds is severely retarded, producing tufts of pale yellow leaves. These tufts often turn brown and die, causing the affected buds to stand out in sharp contrast to normal-appearing leaves on the same branch or in other parts of the tree. With the death of the tufts, affected shoots have relatively sparse foliage, giving diseased trees an open, thin appearance. The virus moves slowly upward so that, after several years, fruit is produced only at the branch extremities. Symptoms may vary somewhat, depending on the age of the tree or the duration of the infection and on the climatic changes from year to year.

GRAPEVINE DECLINE

TomRSV was first isolated from several *Vitis vinifera* cultivars in California, where the disease was called grape yellow vein (41). Initial symptoms in inoculated grapevines consisted of faint chlorotic mottling and line patterns that were either irregular, ringlike, or in the shape of oak leaves. These patterns were seldom seen in naturally infected vineyards, where the most common leaf symptoms were chrome-yellow flecks or yellow flecks of a pebbly texture along the veins. Vines varied widely in both the severity of leaf symptoms and in the number of leaves displaying symptoms. Fruit clusters on diseased vines were often completely sterile or set a reduced number of normal or seedless berries. Cluster stems on which no berries developed turned brown and died.

TomRSV was later isolated from several infected grapevine cultivars in eastern North America (24, 36). Symptoms in infected plants varied widely, depending primarily on the susceptibility of the various cultivars. Some cultivars showed only slight mottling of the foliage, increased cane size, and a slight to moderate reduction in fruit set. In contrast, other cultivars developed a pronounced chlorotic mottling of the foliage, shortened internodes, severe stunting, and fruit clusters that were partially filled with berries varying greatly in size and maturity.

APPLE UNION NECROSIS AND DECLINE

Apple union necrosis and decline is a disease that is recognized in the apple-growing areas of the United States (94, 128) and Canada (70). It is strongly associated with the apple cultivars propagated on MM106 clonal rootstocks of trees with union incompatibility symptoms. Although the evidence is not conclusive in all cases, the association between TomRSV infection and the occurrence of the disease is sufficiently strong to implicate TomRSV as a major cause of the disease. The most reliable diagnostic symptoms are pitting, invagination, and necrosis in the woody cylinder at the graft union. Symptom development at the graft union is thought to result from differences in rootstock and scion susceptibility to TomRSV, since the virus is often detected in rootstocks, but not in scions of diseased trees (128). Other factors, such as fruiting and environmental stresses, may play significant roles in inducing symptom expression (22). Symptoms may include sparse, small, pale leaves, abnormally heavy flowering and fruit set, small fruit, reddish bark with prominent lenticels, death of lateral leaf and flower buds, and short internodes producing bunchy-type terminal growth. In severely affected trees the trunk may be swollen above the union and there may be partial or complete cleavage of the union.

Contaminated nursery stock is not considered to be an important source of infection. Since the virus is common in dandelion and survives in a proportion of seed from infected plants, TomRSV-infected dandelion seed is thought to be a major source for both interorchard and intraorchard spread. The vector nematodes, which are prevalent in many orchards, may acquire virus from orchard weed hosts and transfer it to apple trees (96).

OTHER DISEASE RECORDS

While diseases in raspberry, strawberry, *Prunus* spp., grapevine, and apple account for nearly all of the virus problems caused by TomRSV, there are a few reports of the virus being isolated from other perennial crops, mainly ornamentals. These include red currant (54), pelargonium (65), gladiolus (11), hydrangea (12), elderberry (136), orchid (38), dogwood (111), birdsfoot-trefoil (92), cucumber (79), and ash (53).

Cherry Leaf Roll

Cherry leaf roll virus (CLRV) is a typical nepovirus in all respects, except that it has not been conclusively demonstrated that the virus can be transmitted by nematodes despite serious attempts to do so in North America (35) and Europe (60). There are several arguments to support the hypothesis that CLRV has evolved to the point where it can survive and spread without the involvement of a nematode vector, assuming that it may have had such a vector at one stage in its evolution. First, the virus infects the pollen and, even though infection of a healthy plant that is pollinated with CLRV-infected pollen has not been demonstrated, it may occur often enough to account for all field spread. Second, each host of CLRV appears to be infected with a distinct serotype of the virus, suggesting a prolonged association between the virus and its specific host. This would be expected when the natural means of virus transmission is exclusively by infected pollen. With nematode transmission, a range of host plants in the same locality would have become infected with the same virus strain. Finally, in contrast to other nepoviruses, which are usually restricted to one geographical area, CLRV is established and spreads in both North America and Europe. This is more likely to happen with a virus that does not require a specific nematode vector for natural spread. The virus is disseminated by man with infected planting material, so that even in the absence of a nematode vector, the virus can spread.

ELM MOSAIC

A mosaic disease of elm was first observed in Ohio as early as 1927 and later in several other eastern states (129). Some leaves on affected trees appear normal, while others are abnormally large and dark green, and still others are small, mottled, and distorted. Those leaves that are smaller develop either a yellow-green mottle, a diffuse mottling along the midribs, or chlorotic ringspot symptoms. In severe infection, leaf buds may fail to expand, leaving the branches bare of foliage for a considerable distance. Such trees have the appearance of having their leaves in bunches or tufts. Trees are not killed, but individual branches may die and some trees show a gradual decline in vigor.

Elm mosaic was found to show one-way protection in plant-protection studies with TomRSV and this, in addition to the similarity in physical properties, host range, and symptoms, led to the suggestion that the virus was probably related to TomRSV (139). In later serological studies, it was found to be unrelated to TomRSV, but instead it was found to be related to CLRV (59). The natural vectors of elm mosaic have not been demonstrated; Fulton and Fulton (35) found that it was not transmitted by nematodes belonging to the genus *Xiphinema* and Callahan (15) reported that it was transmitted through the pollen.

WALNUT BLACK-LINE DISEASE

Walnut black-line is considered the most important factor limiting walnut production in some regions of California (86). Symptoms are not observed in English walnuts propagated on English walnut seedling rootstock, but it is a widespread disorder associated with the decline of trees propagated on *Juglans hindsii* or Paradox *(J. hindsii* × *J. regia)* seedling rootstocks. Affected trees show poor terminal growth, yellowing and drooping of leaves, and premature defoliation followed by dieback of terminal shoots, general tree decline, and profuse puckering of the rootstock. Positive diagnosis is based on an examination of the union, where affected trees show small holes and cracks in the bark at the scion–rootstock junction, and, upon removal of the bark, a narrow strip of darkened cambium and phloem tissue. Eventually the black-line completely encircles the tree at the scion–rootstock junction, resulting in death of the scion within 2–6 years. Conclusive proof of the pathogenicity of CLRV was obtained when purified virus was applied to the cambium of the scion, resulting in black-line symptoms within 18 months (85).

The causal relationship between the black-line symptoms and CLRV was indicated by the consistent isolation of CLRV from English walnut scions, but not from the *J. hindsii* or Paradox rootstocks (86). The hypothesis is that these rootstocks show resistance because of a hypersensitive reaction to CLRV. It follows that healthy English walnut scions on CLRV-resistant rootstocks could not become infected via the rootstocks, hence an above-ground vector must be involved, probably CLRV-infected pollen. While this mode of transmission has not been demonstrated conclusively in the field, experiments show that when CLRV-infected pollen is used a percentage of the resulting nuts are infected (87).

OTHER DISEASES

In Europe, CLRV causes a disease in cherries, but in North America, the virus has not been found in *Prunus*. It has been recorded from dogwood (141), golden elderberry (44), and hoary alyssum (72).

Cherry Raspleaf

Cherry raspleaf disease, caused by cherry raspleaf virus (CRLV), occurs in western North America (125). Nyland *et al.* (91) demonstrated that the virus was vectored by *X. americanum* in California, and the same nematode was shown to be a vector in British Columbia (45). The virus satisfies the criteria for inclusion in the nepovirus group, except that its protein coat consists of two species of polypeptide molecules of molecular weight 24,000 and 22,500 (125).

The most characteristic symptom of the disease in cherry is prominent enations on the underside of the leaves. The enations take the form of leafy outgrowths between the lateral veins and along the midrib. Affected leaves are narrow, folded, and deformed. The upper leaf surfaces have a rough, pebbly texture with depressions corresponding to the enations on the lower surface. Newly infected trees usually develop enations on the lower leaves first and the virus then spreads within the tree slowly, particularly in mature cherry trees. Affected spurs and branches may die, giving the tree an open, bare appearance. Trees that are affected at an early age remain stunted, produce few fruits, and are occasionally killed by the virus.

Flat apple was a disease of unknown etiology when first described in Oregon and Washington in 1963 (10). In 1977, Parish (93) provided evidence that the flat apple disease was induced by cherry raspleaf virus. Characteristically affected apples on some branches are distinctly flattened; i.e., they are about half as long as they are wide. The calyx lobes are prominent, the calyx basin is broad and open to the core, and the stem cavity is shallow. Affected apples are smaller and have prominent lenticels, particularly in immature fruit. Leaves in the affected region of the tree are long and narrow and appear brittle and dry. Growing points are more numerous, and the growth is short and stubby. After being infected for a few years, diseased trees are dwarfed and have a dense, bushy appearance.

In addition to cherry and apple, the virus has been recovered from peach, dandelion, plantain, and balsamroot (45). More recently, it has been recovered from a few raspberry plants in Quebec (R. Stace-Smith, unpublished). The origin of these plants is uncertain, but it is possible that the initial infected planting material could have originated in western North America.

Peach Rosette Mosaic

Peach rosette mosaic was recognized as a virus disease of peach in Michigan as early as 1917 (66). In addition to *Prunus* spp., peach rosette mosaic virus (PRMV) infects grapevine (26, 102) and blueberry (101). Klos *et al.* (67) reported that *X. americanum* and a *Criconemoides* sp. were vectors and, more recently, *Longidorus diadecturus* has been implicated as a vector in Ontario (2). Evidence suggests that *L. diadecturus* is more important than *X. americanum* as a vector of PRMV in peach in Ontario. The vector potential of *Criconemoides* sp. has not been confirmed. Natural distribution of the virus has been reported by Dias (23), Dias and Cation (26), and Dias and Allen (25).

Affected peach trees show delayed foliation, chlorotic mottling and distortion of the early formed leaves, and shortening of the internodes to produce a rosette appearance. Normal branches are interspersed with affected ones on an infected tree. Leaf chlorosis is evident early in the growing season and the chlorotic areas vary in size, shape, and color intensity. Tissue growth in the chlorotic area of the leaf is retarded, resulting in leaf distortion. Leaves formed later in the season are of near normal size and are darker green than normal leaves.

In grapevine, PRMV infection results in a delayed dormancy breaking, late and uneven bloom, small and uneven clusters of berries, and leaf deformity and mottling. Affected vines are lighter green than normal. Cane growth is typified by short, crooked growth of the first four to six internodes with odd angles of cane branching, giving the affected vine an umbrellalike growth habit. Vines that have been affected for a number of years become unproductive and often die.

Grapevine Fanleaf

Grapevine fanleaf virus (GFLV) was the first plant virus shown to be transmitted by a plant parasitic nematode. The nematode *X. index* was demonstrated to be the vector in California vineyards in experiments reported in 1958 (51). The virus is serologically related to arabis mosaic virus and in all respects is a typical nepovirus (99). It is known to occur throughout the temperate regions of the world wherever *Vitis vinifera* and hybrid rootstocks of grapevine are grown. The natural host range of the virus is restricted to *Vitis* spp., probably because of the high degree of vector specificity and the narrow host range of the virus. Like other nepoviruses, GFLV can be transmitted experimentally to numerous herbaceous species.

Symptoms in recently infected grapevines include ring and line patterns and spots on some leaves. These symptoms are transitory and are replaced by green and yellow mosaic. The degree of mosaic depends upon the strain of the virus involved and the level of susceptibility of particular cultivars. Sometimes the mosaic symptoms may be entirely lacking. In chronically infected plants, the virus creates a characteristic malformation of the leaves, in which the veins are abnormally spread to give the leaf blade the appearance of a fan. Less susceptible cultivars do not show such malformation, but may show a slight asymmetry and a reduction in the size of the leaf. Affected vines may show double nodes, short internodes, or abnormal branches. The number and size of fruit bunches are smaller than on normal plants, and the berries often fail to develop or remain small and seedless. Infection can cause a quick destruction of the plant or result in a decline over several years.

Blueberry Leaf Mottle

Blueberry leaf mottle virus was isolated from high bush blueberry in Michigan in 1977 (104). Serological tests showed that this previously undescribed nepovirus was distantly related to grapevine Bulgarian latent virus (99). The vector remains unknown, but the pattern of field spread suggests pollen transmission. The virus in blueberries is not known to occur outside of Michigan, but a closely related strain has been isolated from grapevine in New York State (105, 138). In grapevine, the diseased plant showed delayed budbreak, differential growth of shoots, and pale green leaves. By late summer, the affected vine appeared almost normal in growth and foliage color but the fruit clusters were small and contained many aborted berries.

Tomato Top Necrosis

In 1937, Imle and Samson (57) described a tomato disease characterized by intricate patterns of brown necrotic rings and lines on young leaves, broad, sunken necrotic streaks on petioles and stems of young shoots, necrosis of shoot terminals, and corky brown necrotic rings on green and ripe fruits. They noted that these symptoms suggested the name tomato ringspot. The virus, common in many canning-tomato fields in the southern part of Indiana, was also isolated from diseased tomatoes in Missouri, and from a single collection in Illinois (116).

Symptoms produced on tomato, jimson weed, and Turkish tobacco were strikingly different from those produced by the tomato ringspot virus described by Price (98). Serological evidence that the two viruses were distinct was obtained by Bancroft (5), who suggested the name tomato top necrosis virus for a virus isolated more recently from tomatoes in Illinois, and believed to be the same virus as the one described earlier (116). Nematode transmission has not been demonstrated, but the characteristics of TTNV are typical of nepoviruses.

Eucharis Mottle

Eucharis mottle virus was originally described as a new strain of tobacco ringspot virus isolated from *Eucharis candida* imported from Peru into the United States (63). Plants that were grown in quarantine showed a conspicuous mottle characteristic of a virus disease. The virus has only been isolated from the one shipment of bulbs, and nothing is known about its prevalence in Peru. It has been considered to be a distinct strain of TobRSV based on symptom differences in several hosts and on the formation of spurs in agar gel diffusion tests against known isolates of tobacco ringspot virus. More recently, Salazar and Harrison (115) carried out plant-protection and serological tests on several tobacco ringspot and allied viruses. Their results, together with previous results (42, 63), suggested that eucharis mottle should be considered a separate virus, but that further work would be required to test this proposal.

Andean Potato Calico

A potato virus that reacted in immunodiffusion tests against tobacco antiserum was first reported by Fribourg and Salazar in 1972 (32). After further studies on the virus, Fribourg (31) applied the designation "Andean potato calico strain of tobacco ringspot virus," noting that the virus was serologically distinct from eucharis mottle virus and North American isolates of TobRSV. Salazar and Harrison (114) applied the designation "potato black ringspot" to a different isolate of Andean potato calico. They concluded that potato black ringspot was sufficiently distinct from tobacco ringspot that it should be considered a separate virus. The criteria they used were antigenic affinity, behavior in plant-protection tests, and ability to form pseudorecombinants. Further work on the properties of the virus have been reported by Salazar and Harrison (115).

The disease was frequently found in the cultivar Ticahuasi in the irrigated desert plantings of the central coastline of Peru. It was also found in centers for seed-potato production in the highlands. Initial symptoms are yellow areas, which start at the leaf margins and progress until most of the leaf is affected. The virus does not appear to stunt or deform the leaves.

Arracacha Virus A

Plants of arracacha showing a pronounced yellow mosaic were found growing among potatoes during a survey of virus diseases in the Andean highlands of Peru. Samples revealed two polyhedral viruses, one of which had properties of a nepovirus. When serological tests failed to show any relationships with a range of nepoviruses, it was considered a new nepovirus (61). No nematode vector has been demonstrated and little is known about the natural host range and distribution of the virus in Peru.

Potato Virus U

Potato virus U is the name proposed for a virus isolated from a single
potato plant in central Peru (62). The plant (cultivar unknown) had bright
yellow markings and, although several viruses are known to cause cali-
colike symptoms in potato and herbaceous test plants, this virus was unrelated to others commonly isolated in Peru. From physical, chemical,
and biological properties, the virus was considered to be a new member
of the nepovirus group. The virus is readily seed-transmitted and, in preliminary experiments, it was transmitted by an unknown species of the
nematode *Longidorus*.

Ecology and Epidemiology

The Vectors

Trudgill *et al.* (135) defined some strict criteria by which experimental
nematode transmission can be judged as valid. Central to these criteria
is the requirement that the above-soil portion of bait plants, i.e., the healthy
test plants to which viruliferous test nematodes are introduced for possible
transmission of virus, must be shown to contain virus. Fulfillment of this
requirement proves that transmission did not occur as a result of root
contamination with viruliferous nematodes or their feces. The most recent
publications listing definitive and tentative nepoviruses and their vectors
are those of Murant (89) and Taylor and Brown (131). Four species of the
genus *Xiphinema (X. americanum sensu stricto, diversicaudatum, index,*
and *italiae)* as well as six species of the genus *Longidorus (L. apulus,
attenuatus, diadecturus, elongatus, macrosoma,* and *martini)* qualify as
valid vectors. In the Americas, the only species of *Longidorus* reported
as a vector is that of *L. diadecturus* for peach rosette mosaic virus (3).
Longidorus elongatus has been found associated with raspberry in British
Columbia, but a vector association has not been shown (77). In this same
report, it was noted that *X. bakeri* is also associated with raspberry, but
repeated attempts to experimentally transmit tomato ringspot virus with
this species have failed. Until the publication in 1979 by Lamberti and
Bleve-Zacheo (69) revising the taxonomy of nematodes that historically
were placed in *Xiphinema americanum*, all other American nepoviruses
were reported as being vectored by *X. americanum sensu lato* and *X.
index*. The aim of Lamberti and Bleve-Zacheo's paper was to clarify the
identity of several populations of "so-called" *X. americanum* from various
locations of the world.

   In this article, the authors place 15 new species of *Xiphinema* into six
groups, based on the morphobiometrics of the specimens examined. The
groups that are probably significant from the standpoint of nepovirus vector

species in the Americas are as follows: (1) *Xiphinema americanum sensu stricto,* which seems to be limited in its distribution to the eastern part of the United States and Canada. (2) *Xiphinema pachtaicum,* which has been found in California, and *X. californicum,* which has been found along the western seaboard of North America [recent reports of TomRSV transmission by *X. californicum* (56) suggest that previous reports of transmission of cherry raspleaf virus (91), TobRSV (133), and strains of TomRSV (134) by *X. americanum* in western North America were in fact done with *X. californicum*]. (3) No important vector species. (4) *Xiphinema rivesi,* which was recently reported in the eastern United States (88). (5, 6) No important vector species. The effect of Lamberti and Bleve-Zacheo's work has been to cast doubt as to the true species of *Xiphinema* that have been reported as vectors of American nepoviruses.

Blueberry leaf mottle and CLRV are examples of nepoviruses that have evolved a pollen-borne mode of transmission as an effective adaptation to the agricultural setting in which they primarily operate. It is common for virus particles of most nepoviruses to be found in pollen and seed of both naturally and experimentally infected host plants, but most, if not all, nepoviruses with proven nematode vectors are ultimately spread by their nematode vector.

Specificity of Transmission

Some classic experiments have been done involving raspberry ringspot virus transmission, using two serologically different strains of the virus and the two different vector species *L. elongatus* and *L. macrosoma* (50). Pseudorecombinant strains were made by combining the two RNA species from purified virions of the English and Scottish strains. The English strain is transmitted by *L. macrosoma* and the Scottish by *L. elongatus.* By combining the RNA-1 from one strain with the RNA-2 from the other, the smaller of the two RNA species, RNA-2, was identified as the genome piece that codes for the protein coat and nematode transmissibility of the virus. This finding led to the hypothesis that specificity of nepovirus transmission is linked to the interaction between the vector and the viral protein coat. It has been postulated that bound virus is released from the specific site in the nematode as a result of a pH change when the nematode salivates (132). Specificity of nepovirus transmission seems to vary greatly; for example, peach rosette mosaic virus has been experimentally transmitted with similar efficiency by both *L. diadecturus* and *X. americanum* (3). *Xiphinema americanum sensu stricto* transmits TobRSV and PRMV, while *X. americanum, X. riversi,* and *X. californicum* transmit TomRSV (88, 56). It is possible that some reports of nematodes as vectors of viruses are the result of contamination of the roots of test plants by viruliferous nematodes or their feces.

Population Dynamics

A review (143) of the ecology of *Xiphinema* and *Longidorus* covers the effects of soil, plant, climate, and seasonal factors on the population dynamics of the nematodes. That review was for the most part based on European experience, but many of the principles embodied in the article should be applicable in temperate climates of North America. A more recent short, but useful review of the ecology of virus vector nematodes was written by McGuire (78).

Xiphinema index has a much narrower host range than *X. americanum* and other *Xiphinema* spp. It is primarily found associated with grapevine and fig, but in Israel it has been found in association with sour orange, dwarf nettle, strawberry, and marigold (18). *Xiphinema index* can complete its life cycle in 22–27 days in California (100), but in Israel the cycle was found to take between 7 and 9 months. An Australian study (47) revealed that ovogenesis was restricted to the spring and summer months, which correlated with periods of active vine growth. The life cycle was observed to vary from 3 months in the summer to 9 months in the autumn, winter, and spring. In Israel, it was found that *X. index* was favored by a heavy soil with its better water-holding capacity, rather than a light soil, in which the nematodes dried out (18). Similar results were observed in Australia (47), where wet soils permit the nematode to survive as long as there is some air in the soil pores. In general the nematode populations were higher during the fall and winter in sandy soil than in heavy soil from other portions of the same vineyard. Populations were similar between the two soil types during spring and summer; this was attributed to the beneficial effects of the water retention in the heavier soil. Nematode populations were greatest at 15 cm and least at 65 cm under grapevines in Australia. Under California conditions, *X. index* was found to a depth of 185 cm.

While *X. index* has a restricted host range, *X. americanum sensu stricto* has a broad one, with woody plants as well as biennial and perennial herbaceous plants being favored (83, 95). *Xiphinema index* is relatively easy to maintain in captivity, while *X. americanum* does better under field conditions than in captivity. In general, *X. index* is more tolerant of adverse conditions than is *X. americanum* or most other *Xiphinema* spp.

A few population dynamics studies have been done with *Xiphinema* spp. in North America. Griffin and Darling (43) researched population levels of *X. americanum sensu stricto* in an ornamental spruce nursery in Wisconsin. They observed two seasonal population peaks, in which one cycle extended from April to August and the other from September to January. They believed that only eggs and fourth-stage larvae were able to survive the frozen winter conditions. During the winter, adults retreated to a depth of 50 cm, even though the soil was frozen to only 33 cm. Bird and Ramsdell (9) studied *X. americanum sensu stricto* population dynamics associated with grapevines in Michigan. This work was done primarily as part of a soil fumigation experiment to control the vector of

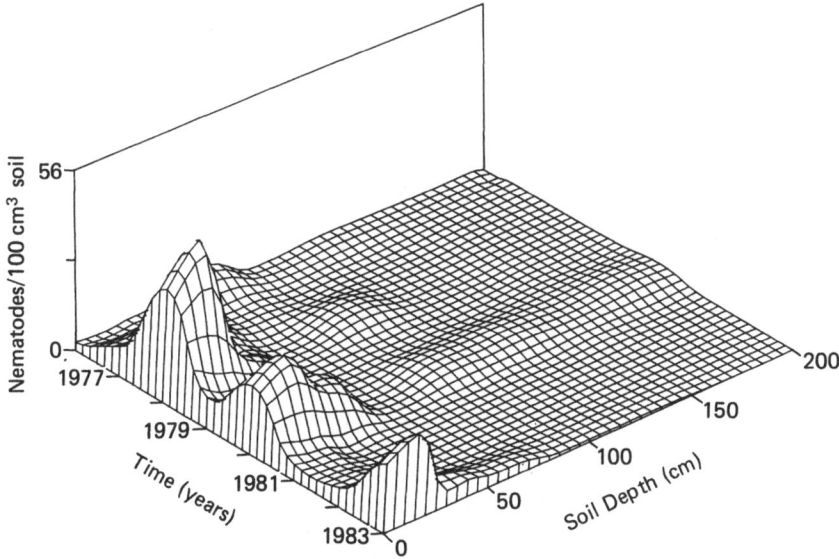

FIGURE 5.1. Population profile of *Xiphinema americanum sensu stricto* taken annually during July or August in a sandy soil in Michigan.

PRMV and infection of replanted healthy vines. Figure 5.1 shows 7 years of population profiles taken annually during July or August to a depth of 200 cm in a sandy soil. It is apparent from the three-dimensional diagram that >90% of the nematodes were in the top 50 cm of soil where grapevine feeder roots were primarily located. Grapevine roots went to at least 200 cm and *X. americanum* was found to the same depth as the grapevine roots. Populations fluctuated from year to year, with declines being correlated with low rainfall.

In central California, where the seasonal temperatures are higher than in the midwestern United States, Ferris and McKenry (27) did a population study of *X. americanum* (probably *X. californicum*) associated with grapevine. They found that population levels were higher in the fall and winter than in the spring and summer. As soil moisture increased, so did population levels. About 90% of the population was found in the upper 30 cm of soil, and gradually fewer nematodes were found as the soil depths increased to 120 cm. Since the ground does not freeze during the winter, populations in the upper 30 cm stayed fairly constant throughout the year. None of the North American studies determined the length of the life cycle of *X. americanum*. Studies in England (143) have shown that *Xiphinema* spp. may take from 2 to 3 years to complete their cycle, while in the midwestern and eastern United States the cycle may take at least 1 year (G.W. Bird, personal communication).

Since *Longidorus* is relatively unimportant as a virus vector in the Americas, we will not discuss its ecology and population dynamics.

Prevention and Control

Detection of Viruses

Before one can proceed with control measures directed toward prevention of nepovirus infection on a preplant basis in new plantings or a replant basis in established plantings, thorough detection of virus in the host plants or weeds is an essential first step. The use of the enzyme-linked immunosorbent assay (ELISA) has been a great step toward making detection of nepoviruses in crop plants and weeds a practical reality. A review of immunosorbent assays in plant pathology (17) discusses the variations of ELISA and other types of ultrasensitive serological tests relative to plant virus detection. The reader is referred to this comprehensive review for more details than will be given here. Although immunosorbent assays are effective, they are limited by the number and quality of antisera available. The possible presence of several viruses can be detected by sap-inoculation to herbaceous indicator plants, followed by serological identification. If several strains of a certain virus are involved, double-antibody sandwich ELISA is too strain-specific and modifications must be made to broaden its scope (6, 17, 68, 84).

Immunosorbent electron microscopy with the added modification of particle "decorating" has been developed for broad use in plant virus detection (84). It has been tried as a method for nepovirus detection in grapevines (113), but it has not been completely reliable. Radioimmunoassay was shown to be more sensitive and reliable than ELISA for the detection of TomRSV in cambium tissues from bark core samples at the bud union of stem-pitting diseased stone fruit trees (55).

ELISA can be used reliably to detect TomRSV in apple trees affected with apple union necrosis and decline (96), although the relative ability of TomRSV to multiply in the scion and/or rootstock will dictate in which tissue detection of the virus is more likely (22). For field surveys, the best sampling technique is to take cambium samples from near the graft union with a cork borer. On stone fruit crops, detection of TomRSV by ELISA is more difficult, and in Montmorency sour cherry it is unreliable using the cork borer technique (D.C. Ramsdell and S. Morrissey, unpublished data). Barrat *et al.* (8) found that even though TomRSV was easily detected in dandelions near affected peach trees, ELISA detection of TomRSV was not reliable when using graft-union cambium samples from the peach trees for testing. In general, with stone fruit trees and with mature apple trees, detection of TomRSV in the above-ground portions of the tree is difficult and unreliable.

Detection of TomRSV in grapevines by ELISA was found to be comparable with biological indexing on *Chenopodium quinoa* (39). Similar comparisons for peach rosette mosaic virus showed that *C. quinoa* indexing was more reliable (106). In bush berries and small fruit crops, ELISA works well when above-ground tissue samples are taken (19). Even though ELISA (58) and *C. quinoa* indexing (137) work well for detection of GFLV, graft indexing onto indicator vines is used in various parts of the world. The reasons for using this indexing method are that other viruslike agents can be detected and in some cases, equipment and greenhouse facilities are not available.

In general, there are few problems associated with ELISA detection of nepoviruses in the weeds they infect. For example, ELISA has been used to detect TomRSV, TobRSV, and PRMV in weeds in orchards (96) and vineyards (103). ELISA is facilitating some excellent epidemiological studies in orchard and vineyard plantings, which will lead to refining future control strategies.

Detection of Vector Nematodes

Determination of potential nematode vectors, their populations at various soil depths, and their field distribution is of paramount importance as part of a scheme to control nepoviruses, especially in woody perennial crops. This is true whether one is sampling prior to planting a new field or replanting an old field. Nepovirus vectors are ectoparasitic and lend themselves to easier sampling than if they were endoparasitic. Factors such as time of year, soil type and texture, root systems of crop and weed plants, and soil drainage have a bearing on how to sample to get an accurate assessment of the nematode population. In fruit crops in North America, where potential virus-vector nematodes may be harbored deep in the soil, samples down to 150 cm are required.

A high percentage of orchards and vineyards in North America have been planted without any prior nematode sampling. As nepovirus diseases become evident in woody crop plantings, growers are becoming aware of the requirement to sample for nematodes prior to planting. Agricultural research and advisory groups should recommend that growers sample to depths of 45 cm in a few locations in addition to the usual 15- to 20-cm sampling depths. Vector nematodes can be abundant at those depths and, if they are, fumigation treatments can be adjusted accordingly.

Weed Management

Recently much attention has been paid to the possible role of weeds in the epidemiology of spread of nepoviruses in North America. In a study of PRMV spread in Michigan vineyards, it was found that three of 16 weed species associated with diseased vines were infected (103). Two important virus sources were dandelion and grape seedlings. Dandelion seeds are windblown and may disseminate PRMV over considerable distances; grape pomace is returned from processors to growers on a random basis as a soil supplement and germinating seeds could introduce virus into vineyards that were previously free from PRMV. Figure 5.2 shows the pattern of spread of PRMV in such a vineyard where dandelions were

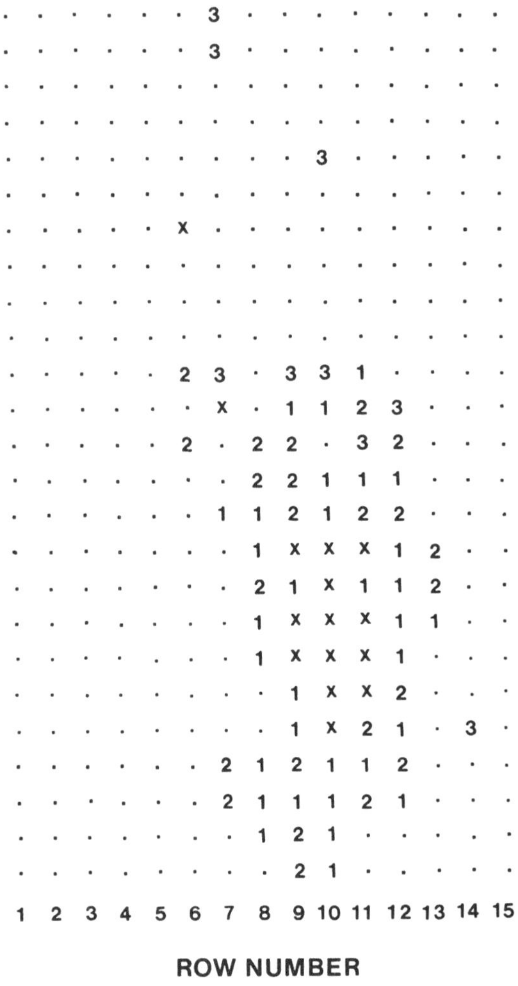

```
·  ·  ·  ·  ·  ·  3  ·  ·  ·  ·  ·  ·  ·  ·
·  ·  ·  ·  ·  ·  3  ·  ·  ·  ·  ·  ·  ·  ·
·  ·  ·  ·  ·  ·  ·  ·  ·  ·  ·  ·  ·  ·  ·
·  ·  ·  ·  ·  ·  ·  ·  ·  ·  ·  ·  ·  ·  ·
·  ·  ·  ·  ·  ·  ·  ·  ·  3  ·  ·  ·  ·  ·
·  ·  ·  ·  ·  ·  ·  ·  ·  ·  ·  ·  ·  ·  ·
·  ·  ·  ·  ·  x  ·  ·  ·  ·  ·  ·  ·  ·  ·
·  ·  ·  ·  ·  ·  ·  ·  ·  ·  ·  ·  ·  ·  ·
·  ·  ·  ·  ·  ·  ·  ·  ·  ·  ·  ·  ·  ·  ·
·  ·  ·  ·  ·  ·  ·  ·  ·  ·  ·  ·  ·  ·  ·
·  ·  ·  ·  2  3  ·  3  3  1  ·  ·  ·  ·  ·
·  ·  ·  ·  ·  ·  x  ·  1  1  2  3  ·  ·  ·
·  ·  ·  ·  ·  2  ·  2  2  ·  3  2  ·  ·  ·
·  ·  ·  ·  ·  ·  2  2  1  1  1  ·  ·  ·  ·
·  ·  ·  ·  ·  1  1  2  1  2  2  ·  ·  ·  ·
·  ·  ·  ·  ·  ·  1  x  x  x  1  2  ·  ·  ·
·  ·  ·  ·  ·  ·  2  1  x  1  1  2  ·  ·  ·
·  ·  ·  ·  ·  ·  1  x  x  x  1  1  ·  ·  ·
·  ·  ·  ·  ·  ·  1  x  x  x  1  ·  ·  ·  ·
·  ·  ·  ·  ·  ·  ·  1  x  x  2  ·  ·  ·  ·
·  ·  ·  ·  ·  ·  ·  1  x  2  1  ·  3  ·  ·
·  ·  ·  ·  ·  2  1  2  1  1  2  ·  ·  ·  ·
·  ·  ·  ·  ·  2  1  1  1  2  1  ·  ·  ·  ·
·  ·  ·  ·  ·  ·  1  2  1  ·  ·  ·  ·  ·  ·
·  ·  ·  ·  ·  ·  2  1  ·  ·  ·  ·  ·  ·  ·
 1  2  3  4  5  6  7  8  9 10 11 12 13 14 15
```

ROW NUMBER

FIGURE 5.2. Pattern of spread of peach rosette mosaic virus (PRMV) in a Concord grape vineyard in southwestern Michigan. (·) Healthy; (x) missing vine; (1–3) vines indexed positive for PRMV in 1981, 1982, and 1983, respectively.

present and where pomace had been spread. It is likely that weed-borne spread of PRMV accounts for a substantial proportion of the infection foci in more than 100 vineyards in Michigan.

A cooperative effort was recently made to assess the extent of orchard weeds that were infected with TomRSV and TobRSV in association with virus-diseased apple and peach orchards in the eastern United States (28, 96). Twenty-one species of annual and perennial weeds were found to be infected with TomRSV and eight species with TobRSV (96). Powell *et al.* (97) surveyed the possible origin of infection sites in Pennsylvania orchards and found that both *Xiphinema* and dandelions were ubiquitous and plentiful. Infected nursery stock provides the initial introduction of TomRSV into the fruit production areas. A change from clean cultivation by discing to minimum tillage has facilitated transmission of the virus by nematodes into dandelions and other weeds. As the number of TomRSV-infected dandelions has increased, the probability of virus transmission to healthy trees has also risen. It appears that TomRSV is endemic in dandelions and other weeds and their deep-rooted nature provides reservoirs for nepoviruses.

Chemical Control

USE OF FUMIGANTS

The arsenal of effective soil fumigants is dwindling. Ethylene dibromide has been declared a powerful carcinogen and its future availability is doubtful. Another previously used fumigant, DBCP (1,2-dibromo-3-chloropropane), was withdrawn in 1976. The remaining fumigants that can be used are DD or Telone II® (1,3-D), Vorlex® (methylisothiocyanate + 1,3-D), and methyl bromide or methyl bromide plus chloropicrin. There are also nonfumigant chemicals and these will be discussed later.

Fallowing is not an effective means of disease control because weeds and remaining roots of woody crop plants act as carryover hosts for the nematode vector and the virus. In California vineyards where vines infected with GFLV were removed, *X. index* was recovered for 10 years and GFLV for 5 years after fallowing was begun (108). In one California vineyard site under fallow, *X. index* was found at a depth of 240 cm. Superimposed shallow (S) plus deep (D) fumigation of the soil was done, applying 20% of the fumigant at a depth of 20–25 cm and 80% at a depth of 75–90 cm. Populations of *X. index* were reduced to zero at depths of 2 m for 3 years. At the end of 3 years, no vines had become infected with GFLV, but vines in the unfumigated control plot had. In another California test involving 2,4-D treatment to kill old GFLV-infected vines and their root systems prior to superimposed S plus D fumigation of soil, *X. index* populations stayed at zero and replanted vines remained free of virus disease for more than 6 years (109).

In Michigan, similar superimposed S plus D fumigant treatments were tested to control *X. americanum* and PRMV infection of grapevines (107). *Xiphinema americanum* was eradicated to depths of at least 180 cm for 8 years (Figure 5.3). Replacement vines have remained free of PRMV.

Soils that contain a considerable amount of clay or high organic matter do not lend themselves to effective soil fumigation. Deep, sandy soils that are in good tilth, free of plant debris, at a temperature of 13–25°C, and have a soil moisture content of 7–10% are ideal for successful soil fumigation.

There are problems associated with soil fumigation and plant health. Methyl bromide can destroy all mycorrhizal fungi and cause nutrient deficiencies to occur unless mycorrhizal fungi are inoculated to nursery stock at planting time. Phytotoxicity may occur if soils are not sufficiently aerated between the time of fumigation and planting.

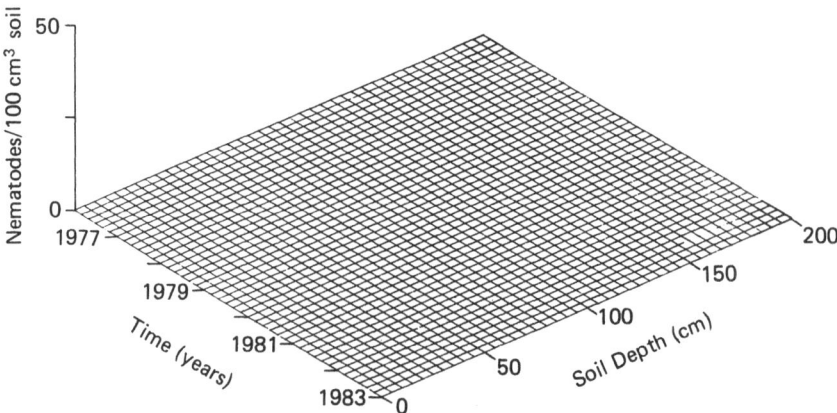

FIGURE 5.3. Population profile of *Xiphinema americanum sensu stricto* taken annually during July or August for 7 years following shallow plus deep fumigation of a sandy soil in Michigan (compare with Fig. 5.1).

The actual cost of a shallow preplant fumigation, where nematodes are present only in the upper zones of the soil (not more than 40–50 cm deep), is not excessive. The chemicals required would cost about $700–$1000/ha treated. Application costs would be extra. Such costs, if amortized over a 10-year period or longer, are economically feasible if new plantings are being put into a field with known nematode-vector and virus infestation.

In replant situations where virus and vector nematodes are present in a field with infection foci, the cost per unit area will be high. It has been estimated that a superimposed S plus D application of Telone II® at 1400 L/ha would cost approximately $3500/ha under California conditions, assuming a 15-ha treatment block (110). This may seem exhorbitant, but in fact this may not be the case. In Michigan, for example, a 20-ha vineyard usually has one or two infection foci, totaling only 1 or 2 ha in area. In order to stop the further spread of the disease, paying this seemingly high cost may well be worth the benefit derived from ultimately saving the vineyard.

USE OF NONFUMIGANTS

With the loss of DBCP as a fumigant for use in existing plantings of fruit crops, much interest has been generated in the use of systemic organic phosphate and carbamate nematicides. Systemic nematicides are relatively nonphytotoxic, are usually in a granular form, and are applied to the soil. The chemical is then disced into the upper soil layers and irrigated to drive the chemical into the main root zone of most woody crop plants. The organic phosphate and carbamate nematicides are cholinesterase inhibitors and because of this, they are extremely toxic to humans and warm-blooded animals and should be used with care. They disrupt the nervous system of the nematodes, interfering with their ability to feed. Direct death of the nematodes does not usually occur, but due to their inability to feed effectively, they starve to death and the population declines.

Although excellent yield increases have been realized from the use of systemic nematicides to control nematodes, this particular type of nematicide is not particularly useful for the control of virus spread. Although nematode feeding is disrupted, some feeding does occur, and virus transmission may still result. The half-life of these chemicals is only a few weeks, so they have limited persistence. The organic phosphate compound phenamiphos has a somewhat longer residual effect than those in the carbamate group.

Perhaps the best use of this type of nematicide is as a followup treatment to preplant soil fumigation to maintain nematode populations at low levels for a longer time. Annual application to the soil surface would be required to maintain the effectiveness.

Other Methods of Control

Virus-free planting stock is the foundation for the prevention of nepovirus disease. In California millions of grapevine scions and cultivars grafted onto rootstocks are propagated as virus-free as a result of thermotherapy and shoot-tip culture (90). Alternatively, apical meristems can be cultured *in vitro* with or without thermotherapy to eradicate stocks of virus (82). In the late 1800s and early 1900s, grape stocks were imported into California from Europe. A high percentage of these stocks contained GFLV, and the soil with some of the stocks contained *X. index*. This example points out the prime value of quarantine as a method of preventing new diseases from entering a country.

One of the potentially most effective long-term strategies against nepovirus diseases is that of genetic resistance. Ideally, a rootstock should possess resistance to direct feeding damage by the nematode and immunity to the virus (i.e., the virus will not multiply in the rootstock). Alternatively, own-rooted woody plants could be developed to tolerate both nematode feeding and virus infection. In tree fruits, there is good promise that own-rooted 'Stanley' plum will remain free of TomRSV infection when grown in soil infested with viruliferous *Xiphinema* spp. (40). Muscadine grape has recently been shown to possess resistance to GFLV when exposed to viruliferous *X. index* (108). Another potentially valuable approach might be to graft a virus-immune interpiece between the rootstock and scion of woody plants. Jiminez and Goheen (58) tested the susceptibility of 58 grape cultivars to GFLV and their ability to pass GFLV through their tissues to a healthy bud of *Vitis rupestris* St. George, a susceptible indicator cultivar. Twelve scion or rootstock cultivars gave negative readings when tested by ELISA. It is possible that the virus titer was sufficiently low so as to be undetectable by this assay method.

It is clear that prevention and control of nepovirus disease require several strategies to be done in concert. With the recent advent of ultrasensitive immunosorbent assays a much better understanding of the epidemiology of nepovirus diseases is being gained. Better ways of using the standard gaseous soil fumigants have been developed in order to control vector nematodes more thoroughly. The newer systemic nematicides can be used to augment control given by the older fumigant materials. Further efforts toward developing crop resistance or immunity to the viruses will ultimately be the best strategy.

Concluding Remarks

Our initial intention was to review the nepoviruses of the world, but with the vast amount of literature on the subject, we realized that it would be impossible to review such a topic without restricting the content to a few specific aspects. We decided to confine our coverage to the nepoviruses of the Americas, but there was still an abundance of references and a danger of producing a superficial review unless the scope was restricted further. We have therefore concentrated on the pathological aspects of the nepoviruses, omitting most of the virological aspects that have been published. We took this course deliberately, recognizing that there is nothing particularly unique about the virology of the American members of the group, but that each virus has its special significance from the point of view of its associated diseases. In this review we have attempted to give the reader an overview of the diseases attributed to individual nepoviruses, the epidemiology of the diseases, and strategies for control. Even in restricting the scope, we have had to omit many references, but we feel that all of the major studies have been noted.

After reviewing the literature on nepoviruses as disease agents in the Americas, we are left with the inevitable conclusion that the nepoviruses are extremely important in agricultural crops in the United States and Canada, but are of little significance in Central and South America. This conclusion is based on the extent of the published work, and one wonders if we are not ignorant of the true state of affairs. On a worldwide basis, some 30 known or suspected nepoviruses have been described and new ones are continually being discovered. The known members almost certainly represent only a fraction of those that exist in nature. Most of the effort to date has been directed toward an understanding of the identity and mode of transmission of viruses that cause serious diseases in commercial crops. Minimum effort has been expended in searching for viruses in plants that are symptomlessly infected, as is so common with nepoviruses, or in sampling the many native species that are of little agricultural interest. We predict that several new nepoviruses will be detected in the Americas over the next decade, particularly in Central and South America, where plant viruses have not been as intensively studied as in North America. Given our understanding of the epidemiology and control strategies, we require considerably more information on the vector potential of the various *Xiphinema* spp., the role of nematodes compared with that of seed and pollen as virus vectors, and the importance of various weed species as alternate hosts of the virus.

References

1. Allen, W.R., Davidson, T.R., and Briscoe, M.R., 1970, Properties of a strain of strawberry latent ringspot virus isolated from sweet cherry growing in Ontario, *Phytopathology* **60**:1262–1265.
2. Allen, W.R., Van Schagen, J.G., and Eveleigh, E.S., 1982, Transmission of peach rosette mosaic virus to peach, grape, and cucumber by *Longidorus diadecturus* obtained from diseased orchards in Ontario, *Can. J. Plant Pathol.* **4**:16–18.
3. Allen, W.R., Van Schagen, J.G., and Ebsary, B.A., 1984, Comparative transmission of the peach rosette mosaic virus by Ontario populations of *Longidorus diadecturus* and *Xiphinema americanum* (nematode: *Longidoridae*), *Can. J. Plant Pathol.* **6**:29–32.
4. Auger, J., and Converse, R.H., 1982, Raspberry bushy dwarf and tomato ringspot viruses in Chilean red raspberries, *Acta Hortic.* **129**:9 (abstract).
5. Bancroft, J.B., 1968, Tomato top necrosis virus, *Phytopathology* **58**:1360–1363.
6. Barbara, D.J., and Clark, M.F., 1982, A simple indirect ELISA using F(ab')$_2$ fragments of immunoglobulin, *J. Gen. Virol.* **58**:315–322.
7. Barnett, O.W., and Baxter, L.W., 1976, Arabis mosaic virus from wild dogwood in South Carolina, *Proc. Am. Phytopathol. Soc.* **3**:249 (abstract).
8. Barrat, J.G., Scorza, R., and Otto, B.E., 1984. Detection of tomato ringspot virus in peach orchards, *Plant Dis.* **68**:198–200.
9. Bird, G.W., and Ramsdell, D.C., 1985, Population trends and vertical distribution of plant-parasitic nematodes associated with *Vitis labrusca* L. in Michigan. *J. Nematol.* **17**:100–107.
10. Blodgett, E.C., Aichele, M.D., Coyier, D.L., and Milbrath, J.A., 1963, The flat apple disease, *Plant Dis. Rep.* **47**:769–771.
11. Bozarth, R.F., and Corbett, M.K., 1958, Tomato ringspot virus associated with stub or stub head disease of gladiolus in Florida, *Plant Dis. Rep.* **42**:217–221.
12. Brierley, P., 1954, Symptoms in the florists' hydrangea caused by tomato ringspot virus and an unidentified sap-transmissible virus, *Phytopathology* **44**:696–699.
13. Cadman, C.H., 1963, Biology of soil-borne viruses, *Annu. Rev. Phytopathol.* **1**:143–172.
14. Cadman, C.H., and Lister, R.M., 1961, Relationship between tomato ringspot and peach yellow bud mosaic viruses, *Phytopathology* **51**:29–31.
15. Callahan, K.L., 1957, Pollen transmission of elm mosaic virus, *Phytopathology* **47**:5 (abstract).
16. Chamberlain, G.C., 1938, Yellow blotch-curl: A new virus disease of the red raspberry in Ontario, *Can. J. Res. C*, **16**:118–124.
17. Clark, M.F., 1981, Immunosorbent assays in plant pathology, *Annu. Rev. Phytopathol.* **19**:83–106.

18. Cohen, E., and Mordechai, M., 1970, The influence of some environmental and cultural conditions on rearing populations of *Xiphinema* and *Longidorus*, *Nematologica* **16**:85–93.

19. Converse, R.H., 1979, Recommended virus-indexing procedures for new USDA small fruit and grape cultivars, *Plant Dis. Rep.* **63**:848–851.

20. Converse, R.H., 1981, Infection of cultivated strawberries by tomato ringspot virus, *Phytopathology* **71**:1149–1152.

21. Converse, R.H., and Ramsdell, D.C., 1982, Occurrence of tomato and tobacco ringspot viruses and of dagger and other nematodes associated with cultivated and high bush blueberries in Oregon, *Plant Dis.* **68**:710–712.

22. Cummins, J.N., and Gonsalves, D., 1982, Recovery of tomato ringspot virus from inoculated apple trees, *J. Am. Soc. Hortic. Sci.* **107**:798–800.

23. Dias, H.F., 1975, Peach rosette mosaic virus, CMI/ABB Descriptions of Plant Viruses, No. 150, Commonwealth Mycological Institute/Association of Applied Biologists, Kew, Surrey, England.

24. Dias, H.F., 1977, Incidence and geographic distribution of tomato ringspot virus in De Chaunac vineyards in the Niagara peninsula, *Plant Dis. Rep.* **61**:24–28.

25. Dias, H.F., and Allen, W.R., 1980, Characterization of the single protein and two nucleic acids of peach rosette mosaic virus, *Can. J. Bot.* **58**:1747–1754.

26. Dias, H.F., and Cation, D., 1976, The characterization of a virus responsible for peach rosette mosaic and grape decline in Michigan, *Can. J. Bot.* **54**:1228–1239.

27. Ferris, H., and McKenry, M.V., 1974, Seasonal fluctuations in the spatial distribution of nematode populations in a California vineyard, *J. Nematol.* **6**:203–210.

28. Forer, L.B., and Stouffer, R.F., 1982, *Xiphinema* spp. associated with tomato ringspot virus infection of Pennsylvania fruit crops, *Plant Dis.* **66**:735–736.

29. Francki, R.I.B., and Hatta, T., 1977, Nepovirus (tobacco ringspot virus) group, in: K. Maramorosch (ed.), *The Atlas of Insect and Plant Viruses*, Academic Press, New York, pp. 221–235.

30. Frazier, N.W., Yarwood, C.E., and Gold, A.H., 1961, Yellow-bud virus endemic along California coast, *Plant Dis. Rep.* **45**:649–651.

31. Fribourg, C.E., 1977, Andean potato calico strain of tobacco ringspot virus, *Phytopathology* **67**:174–178.

32. Fribourg, C.E., and Salazar, L.F., 1972, Newly detected potato viruses in Peru, in: E.R. French (ed.), *Prospects for the Potato in the Developing World*, International Potato Center, Lima, Peru, pp. 230–233.

33. Fromme, F.D., Wingard, S.A., and Priode, C.N., 1927, Ringspot of tobacco; an infectious disease of unknown cause, *Phytopathology* **17**:321–328.

34. Fulton, J.P., 1962, Transmission of tobacco ringspot virus by *Xiphinema americanum*, *Phytopathology* **52**:375.

35. Fulton, J.P., and Fulton, R.W., 1970, A comparison of some properties of elm mosaic and tomato ringspot viruses, *Phytopathology* **60**:114–115.

36. Gilmer, R.M., and Uyemoto, J.K., 1972, Tomato ringspot virus in 'Baco Noir' grapevines in New York, *Plant Dis. Rep.* **56**:133–135.

37. Gilmer, R.M., Uyemoto, J.K., and Kelts, J.L., 1970, A new grapevine disease induced by tobacco ringspot virus, *Phytopathology* **60**:619–627.

38. Goff, L.M., and Corbett, M.K., 1977, Association of tomato ringspot virus with a chlorotic leaf streak of cymbidium orchids, *Phytopathology* **67**:1096–1100.

39. Gonsalves, D., 1979, Detection of tomato ringspot virus in grapevines: A comparison of *Chenopodium quinoa* and enzyme-linked immunosorbent assay (ELISA), *Plant Dis. Rep.* **63**:962–965.
40. Gonsalves, D., Cummins, J.N., and Rosenberger, D.A., 1983, NE-14 Annual Report for 1983, New York Agricultural Experiment Station, Cornell University, Geneva, New York (USDA–CSRS Regional Research Report). Unpaged.
41. Gooding, G.V., 1963, Purification and serology of a virus associated with the grape yellow-vein disease, *Phytopathology* **53**:475–480.
42. Gooding, G.V., 1970, Natural serological strains of tobacco ringspot virus, *Phytopathology* **60**:708–713.
43. Griffin, G.D., and Darling, H.M., 1964, An ecological study of *Xiphinema americanum* Cobb in an ornamental spruce nursery, *Nematologica* **10**:471–479.
44. Hansen, A.J., and Stace-Smith, R., 1971, Properties of a virus isolated from golden elderberry, *Sambucus niger aurea, Phytopathology* **61**:1222–1229.
45. Hansen, A.J., Nyland, G., McElroy, F.D., and Stace-Smith, R., 1974, Origin, cause, host range and spread of cherry rasp leaf disease in North America, *Phytopathology* **64**:721–727.
46. Hansen, C.M., and Campbell, R.N., 1979, Strawberry latent ringspot virus from 'Plain' parsley in California, *Plant Dis. Rep.* **63**:142–146.
47. Harris, A.R., 1979, Seasonal populations of *Xiphinema index* in vineyards soils of north-eastern Victoria, Australia, *Nematologica* **25**:336–347.
48. Harrison, B.D., 1977, Ecology and control of viruses with soil-inhabiting vectors, *Annu. Rev. Phytopathol.* **15**:331–360.
49. Harrison, B.D., and Murant, A.F., 1977, Nepovirus group, CMI/AAB Descriptions of Plant Viruses, No. 185, Commonwealth Mycological Institute/Association of Applied Biologists, Kew Surrey, England.
50. Harrison, B.D., Murant, A.F., Mayo, M.A., and Roberts, I.M., 1974, Distribution and determinants for symptom production, host ranges and nematode transmissibility between the two RNA components of raspberry ringspot virus, *J. Gen. Virol.* **22**:233–247.
51. Hewitt, W.B., Raski, D.J., and Goheen, A.C., 1958, Nematode vector of soilborne fanleaf virus of grapevines, *Phytopathology* **48**:586–595.
52. Hibben, C.R., and Bozarth, R.F., 1972, Identification of an ash strain of tobacco ringspot virus, *Phytopathology* **62**:1023–1029.
53. Hibben, C.R., and Reese, J.A., 1983, Identification of tomato ringspot virus and mycoplasma-like organisms in stump sprouts of ash, *Phytopathology* **73**:367 (abstract).
54. Hildebrand, E.M., 1939, Tomato ringspot on currant, *Am. J. Bot.* **29**:362.
55. Hoy, J.W., Mircetich, S.M., and Shepherd, R.J., 1981, Immunosorbent assays for tomato ringspot virus in stone fruit trees, *Phytopathology* **71**:227 (abstract).
56. Hoy, J.W., Mircetich, S.M., and Lownsbery, B.F., 1984, Differential transmission of Prunus tomato ringspot virus strains by *Xiphinema californicum, Phytopathology* **74**:332–335.
57. Imle, E.P., and Samson, R.W., 1937, Studies on a ringspot type of virus of tomato, *Phytopathology* **27**:132 (abstract).
58. Jiminez, F., and Goheen, A.C., 1980, The use of enzyme-linked immunosorbent assay for detection of grape fanleaf virus, in: A.J. McGinnis (ed.), *Proceedings of the 7th Meeting of the International Council for the Study of*

Viruses and Virus-like Diseases of the Grapevine, Agriculture Canada, Research Branch, pp. 283–292.

59. Jones, A.T., and Murant, A.F., 1971, Serological relationship between cherry leaf roll, elm mosaic and golden elderberry mosaic viruses, *Ann. Appl. Biol.* **69**:11–15.

60. Jones, A.T., McElroy, F.D., and Brown, D.J.F., 1981, Tests for transmission of cherry leafroll virus using *Longidorus, Paralongidorus* and *Xiphinema* nematodes, *Ann. Appl. Biol.* **99**:143–150.

61. Jones, R.A.C., and Kenton, R.H., 1978, Arracacha virus A, a newly recognized virus infecting arracacha (*Arracacha xanthorrhiza* Umbelliferae) in the Peruvian Andes, *Ann. Appl. Biol.* **90**:85–91.

62. Jones, R.A.C., Fribourg, C.E., and Koenig, R., 1983, A previously undescribed nepovirus isolated from potato in Peru, *Phytopathology* **73**:195–198.

63. Kahn, R.P., Scott, H.A., and Monroe, R.L., 1962, Eucharis mottle strain of tobacco ringspot, *Phytopathology* **52**:1211–1216.

64. Kemp, W.G., 1967, Natural occurrence of tobacco ringspot virus in pelargonium in Ontario, *Can. J. Plant Sci.* **47**:295–300.

65. Kemp, W.G., 1969, Detection of tomato ringspot virus in pelargonium in Ontario, *Can. Plant Dis. Surv.* **49**:1–4.

66. Klos, E.J., 1976, Rosette mosaic, in: *Virus Disease and Noninfectious Disorders of Stone Fruits in North America*, Agriculture Handbook 437, U.S. Department of Agriculture, Washington, D.C., pp. 135–138.

67. Klos, E.J., Finck, F., Knierin, J.A., and Cation, D., 1967, Peach rosette mosaic transmission and control studies, *Mich. Agric. Exp. Stn. Q. Bull.* **49**:287–293.

68. Koenig, R., 1981, Indirect ELISA methods for the broad specificity detection of plant viruses, *J. Gen. Virol.* **55**:53–62.

69. Lamberti, F., and Bleve-Zacheo, T., 1979, Studies on *Xiphinema americanum sensu lato* with descriptions of fifteen new species (Nematoda, Longidoridae), *Nematol. Medit.* **7**:51–106.

70. Lana, A.F., Peterson, J.F., Rouselle, G.L., and Vrain, T.C., 1983, Association of tobacco ringspot virus with a union incompatibility of apple, *Phytopathol. Z.* **106**:141–148.

71. Lister, R.M., Rainere, L.C., and Varney, E.H., 1963, Relationships of viruses associated with ringspot diseases of blueberry, *Phytopathology* **53**:1031–1035.

72. Lockhart, B.E.L., 1977, Berteroa ringspot virus, a new strain of cherry leafroll virus, *Proc. Am. Phytopathol. Soc.* **4**:90 (abstract).

73. Lucas, G.B., 1975, Ringspot, in: G.B. Lucas (ed.), *Diseases of Tobacco*, 3rd ed., Biological Consulting Associates, Raleigh, North Carolina, pp. 523–530.

74. Martelli, G.P., 1975, Some features of nematode-borne viruses and their relationship with the host plants, in: F. Lamberti, C.E. Taylor, and J.W. Seinhorst (eds.), *Nematode Vectors of Plant Viruses*, Plenum Press, New York, pp. 223–252.

75. Matthews, R.E.F., 1982, Classification and nomenclature of viruses, *Intervirology* **17**:1–199.

76. Mayo, M.A., Murant, A.F., and Harrison, B.D., 1971, New evidence on the structure of nepoviruses, *J. Gen. Virol.* **12**:175–178.

77. McElroy, F.D., 1975, Nematode transmitted viruses in British Columbia, Canada, in: F. Lamberti, C.E. Taylor, and J.W. Seinhorst (eds.), *Nematode Vectors of Plant Viruses*, Plenum Press, New York, pp. 287–288.

78. McGuire, J.M., Epidemiology of nematode-borne viruses and their vectors, *Proc. Am. Phytopathol. Soc.* **4**:42–49.
79. McKeen, C.D., 1960, An occurrence of tomato ringspot virus on greenhouse cucumber in Ontario, *Proc. Can. Phytopathol. Soc.* **27**:14–15.
80. McLean, D.M., and Meyer, H.M., 1961, A survey of cucurbit viruses in the lower Rio Grande Valley of Texas: Preliminary report, *Plant Dis. Rep.* **45**:137–139.
81. Mellor, F.C., and Stace-Smith, R., 1963, Reaction of strawberry to a ringspot virus from raspberry, *Can. J. Bot.* **41**:865–870.
82. Mellor, F.C., and Stace-Smith, R. 1977, Virus-free potatoes by tissue culture, in: J. Reinert and Y.P.S. Bajaj (eds.), *Applied and Fundamental Aspects of Plant Cell, Tissue and Organ Culture,* Springer-Verlag, Berlin, pp. 616–635.
83. Miller, P.M., 1980, Reproduction and survival of *Xiphinema americanum* on selected woody plants, crops and weeds, *Plant Dis.* **64**:174–175.
84. Milne, R.G., and Luisoni, E., 1977, Rapid immune electron microscopy of virus preparations, in: Maramorosch and H. Koprowsky (eds.), *Methods in Virology,* Vol. 6, Academic Press, New York, pp. 265–281.
85. Mircetich, S.M., and Rowhani, A., 1982, The causal relationship of cherry leafroll virus and the blackline disease of English walnut trees, *Phytopathology* **72**:988 (abstract).
86. Mircetich, S.M., Sandborn, R.R., and Ramos, D.E., 1980, Natural spread, graft transmission, and possible etiology of walnut blackline disease, *Phytopathology* **70**:962–968.
87. Mircetich, S., Rowhani, A., and Cucuzza, J., 1982, Seed and pollen transmission of cherry leafroll virus (CLRV-W), the causal agent of the blackline disease (BL) of English walnut trees, *Phytopathology* **72**:988 (abstract).
88. Mountain, W.L., Powell, C.A., Forer, L.B., and Stouffer, R.F., 1983, Transmission of tomato ringspot virus from dandelion via seed and dagger nematodes, *Plant Dis.* **67**:867–868.
89. Murant, A.F., 1981, Nepoviruses, in: E. Kurstak (ed.), *Handbook of Plant Virus Infections and Comparative Diagnosis,* Elsevier, Amsterdam, pp. 197–238.
90. Nyland, G., and Goheen, A.C., 1969, Heat therapy of virus diseases of perennial plants, *Annu. Rev. Phytopathol.* **7**:331–354.
91. Nyland, G., Lownsbery, B.F., Lowe, S.K., and Mitchell, J.F., 1969, The transmission of cherry rasp leaf virus by *Xiphinema americanum. Phytopathology* **59**:1111–1112.
92. Ostazeski, S.A., and Scott, H.A., 1966, Natural occurrence of tomato ringspotvirus in birdsfoot-trefoil, *Phytopathology* **56**:585–586 (abstract).
93. Parish, C.L., 1977, A relationship between flat apple disease and cherry rasp leaf disease, *Phytopathology* **67**:982–984.
94. Parish, C.L., and Converse, R.H., 1981, Tomato ringspot virus associated with apple union necrosis and decline in western United States, *Plant Dis.* **65**:261–263.
95. Powell, C.A., Forer, L.B., and Stouffer, R.F., 1982, Reservoirs of tomato ringspot virus in fruit orchards, *Plant Dis.* **66**:583–584.
96. Powell, C.A., Forer, L.B., Stouffer, R.F., Cummins, J.N., Gonsalves, D., Rosenberger, D.A., Hoffman, J., and Lister, R.M., 1984a, Orchard weeds as hosts of tomato ringspot and tobacco ringspot viruses, *Plant Dis.* **68**:242–244.

97. Powell, C.A., Mountain, W.L., Dick, T., Forer, L.B., Derr, M.A., Lathrop, L.D., and Stouffer, R.F., 1984b, Distribution of tomato ringspot virus in dandelion in Pennsylvania, *Plant Dis.* **68**:796–798.

98. Price, W.C., 1936, Specificity of acquired immunity from tobacco-ring-spot diseases, *Phytopathology* **26**:665–675.

99. Quacquarelli, A., Gallitelli, D., Savino, V., and Martelli, G.P., 1976, Properties of grapevine fanleaf virus, *J. Gen. Virol.* **32**:349–360.

100. Radewald, J.D., and Raski, D.J., 1962, A study of the life cycle of *Xiphinema index, Phytopathology* **52**:748 (abstract).

101. Ramsdell, D.C., and Gillett, J.M., 1981, Peach rosette mosaic virus in highbush blueberry, *Plant Dis.* **65**:757–758.

102. Ramsdell, D.M., and Myers, R.L., 1974, Peach rosette mosaic virus, symptomatology and nematodes associated with grapevine 'degeneration' in Michigan, *Phytopathology* **64**:1174–1178.

103. Ramsdell, D.C., and Myers, R.L., 1978, Epidemiology of peach rosette mosaic virus in a Concord grape vineyard, *Phytopathology* **68**:447–450.

104. Ramsdell, D.C., and Stace-Smith, R., 1979, Blueberry leaf mottle, a new disease of highbush blueberry, *Acta Hortic.* **95**:37–48.

105. Ramsdell, D.C., and Stace-Smith, R., 1981, Physical and chemical properties of the particles and ribonucleic acid of blueberry leaf mottle virus, *Phytopathology* **71**:468–472.

106. Ramsdell, D.C., Andrews, R.W., Gillett, J.M., and Morris, C.E., 1979, A comparison between enzyme-linked immunosorbent assay (ELISA) and *Chenopodium quinoa* for detection of peach rosette mosaic virus in 'Concord' grapevines, *Plant Dis. Rep.* **63**:74–78.

107. Ramsdell, D.C., Bird, G.W., Gillett, J.M., and Rose, L.M., 1983, Superimposed shallow and deep soil fumigation to control *Xiphinema americanum* and peach rosette mosaic virus reinfection in a Concord vineyard, *Plant Dis.* **67**:625–627.

108. Raski, D.J., Hewitt, W.B., Goheen, A.C., Taylor, C.E., and Taylor, R.H., 1965, Survival of *Xiphinema index* and reservoirs of fanleaf virus in fallowed vineyard soils, *Nematologica* **11**:349–352.

109. Raski, D.J., Hewitt, W.B., and Schmitt, R.V., 1971, Controlling fanleaf virus-dagger nematode disease complex in vineyards by soil fumigation, *Calif. Agric.* **25**(4):11–14.

110. Raski, D.J., Goheen, A.C., Lider, L.A., and Meredith, C.P., 1983, Strategies against grapevine fanleaf virus and its nematode vector, *Plant Dis.* **67**:335–339.

111. Reddick, B.B., Barnett, O.W., and Baxter, L.W., 1979, Isolation of cherry leafroll, tobacco ringspot, and tomato ringspot viruses from dogwood trees in South Carolina, *Plant Dis. Rep.* **63**:529–532.

112. Rush, M.C., and Gooding, G.V., 1970, The occurrence of tobacco ringspot virus strains and tomato ringspot virus in host indigenous to North Carolina, *Phytopathology* **60**:1756–1760.

113. Russo, M., Martelli, G.P., and Savino, V., 1980, Immunosorbent electron microscopy for detecting sap-transmissible viruses of grapevine, in: A.J. McGinnis (ed.), *Proceedings of the 7th Meeting of the International Council for the Study of Viruses and Virus-like Diseases of the Grapevine,* Agriculture Canada, Research Branch, pp. 251–257.

114. Salazar, L.F., and Harrison, B.D., 1978a, Host range and properties of potato black ringspot virus, *Ann. Appl. Biol.* **90:**375–386.
115. Salazar, L.F., and Harrison, B.D., 1978a, The relationship of potato black ringspot virus to tobacco ringspot and allied viruses. *Ann. Appl. Biol.* **90:**387–394.
116. Samson, R.W., and Imle, E.P., 1942, A ring-spot type of virus disease of tomato, *Phytopathology* **32:**1037–1047.
117. Schlocker, A., and Traylor, J.A., 1976, Yellow bud mosaic, in: *Virus Disease and Noninfectious Disorders of Stone Fruits in North America,* Agriculture Handbook 437, U.S. Department of Agriculture, Washington, D.C., pp. 156–165.
118. Sinclair, J.B., and Shurtleff, M.C. (eds.), 1975, *Compendium of Soybean Diseases,* American Phytopathological Society, St. Paul, Minnesota. 69 pp.
119. Sinclair, J.R., and Walker, J.C., 1956, A survey of ringspot on cucumber in Wisconsin, *Plant Dis. Rep.* **40:**19–20.
120. Smith, S.H., and Traylor, J.A., 1969, Stem pitting of yellow bud mosaic virus-infected peaches, *Plant Dis. Rep.* **53:**666–667.
121. Smith, S.H., Stouffer, R.F., and Soulen, D.M., 1973, Induction of stem pitting in peaches by mechanical inoculation with tomato ringspot virus, *Phytopathology* **63:**1404–1406.
122. Stace-Smith, R., 1962, Studies on Rubus virus diseases in British Columbia. IX. Ringspot disease of red raspberry, *Can. J. Bot.* **39:**559–565.
123. Stace-Smith, R., 1984, Red raspberry virus diseases in North America, *Plant Dis.* **68:**274–279.
124. Stace-Smith, R., and Hansen, A.J., 1974, Occurrence of tobacco ringspot virus in sweet cherry, *Can. J. Bot.* **52:**1647–1651.
125. Stace-Smith, R., and Hansen, A.J., 1976, Cherry raspleaf, CMI/AAB Descriptions of Plant Viruses, No. 159, Commonwealth Mycological Institute/Association of Applied Biologists, Kew, Surrey, England.
126. Stobbs, L.W., and Van Schagen, J.G., 1985, Relationship between grapevine Joannes-Seyve virus and tomato blackring virus, *Can. J. Plant Pathol.* **7:**37–40.
127. Stone, W.J., Mink, G.I., and Bergeson, G.B., 1962, A new disease of American spearmint caused by tobacco ringspot virus, *Plant Dis. Rep.* **46:**623–624.
128. Stouffer, R.F., Hickey, K.D., and Welsh, M.F., 1977, Apple union necrosis and decline, *Plant Dis. Rep.* **61:**20–24.
129. Swingle, R.U., Tilford, P.E., and Irish, C.F., 1943, A graft transmissible mosaic of American elm, *Phytopathology* **33:**1196–2000.
130. Tall, M.G., Price, W.C., and Wertman, K., 1949, Differentiation of tobacco and tomato ringspot viruses by cross immunization and complement fixation, *Phytopathology* **39:**288–299.
131. Taylor, C.E., and Brown, D.J.F., 1981, Nematode–virus interactions, in: B.M. Zuckerman and R.A. Rohde (eds.), *Plant Parasitic Nematodes,* Academic Press, New York, pp. 281–301.
132. Taylor, C.E., and Robertson, W.M., 1977, Virus vector relationships and mechanics of transmission, *Proc. Am. Phytopathol. Soc.* **4:**20–29.
133. Téliz, D., 1967, Effects of nematode extraction method, soil moisture and nematode numbers on the transmission of tobacco ringspot virus by *Xiphinema americanum, Nematologica* **13:**177–185.

134. Téliz, D., Grogan, R.G., and Lownsbery, B.F., 1966, Transmission of tomato ringspot, peach yellow bud mosaic, and grape yellow vein diseases by *Xiphinema americanum, Phytopathology* **56:**658–663.
135. Trudgill, D.L., Brown, D.J.F., and McNamara, D.G., 1983, Methods and criteria for assessing the transmission of plant viruses by Longidorid nematodes, *Rev. Nematol.* **6:**133–141.
136. Uyemoto, J.K., 1970, Symptomatically distinct strains of tomato ringspot virus isolated from grape and elderberry, *Phytopathology* **60:**1838–1841.
137. Uyemoto, J.K., Goheen, A.C., Luhn, C.F., and Petersen, L.J., 1976, Use of *Chenopodium quinoa* in indexing for grapevine fanleaf virus, *Plant Dis. Rep.* **60:**536–538.
138. Uyemoto, J.K., Taschenberg, E.F., and Hummer, D.K., 1977, Isolation and identification of a strain of grapevine Bulgarian latent virus in Concord grapevine in New York State, *Plant Dis. Rep.* **61:**949–953.
139. Varney, E.H., and Moore, J.D., 1952, Strain of tomato ringspot virus from American elm, *Phytopathology* **42:**476–477 (abstract).
140. Vaughan, E.K., Johnson, F., Fitzpatrick, R.E., and Stace-Smith, R., 1951, Diseases observed on bramble fruits in the Pacific Northwest, *Plant Dis. Rep.* **35:**34–37.
141. Waterworth, H.E., and Lawson, R.H., 1973, Purification, electron microscopy, and serology of the dogwood strain of cherry leafroll virus, *Phytopathology* **63:**141–146.
142. Waterworth, H.E., and Povish, W.R., 1972, Tobacco ringspot virus from naturally infected dogwood, autumn crocus and forsythia, *Plant Dis. Rep.* **56:**336–337.
143. Weischer, B., 1975, Ecology of *Xiphinema* and *Longidorus,* in: F. Lamberti, C.E. Taylor, and J.W. Seinhorst (eds.), *Nematode Vectors of Plant Viruses,* Plenum Press, New York, pp. 291–307.
144. Wilkinson, R.E., 1952, Woody plant hosts of the tobacco ringspot virus, *Phytopathology* **42:**478 (abstract).
145. Zeller, S.M., and Braun, A.J., 1943, Decline disease of raspberry, *Phytopathology* **33:**156–161.

6
Viral Replication, Translation, and Assembly of Nepoviruses

Donald C. Ramsdell

Multicomponent Nature of Nepoviruses

The first recognition that nepoviruses are multicomponent was by Stace-Smith *et al.* (40). Partially purified tobacco ringspot virus (TRSV) was shown to have three zones in sucrose density gradients (SDG), with S values of 53, 94, and 128, respectively, for top (T), middle (M), and bottom (B) components. These components were shown to be serologically identical. Electron microscopy revealed that all components were spherical, ~28 mm in diameter, and that the T particles were empty shells. Further, the B component was shown to contain 42% RNA and was infectious. Then, in 1966, Diener and Schneider (6) were the first to recognize the existence of two species of RNA in TRSV virions (RNA-1 and RNA-2), by separating them in SDG. They found that the single RNA species from M was noninfectious and that the two RNA species from the B component contained an infectious RNA (32S) and a noninfectious RNA (24S) (RNA-1 and RNA-2), with the noninfectious RNA being in much greater proportion than the infectious RNA. The noninfectious RNA from B sedimented at the same rate as that from the M component. They hypothesized that two smaller, noninfectious RNAs might join to form the longer, infectious RNA. In 1972, Murant *et al.* (20) demonstrated that raspberry ringspot virus (RRV) (a nepovirus of European origin) had T, M, and B components with S values of 52, 92, and 130, respectively, and containing 0, 30, and 44% single-stranded RNA. There were two species of RNA; RNA-1, with a molecular weight of 2.4×10^6 daltons, and RNA-2, with a molecular weight of 1.4×10^6 daltons. They postulated that M particles contained one molecule of RNA-2 and that B particles contained either one molecule of RNA-1 or probably two of RNA-2. In a companion publication, Harrison *et al.* (13) reported that only B particles of RRV were infectious alone, but infectivity was increased by adding excess M particles

Donald C. Ramsdell, Department of Botany and Plant Pathology, Michigan State University, East Lansing, Michigan 48824, USA.

to B particles. Furthermore, SDG and polyacrylamide gel electrophoresis (PAGE)-purified RNA-1 was infective, while RNA-2 was not. Infectivity of RNA-1 was greatly increased by the addition of RNA-2. In addition, RNA-2 from M and B particles behaved simiarly. The authors suggested that RNA-1 and RNA-2 of RRV each carry different pieces of genetic information, or, in other words, that RRV possesses a split genome. Later, in 1974, Schneider *et al.* (36) demonstrated that tomato ringspot virus (TmRSV) contains two nucleoproteins, with S values of 119 and 127. These had buoyant densities of 1.495 and 1.51 g/cm^3, respectively, in cesium chloride buoyant density gradients. Two RNAs were extracted with S values of 30.9 and 32.6 in SDG and it was presumed in this case that the smaller RNA came from the M and the larger from the B component. They found that B (referred to as B2) was relatively more infectious than M (referred to as B1) after SDG separation of the components. They stated that because of enhancement of infectivity of a partially separated component by the addition of the second component, this fits with the possibility of a multipartite genome in TmRSV.

Replication Studies

Rezaian and co-workers have done some excellent and definitive work dealing with the time course of TRSV replication (26–28). They purified an RNA-dependent RNA polymerase from TRSV-infected cucumber (26), which was previously shown by Peden *et al.* (23) to be elicited as a result of TRSV infecting cucumber. Rezaian *et al.* (27) also extracted double-stranded (ds) RNA from infected tissues using phenol extraction treatments with 2-methoxy ethanol, salt fractionation, and digestion by DNase and RNase. They hybridized the resulting melted dsRNA to ^{14}C-labeled TRSV ssRNA on filters to detect base sequences complementary between viral RNA and the dsRNA extracts from infected cucumber. They found that there was a rapid increase in both the amount of virus-specific dsRNA and polymerase activity just before and during rapid synthesis of virus. As soon as virus synthesis declined, the level of polymerase activity fell concomitantly. During this time (the first 3 days after inoculation) encapsulated TRSV titer rose rapidly. By 5 days, accumulated virus was at the maximum. Although it was expected that the dsRNA would be twice the molecular weight of each of the two species of viral ssRNA, thus proving that the dsRNA was indeed a replication intermediate, this did not occur. Rather, the dsRNA was found to be heterogeneous and of lower molecular weight than expected. Experiments were then done using TRSV and a bean strain of TMV to inoculate bean plants. Although the same results occurred as before with TRSV dsRNA being again heterogeneous and of lower molecular weight than expected, the dsRNA of TMV was about twice the molecular weight of ssRNA of TMV. Later, Schneider *et al.* (37) reported the presence of a high-molecular weight dsRNA recovered from bean plants infected with TRSV. In 1974, Rezaian and Francki (27) used the polydisperse dsRNA resulting from TRSV infection of cucumber in RNA–RNA hybridization experiments to determine what proportion of the nucleotide sequences of the viral RNA were complementary to those of the dsRNA. Using saturation hybridization of ^{14}C-labeled TRSV RNA with virus-specific dsRNA, they found that the maximum amount of dsRNA used for annealing rendered ~88% of the viral RNA resistant to RNase. This was similar to a saturation curve of a dsRNA annealing to its homologous ssRNA. These results indicated that all of the nucleotide sequences of the viral ssRNA are present and in the same order in the complementary sequences of the dsRNA. Further experiments were done employing competition hybridization. Competition hybridization was done using TRSV-specific dsRNA with isolated ^{14}C-labeled TRSV RNA-2 in the presence of either unlabeled RNA-2 (from M component) or RNA-1 or the homologous RNA (RNA-2). RNA-2 and the RNA from the purified M component competed equally with labeled RNA-2 for sequences in the melted dsRNA. However, RNA-1 and RNA-2 competed against each other much less effectively. The reciprocal experiment was also run, where TRSV-specific dsRNA was tested with isolated ^{14}C-labeled RNA-1 in the presence of competing RNA-2 (from M component) or the homologous RNA (RNA-1). The results were similar to the first experiment; each heterologous competing RNA was only about 14% as efficient as the homologous RNA. Rezaian and Francki stated that, "the fact that RNA-1 and RNA-2 of TRSV have largely distinct nucleotide sequences suggests that the virus has a divided genome although, at present, there is no conclusive evidence that both RNAs are functional."

Cellular Sites of Replication and Synthesis

Rezaian *et al.* (28) in 1976 investigated the site of TRSV synthesis in infected cucumber cotyledon cells, using ^{14}C-labeled TRSV ssRNA to detect TRSV-specific dsRNA and assays for RNA-dependent RNA polymerase. Low-speed differential centrifugation was used to determine if certain organelles or the supernatants contained radioactivity when probes or assays were used. Using the antibiotics chloramphenicol (which inhibits protein synthesis on 70S ribosomes) and cycloheximide (which inhibits protein synthesis on 80S ribosomes), they were able to gain new information as to the sites of TRSV multiplication. They also used electron microscopy to follow ultrastructural changes over a time course of infection. They found that as with previous work, virus replication is at its peak by 3 days postinoculation. Furthermore, TRSV protein synthesis is sensitive to cycloheximide, but not to chloramphenicol, which indicates that the 80S cytoplasmic ribosomes are involved rather than the 70S chloroplast ribosomes. Electron microscopy failed to reveal any virus-induced changes in the structure of any of the organelles, but did establish increased structural complexity of the cytoplasmic membrane systems in the cells of infected cotyledons.

The Presence of 5′-Genome-Linked Proteins

In 1978, Harrison and Barker (11) reported that a protease-sensitive structure was needed for infectivity of TRSV and tomato black ring virus (TBRV). Both RNA-1 and RNA-2 from these viruses was rendered noninfectious as a result of treatment with proteinase K or pronase. Using radioiodination techniques, Mayo *et al.* (17) showed that TRSV contains a covalently linked genome protein of molecular weight ~4000. In 1982, Mayo *et al.* (18) reported that the genome-linked protein (VPg) is present in several nepoviruses [strawberry latent ringspot (SLRV), TBRV, TRSV, and RRV] and further, that it is present on both RNAs and even on a satellite RNA (RNA-3) of SLRV. The presence of 5′-end VPg proteins is common to picornaviruses, comoviruses, sobemoviruses, and polyviruses as well as nepoviruses (5).

The Presence of 3′-Genomic Polyadenylate

In 1979, Mayo *et al.* (16) reported the presence of polyadenylate [Poly(A)] among five nepoviruses: RRV, SLRV, TRSV, TBRV, and TmRSV. The RNA of these viruses was bound to oligo(dT)-cellulose columns under conditions of high-ionic strength buffer and eluted with buffers of low ionic strength. The polyadenylate was stated to be probably located at the 3′ termini of the RNA molecules. Several other plant virus groups have virus members with 3′-poly(A) present (5). It is interesting to note that the 3′ end of most messenger RNA found in eukaryotic cells contains poly(A) (1).

Little or No Base Sequence Homology Exists Between the Different RNA Species

Replication of TBRV was studied in protoplasts by Robinson *et al.* 1980 (29). They found when using tobacco protoplasts inoculated with purified M particles (only RNA-2) that no replication of TBRV RNA was detected. However, if protoplasts were inoculated with only purified B (only RNA-1), extracts made from the protoplasts 24 or 48 hr later contained RNA that hybridized to cDNA copies of RNA-1 and had the same mobility as RNA-1 in polyacrylamide–agarose gels. In addition, this RNA-1 was capable, when mixed with M component (only RNA-2), of causing local lesions to form in *Chenopodium quinoa* leaves. This indicates that the VPg protein is virus-coded by RNA-1.

In hybridization experiments using cDNA copies of RNA-1 and RNA-2, no sequence homology between the two species of RNA was detected.

Cell-Free Protein Synthesis

Fritsch *et al.* 1980 (9) did *in vitro* translation of genome and satellite RNAs of TBRV into protein products (satellites will be discussed in more detail later), using wheat germ extracts and rabbit reticulocyte lysates. The genome RNA consisted of RNA-1 (2.8×10^6 daltons) and RNA-2 (1.6×10^6 daltons). It should be noted here that these genome molecular weight values are higher than those reported earlier (12) for TBRV, because of more recent findings by co-workers in Scotland (21) that viral RNAs denatured with glyoxol prior to PAGE exhibit a relatively higher molecular weight than with other previously used methods. These positive strand mRNAs translated protein products (using a radiolabeled amino acid) of maximum molecular weight 2.2×10^5 and 1.6×10^5 daltons, respectively. The products represented 80 and 100% of the coding capacity of the respective RNAs. The smaller polypeptide reacted with antiserum to TBRV particles, which is further evidence that RNA-2 codes for nepovirus coat protein formation, as was shown earlier as a result of pseudorecombinant studies with this virus (25).

Gergerich *et al.* 1983 (10) performed *in vitro* translation using cell-free wheat embryo extracts and either unfractionated TRSV RNA or separated RNA-2 of TRSV as message from three isolates of TRSV ranging from severe to intermediate to mild pathogenicity in an attempt to find a protein product(s) that might explain these differences in pathogenicity. RNA from each of the three isolates was used as message to translate protein products. Translation of the RNA of each of the isolates was done with ^{14}C-valine or ^{3}H-valine. The protein products from one isolate labeled with ^{3}H-valine and the products from a second isolate labeled with ^{14}C-valine were electrophoresed in the same gel along with known protein molecular weight markers. Gels were sliced and counted so that the amount of labeled valine incorporated as a result of translation by each of the two isolates could be directly compared for amount of polypeptides synthesized and their molecular weights in the same gel. The tobacco-TRSV isolate of intermediate pathogenicity (serogroup 38 from G. V. Gooding, Jr.) coded for six major bands of protein with a maximum molecular weight of 133,000 daltons and a minimum molecular weight of 10,000 daltons. When the "severe" blueberry isolate protein products were compared with the mild soybean-TRSV or tobacco-TRSV protein products, the range of size of protein products was about the same for those coded for by the tobacco isolate, except that there was a consistent 10,500-dalton product peculiar to the "severe" blueberry isolate not found in the protein products coded for by the other two isolates. Translation with unfractionated "severe" blueberry TRSV RNA and RNA-2 alone indicated that the 10,500-dalton polypeptide was coded for by RNA-1. This peculiar low-molecular weight protein did not coelectrophoresce with TRSV coat protein pretreated in the same manner as was done by Chu and Francki (3). They dissociated TRSV coat protein in the presence of sodium dodecyl sulfate (SDS), urea, and 2-mercaptoethanol (2-ME) and electrophoresced the preparation in polyacrylamide gels and found that 85% of it migrated to a band corresponding to the expected molecular weight of 57,000 daltons, but the remainder formed bands over an estimated molecular weight range of 14,000 to more than 110,000 daltons. They postulated that this represented a polymeric series of 1, 2, 3, 4, and 8. They postulated that the TRSV capsid is constructed from a single polypeptide species with a molecular weight of 12,988 daltons, and that the usually accepted TRSV coat protein subunit molecular weight of 55,000–57,000 daltons is really a tetramer of the smaller entity. More will be said about this later.

Satellite RNAs

Bruening (2) and Murant and Mayo (19) have recently written an excellent review article about satellites of nepoviruses and viruses of other groups as well. To date, satellite RNA has been found in the following nepoviruses: TRSV, TBRV, arabis mosaic virus (AMV), myrobalan latent ringspot virus (MLRV), chicory yellow mottle virus (CYMV), and strawberry latent ringspot virus (SLRV).

Satellite RNA (S-RNA) is of relatively low molecular weight (40,000–130,000 daltons) and requires the presence of the larger RNA of its "helper virus" in order to multiply in host cells. The S-RNA is not needed for the multiplication of the helper virus and possesses no significant amounts of base sequence homology with the helper virus. Nepovirus satellite RNAs are encapsulated with the helper virus RNA, which is coded for by the helper virus, and such virions are of about the same diameter (as measured by electron microscopy) as virions devoid of S-RNA; however, S values of such virions are variable. In the case of tobacco necrosis virus (TNV) S-RNA, it codes for its own protein separately from that of the helper virus RNA. The satellite virus (SV) TNV is serologically unrelated to the helper TNV virions. The virions of SV-TNV have a diameter of ~17 nm and an A value of 50, while TNV virions are ~30 nm in diameter and have an S value of about 118.

Since TRSV is a nepovirus of American origin, it will be used as an example of a nepovirus with S-RNA for the purposes of further discussion of the subject. Schneider 1971 (31) discovered virus particles associated with TRSV that when analyzed by SDG centrifugation were polydisperse, i.e., they overlapped the usual M- and B-component zones. A single RNA component was extracted from these particles and it had an S value of about 7.3. This small RNA multiplied in the presence of its companion TRSV RNA, but it was not infective when inoculated alone or with the

RNA of other nepoviruses, including Eucharis mottle virus, a close relative of TRSV (30). The satellite of TRSV (S-TRSV) could be eliminated from cultures and it was demonstrated that when S-TRSV was inoculated in connection with two different strains of TRSV, the coat protein of the S-TRSV was determined by the strain of TRSV from which it originally came (34). Therefore, the S-TRSV capsid must be specified by its helper TRSV genome.

The S-TRSV RNA was shown to have little base sequence homology with helper TRSV RNA (32), which qualifies it as a true satellite. Sogo *et al.* (39) calculated the molecular weight of S-TRSV RNA to be 115,000–125,000 daltons based upon electron microscopy of the RNA component compared to QB RNA and carnation mottle virus RNA. The overlapping of M and B components by S-TRSV RNA-containing virions mentioned earlier is due to the packaging of nucleoprotein particles containing from 12 to 15 S-RNA molecules (35). It was demonstrated (33) that a dsRNA resulted in cucumber tissue following inoculation of a S-TRSV RNA. The molecular weight of the dsRNA was slightly greater than twice that of the S-RNA (38). The dsRNA was found to be infectious following strand separation by heating. Electron microscopy of the dsRNA revealed that 91% of the population was linear, varying in length from 40 to 3000 nm; 6% consisted of relaxed circles, while the remaining 3% were "racketlike" structures. The linear dsRNA molecules, although variable in length, fell into a preferential length of 130 nm, not ~68 nm, the length of ssTRSV RNA. The circular forms were also variable in size; the shortest circles were 130 nm and the increment of each longer circle was ~130 nm. Also, the circular portion of the "racketlike" structure was ~130 nm with a variable-length linear tail. Kiefer *et al.* (14) referred to these as "multimeric forms" of TRSV S-RNA. It is of interest to note that they found TRSV S-RNA devoid of 3' VPg and 5' polyadenylate. The authors speculated that the TRSV S-RNA linear, circular, and "racketlike" dsRNA was perhaps due to replication akin to Owen and Diener's proposed "rolling circle" model for viroids (22).

One must wonder about the role of a satellite virus in virus-caused disease development. The only well-documented role is that of the satellite RNA of AMV in AMV(H), strain which is probably responsible for the peculiar etiology associated with hop nettlehead disease (4).

Protein Subunits and Reassembly

Mayo *et al.* (15), based upon work with RRV, have proposed that nepoviruses are icosahedral and consist of 60 asymmetric units, each of which may contain one or more chemical subunits. They demonstrated that the protein subunits of RRV have a molecular weight of 54,000 ± 2000, using 7.5% polyacrylamide gels. Even after treating the coat protein with various additives to further break it down, it still electrophoresced at an R_f indicating the same molecular weight as above. They concluded that the coat protein polypeptide is not composed of smaller polypeptides formed by disulfide bridges. They calculated the ratio of the total weight of the protein shell to that of the polypeptide molecule, using partial specific volume of the virions as a basis, and arrived at a probable range of 57–69 structural subunits consisting of a single polypeptide molecule. This closely fits the now accepted simple icosahedron of $T = 1$ structure peculiar to nepoviruses and satellite viruses. The 12,988 dalton protein subunit size of TRSV proposed by Chu and Francki (3) mentioned earlier has not been readily accepted. The protein coat subunit size for SLRV and cherry raspleaf virus (CRLV) are at variance with that of the other nepoviruses. Both of these viruses have two different sizes of subunits; those of SLRV are 29,000 and 44,000, while those of CRLV are 24,000 and 22,500 (12). Both of these viruses have a reported nematode vector and other properties common to nepoviruses, so they are retained as members of the group. Although broad bean wilt virus (7) is a member of the comovirus group (beetle-transmitted), it has other properties in common with SLRV. It would be an interesting experiment to try to form pseudorecombinants between strains of these viruses.

Post-translational cleavage of protein products has been demonstrated with cowpea mosaic virus *in vitro* (24). A 130,000-dalton protein product of TRSV RNA-2 message resulted from *in vitro* translation studies by Gergerich *et al.* (10) and a 200,000-dalton protein product of unfractionated RNA message of TBRV resulted from *in vitro* synthesis (8). Since there are affinities between the comovirus and nepovirus groups, it would be interesting to investigate this possibility further with nepoviruses by the addition of protease inhibitors and/or amino acid analogs to *in vitro* translation systems.

References*

1. Brawerman, G., 1974, Eukaryotic messenger RNA, *Annu. Rev. Biochem.* **43**:621–642.
2. Bruening, G., 1977, Plant covirus systems: Two component systems, in: H. Fraenkel-Conrat and R. R. Wagner (eds.), *Comprehensive Virology,* Vol. II, Plenum Press, New York, pp. 55–141.
3. Chu, P.W.G., and Francki, R.I.B., 1979, The chemical subunit of tobacco ringspot virus coat protein, *Virology* **93**:398–412.
4. Davies, D.L., and Clark, M.F., 1983, A satellite-like nucleic acid of arabis mosaic virus associated with hop nettlehead disease, *Ann. Appl. Biol.* **103**:439–448.
5. Davis, J.W., and Hull, R., 1982, Genome expression of plant positive-strand RNA viruses, *J. Gen. Virol.* **61**:1–14.

*Literature review completed December 1983.

6. Diener, T.O., and Schneider, I.R., 1966, The two components of tobacco ringspot virus nucleic acid: Origin and properties. *Virology* **29**:100–105.

7. Doel, T.R., 1975, Comparative properties of type, nasturtium ringspot and petunia ringspot strains of broad bean wilt virus, *J. Gen. Virol.* **26**:95–108.

8. Fritsch, C., Mayo, M.A., and Murant, A.F., 1978, Translation of satellite RNA of tomato black ring virus *in vitro* and in tobacco protoplasts, *J. Gen. Virol.* **40**:587–593.

9. Fritsch, C., Mayo, M.A., and Murant, A.F., 1980, Translation products of genome and satellite RNAs of tomato black ring virus, *J. Gen. Virol.* **46**:381–390.

10. Gergerich, R.C., Asher, Jr., J.H., and Ramsdell, D.C., 1983, A comparison of some serological and biological properties of seven isolates of tobacco ringspot virus, *Phytopathol. Z.* **107**:292–300.

11. Harrison, B.D., and Barker, H., 1978, Protease-sensitive structure needed for infectivity of nepovirus RNA, *J. Gen. Virol.* **40**:711–715.

12. Harrison, B.D., and Murant, A.F., 1977, Nepovirus group, CMI/AAB Descriptions of Plant Viruses, No. 185, Commonwealth Mycological Institute/Association of Applied Biologists, Kew, Surrey, England.

13. Harrison, B.D., Murant, A.F., and Mayo, M.A., 1972, Evidence for two functional RNA species in raspberry ringspot virus, *J. Gen. Virol.* **16**:339–348.

14. Kiefer, M.C., Daubert, S.D., Schneider, I.R., and Bruening, G., 1982, Multimeric forms of satellite of tobacco ringspot virus RNA, *Virology* **121**:262–273.

15. Mayo, M.A., Murant, A.F., and Harrison, B.D., 1971, New evidence for the structure of nepoviruses, *J. Gen. Virol.* **12**:175–178.

16. Mayo, M.A., Barker, H., and Harrison, B.D., 1979a, Polyadenylate in the RNA of five nepoviruses, *J. Gen. Virol.* **43**:603–610.

17. Mayo, M.A., Barker, H., and Harrison, B.D., 1979b, Evidence for a protein co-valently linked to tobacco ringspot virus RNA, *J. Gen. Virol.* **43**:735–740.

18. Mayo, M.A., Barker, H., and Harrison, B.D., 1982, Specificity and properties of the genome-linked proteins of nepoviruses, *J. Gen. Virol.* **59**:149–162.

19. Murant, A.F., and Mayo, M.A., 1982, Satellites of plant viruses, *Annu. Rev. Phytopathol.* **20**:49–70.

20. Murant, A.F., Mayo, M.A., Harrison, B.D., and Goold, R.A., 1972, Properties of virus and RNA components of raspberry ringspot virus, *J. Gen. Virol.* **16**:327–338.

21. Murant, A.F., Taylor, M., Duncan, G.H., and Raschke, J.H., 1981, Improved estimates of molecular weight of plant virus RNA by agarose gel electrophoresis and electron microscopy after denaturation with glyoxol, *J. Gen. Virol.* **53**:321–332.

22. Owens, R.A., and Diener, T.O., 1982, RNA intermediates in potato spindle tuber viroid replication, *Proc. Natl. Acad. Sci. USA Cell Biol.* **79**:113–117.

23. Peden, K.W.C., May, J.T., and Symons, R.H., 1972, A comparison of two plant virus-induced RNA polymerases, *Virology* **47**:498–501.

24. Pelham, H., 1979, Synthesis and proteolytic processing of cowpea mosaic virus proteins in reticulocyte lysates, *Virology* **96**:463–477.

25. Randles, J.W., Harrison, B.D., Murant, A.F., and Mayo, M.A., 1977, Packaging and biological activity of the two essential RNA species of tomato black ring virus, *J. Gen. Virol.* **36**:187–194.

26. Rezaian, M.A., and Francki, R.I.B., 1973, Replication of tobacco ringspot virus. I. Detection of a low molecular weight double stranded RNA from infected plants, *Virology* **56**:238–249.
27. Rezaian, M.A., and Francki, R.I.B., 1974, Replication of tobacco ringspot virus. II. Differences in nucleotide sequences between the viral RNA components, *Virology* **59**:275–280.
28. Rezaian, M.A., Francki, R.I.B., Chu, P.W.G., and Hatta, T., 1976, Replication of tobacco ringspot virus. III. Site of virus synthesis in cucumber cotyledons, *Virology* **74**:481–488.
29. Robinson, D.J., Basker, H., Harrison, B.D., and Mayo, M.A., 1980, Replication of RNA-1 of tomato black ring virus independently of RNA-2, *J. Gen. Virol.* **51**:317–326.
30. Schneider, I.R., 1970, Interaction between several viruses and the satellite-like virus of tobacco ringspot virus, *Phytopathology* **60**:1312 (abstract).
31. Schneider, I.R., 1971, Characteristics of a satellite-like virus of tobacco ringspot virus, *Virology* **45**:108–122.
32. Schneider, I.R., 1978, Defective plant viruses, *in:* J.A. Romberger *et al.* (eds.), *Beltsville Symposia in Agricultural Research,* Allanheld, Osmun, Montclair, New Jersey, pp. 201–219.
33. Schneider, I.R., and Thompson, S.M., 1977, Double-stranded nucleic acids found in tissue infected with the satellite of tobacco ringspot virus, *Virology* **78**:453–462.
34. Schneider, I.R., and White, R.M., 1976, Tobacco ringspot virus codes for the coat protein of its satellite, *Virology* **70**:244–246.
35. Schneider, I.R., Hull, R., and Markham, R., 1972, Multidense satellite of tobacco ringspot virus: A regular series of components of different densities, *Virology* **47**:320–330.
36. Schneider, I.R., White, R.M., and Civerolo, E.R., 1974a, Two nucleic acid-containing components of tomato ringspot virus, *Virology* **57**:139–146.
37. Schneider, I.R., White, R.M., and Thompson, S.M., 1974, High molecular weight double-stranded nucleic acids from tobacco ringspot virus infected plants, *Proc. Am. Phytopathol. Soc.* **1**:82.
38. Sogo, J.M., and Schneider, I.R., 1982, Electron microscopy of double stranded nucleic acids in tissue infected with the satellite tobacco ringspot virus, *Virology* **117**:401–415.
39. Sogo, J.M., Schneider, I.R., and Koller, T., 1974, Size determination by electron microscopy of the RNA of tobacco ringspot satellite virus, *Virology* **57**:459–466.
40. Stace-Smith, R., Reichmann, M.E., and Wright, N.S., 1965, Purification and properties of tobacco ringspot virus and two RNA-deficient components, *Virology* **25**:487–494.

7
Soil-Borne Viruses of Plants

Chuji Hiruki and David S. Teakle

Introduction

A virus is soil-borne if it infects plants via soil. This may occur either
through the activity of a soil-borne vector or by mechanical means, and
usually this implies virus transmission to the underground parts of plants.
However, if leaves or other aerial parts were infected following contact
with soil (2), this would also be included. Viruses excluded from this definition are those transmitted by direct root-to-root contact, such as potato
virus X (111) and cucumber mosaic virus (147), or root grafts, such as
apple mosaic virus (61). Thus, to be soil-borne, a virus should have an
existence in soil outside of a living host plant.

Perhaps a quarter of a plant is made up of roots and other subterranean
structures, which carry out the important functions of anchoring the plant,
absorbing minerals and water, and synthesizing and storing nutrients. Like
the aerial parts, the roots can be divided into those parts that are highly
active metabolically, particularly the root-tips, and those that are more
supportive, such as the larger, older roots. Infection of roots by soil-borne
viruses usually occurs in the metabolically active root tips, for several
reasons. First, young roots lack the protective covering that may develop
on older roots. Second, young, actively growing tissues are more susceptible to infection than older, mature tissues. Third, nematode and fungal
vectors of viruses are attracted to the root tips, probably by their abundant
exudates, and their attack results in virus transmission (128). Natural
wounds, which probably occur frequently in roots of some plants as a
result of growth, changes in turgor, and the buffeting effects of wind, are
probably sites of infection by mechanically transmitted soil-borne viruses.

Chuji Hiruki, Department of Plant Science, University of Alberta, Edmonton, Alberta
T6G 2P5, Canada.
David S. Teakle, Department of Microbiology, University of Queensland, St. Lucia,
Brisbane, Queensland, Australia 4067.

The number of soil-borne viruses has grown steadily, but is still far lower than the number transmitted by aerial means. Of 248 viruses described in CMI/AAB Descriptions of Plant Viruses to 1984, 41 (one-sixth) had some claim to being soil-borne. The viruses range greatly in their dependence on the soil for transmission. For instance, some of the fungus-transmitted viruses, such as soil-borne wheat mosaic, have no known alternative means of natural spread. Others, such as tobacco mosaic virus, are mainly transmitted mechanically between tops, although it is also mechanically transmitted from debris to roots (15), while bean mild mosaic virus has at least five beetle vectors (167) as well as being transmitted mechanically in soil (34).

This review is concerned mainly with those properties of the viruses that are relevant to their being soil-borne. Such properties as particle stability, the possession of a divided genome, and the ability to be acquired, retained, and transmitted by a vector will be discussed. Properties of the vectors that have been reviewed elsewhere (21, 128, 137, 138) will be treated relatively briefly.

Mechanically Transmitted Soil-Borne Viruses

Properties

Although mechanical transmission of tobacco mosaic virus in soil was demonstrated more than 80 years ago (7), soil transmission of this and other mechanically transmitted viruses has received little attention. The reasons for this include, first, that relatively few viruses have been shown to be commonly transmitted mechanically in soil; second, that such viruses may have more efficient and obvious methods of aerial transmission; and, third, that sometimes infection of roots occurs, but systemic movement of virus is erratic and delayed, and therefore infection is easily overlooked. Despite these facts, the mechanical transmission of viruses in soil may be agriculturally important. For instance, if virus-free plants of tomato are established in tomato mosaic virus-infested soil, the roots and later the tops of some of the plants become infected and the virus is then transmitted readily between tops of the plants (15). A similar situation could exist with carnation and carnation mottle virus, which can be transmitted mechanically in soil (141).

Viruses of three major particle shapes have been reported to be transmitted mechanically in soil (Table 7.1). These are the viruses with straight tubular particles (Tobamovirus and Tobravirus groups), with filamentous particles (Potyvirus group) or with small isometric particles (Dianthovirus, Sobemovirus, Tombusvirus, and other groups). The viruses whose mechanical transmission in soil is most frequent, important, and best established are those tobamoviruses that have highly stable particles and lack a fungal vector, and some of the isometric viruses, such as the tombus-

TABLE 7.1. Properties of some viruses reported to be transmitted mechanically in soil.[a]

Particle shape and virus group	Virus	Particle size (nm)	Genome components (mol. wt. × 10^6)	Coat protein (mol. wt. × 10^3)	Other transmission methods reported	References[b]
Straight tubular						
Tobamovirus	Cucumber green mottle mosaic	300 × 18	—	17	Leaf rubbing, seed	58(154)
Tobamovirus	Tobacco mosaic	300 × 18	2.0	18	Leaf rubbing	173a (151)
Tobamovirus	Tomato mosaic	300 × 18	2.0	18	Leaf rubbing, seed	52a (156)
Tobravirus	Tobacco rattle	190(L) + 45 – 115(S)[d] × 23	2.5 + 0.7–1.3	24	Nematodes	37a (12)
Filamentous						
Potyvirus	Sugarcane mosaic	750 × 13	—	—	Aphids, seed	104a (88)
Isometric						
Dianthovirus	Red clover necrotic mosaic	27	1.5 + 0.5	40	Fungus	57a (181)
Dianthovirus	Sweet clover necrotic mosaic	35	1.4 + 0.6	38	—	52
Sobemovirus	Southern bean mosaic	30	1.4	28	Beetles, seed	141, 153
Sobemovirus	Sowbane mosaic	26	1.3	31	Seed, leaf rubbing	69a (64), 141
Necrovirus	Tobacco necrosis	26	1.3 + 1.6	22.6, 33.5	Fungus	69b (14), 151
Tombusvirus	Carnation mottle[c]	28	—	—	Leaf rubbing	55 (7), 141
Tombusvirus	Cymbidium ringspot[c]	30	1.7	43	Leaf rubbing	57 (178)
Tombusvirus	Galinsoga mosaic[c]	34	—	—	Leaf rubbing	6a (252)
Tombusvirus	Tomato bushy stunt	30	1.5	38, 42	—	90a (69)
Ungrouped	Bean mild mosaic	28	1.3	—	Beetles	167 (231)

[a] Data from CMI/AAB Descriptions of Plant Viruses.
[b] Numbers in parentheses refer to CMI/AAB Descriptions of Plant Viruses.
[c] Possible member of group indicated.
[d] L, long particles; S, short particles.

viruses. The viruses will be discussed individually, and then general comments will be made on factors affecting their soil transmission.

CUCUMBER GREEN MOTTLE MOSAIC VIRUS

This can cause an important disease of cucurbitaceous crops in Europe, India, and Japan (58). Although it is often transmitted by foliage contact and handling of plants during cultivation, and distributed through infected cucurbit seed, infection also occurs via the roots. When melon seedlings were planted in soil contaminated by infective soil debris, the virus was detected in roots within 10–15 days (78). The virus survived in soil air-dried for 47 days, but soil transmission was prevented by soil fumigation with methyl bromide or chloropicrin, or steam sterilization (78).

TOBACCO MOSAIC VIRUS (TMV)

Beijerinck (7) reported that tobacco plants became infected when transplanted into pots of soil that months earlier had supported the growth of a TMV-infected tobacco plant. Since the soil had been stirred with a piece of wood during the growth of the test plants, root infection through inflicted wounds was considered possible. In a later experiment he placed soil from around infected roots near the stems of healthy seedlings, and dug some of the soil under while attempting to avoid injury to the roots. Since the plants became infected within 4 weeks, he concluded that normal roots can probably take up the virus from soil through their intact surfaces. However, some doubt exists as to whether the infested soil was dug in without any root injury.

Soil transmission was controversial for some time, but it has now been generally accepted that TMV may survive in soil between crops (69).

In addition to the normal tobacco strains, the serologically unrelated, highly necrotic Rotterdam-B strain in Sumatra is known to be soil-borne. When tobacco seedlings were planted in soil from around infected plants that had been supplemented with virus-containing parts of diseased plants, symptoms developed within 10 days and some plants were dead in 2–3 weeks. Rotating fields where the disease occurred to other crops for 7 years did not eliminate the disease in tobacco (146).

TOMATO MOSAIC VIRUS (ToMV)

Of all of the viruses transmitted mechanically in soil, tomato mosaic virus has been most studied (16). In glasshouse crops, root debris from the previous crop is undoubtedly the most important source of the virus (30), and a similar situation apparently may exist in the field (79).

Infection probably occurs initially in the roots of a small or moderate number of plants, and when their tops become infected, spread to other tops during handling is usually rapid. Since the virus is normally carried down passively in the stream of assimilates, the time for systemic infection following root infection may be long. Broadbent (15) found that the time for systemic infection varied between 3 weeks and 6 months, with an average of 2–3 months when plants had at least five to seven true leaves at time of exposure. It would be expected to be faster for smaller seedlings.

Repeated tillage of the soil and/or use of a cover crop of insusceptible wheat were shown to greatly reduce or eliminate infection in tomato planted in soil infested with ToMV-containing residues (85). Inactivation of ToMV in soil is probably carried out by microorganisms, and adequate moisture and air are probably important in this (17). Powdered leaf debris lost infectivity within 1 month in moist soil, but remained infective for 2 years in dry. Root debris remained infective for 6 months even in moist soil (17). Control of ToMV in soil by heat or chemicals is difficult because of the resistance of the virus to inactivation. Tomato mosaic virus in dead root tissues was inactivated by moist heat within 10 min at 85°C, but in fairly fresh roots it took 20 min at 90°C. This virus was not inactivated within dry debris by dry heat for 20 min at 100°C.

Formaldehyde at 40,000 ppm (five times normal) reduced the virus by 99.7%, but methyl bromide failed to destroy ToMV in roots in soil. Chloropicrin increased the persistence of ToMV in roots.

TOBACCO RATTLE VIRUS (TRV)

This is normally transmitted in soil by trichodorid nematodes (128). However, its mechanical transmission in soil without the assistance of a nematode vector has been reported in Brazil (112). When plants of *Nicotiana tabacum* were infected with TRV, the roots released virus detectable in drainage water. Further, transmission between roots of diseased and healthy plants occurred both when roots were in contact and when they were not. It was considered that some infection of tomato in Brazil resulted from mechanical transmission in soil. These results differed from those of Sol (123) in the Netherlands, who obtained transmission when roots were in contact, but not when healthy roots contacted debris of infected roots.

SUGARCANE MOSAIC VIRUS (SMV)

It is somewhat surprising that this virus, a member of the Potyvirus group, has been reported to be transmitted in soil, because it is only moderately stable and is normally aphid-borne (Table 7.1). However, Bond and Pirone (10) found that 0–10% of healthy sorghum plants receiving drainage water from infected sorghum plants became infected with sugarcane mosaic virus. However, no infection occurred when seeds were planted in soil containing root debris of infected plants.

BEAN MILD MOSAIC VIRUS

This isometric virus from Central and South America is readily transmitted by beetles (168). However, glasshouse tests showed that the virus was transmitted following contact between roots, when healthy plants were grown in infested soil from which most of the infected bean roots had been removed, and when healthy plants were grown in noncleansed, infested containers. When gel serology was used, the time to detect soil transmission ranged from 54 to 90 days (34). Since plants were often infected without showing symptoms, glasshouse spread was difficult to monitor.

CARNATION MOTTLE VIRUS (CaMV)

This virus is apparently worldwide in distribution in carnation (55) and has also been recorded from lettuce in England (148). Infection occurred when carnation seedlings were grown in soil containing debris of infected carnation plants (141). Although soil transmission of CaMV under commercial conditions has not yet been reported, it is likely that it occurs. Since transmission during handling occurs readily (53), detection of soil transmission may be difficult.

CYMBIDIUM RINGSPOT VIRUS

This virus occurs in cymbidium orchids and *Trifolium repens* in England. It is apparently continuously released from growing roots of infected *Nicotiana clevelandii,* and transplanting healthy *N. clevelandii* into infested soil resulted in a high level of infection. Further, when 1 ml of a 0.1 mg/ml preparation of the virus was pipetted around healthy seedlings growing in pots of sterilized soil, 80% of the pots contained one or more infected plants after 5 weeks (59). No natural transmission in the field or in commercial orchids has yet been shown to be caused by soil-borne virus.

GALINSOGA MOSAIC VIRUS (GMV)

This virus is known only from Queensland, Australia, where it may infect the roots and sometimes the tops of the weed *Galinsoga parviflora* (6). Tests have shown that it readily infects the roots of plants grown in naturally infective soil, but the addition of fungicides did not remove the infectivity of soil and attempts to show that the virus was transmitted by *Olpidium brassicae* failed. Since Pasteurized or autoclaved potting mix became infective when mixed with virus-containing roots, tops, leaf extracts, or drainage water, it was concluded that the virus was transmitted mechanically in soil (120).

RED CLOVER NECROTIC MOSAIC VIRUS (RCNMV)

This virus occurs in Europe (57a), Australia (90), and Canada (108a). It has been shown to be exuded into soil from roots of infected *Trifolium pratense* and *Nicotiana clevelandii* and can infect healthy *N. clevelandii* bait seedlings planted in such soil, apparently without a vector (57a). However, mechanical transmission in soil can be erratic, especially if plants are grown under relatively warm conditions (90). The possibility has been raised that *Olpidium radicale* is a vector of RCNMV (14), but evidence is not yet convincing. *Olpidium brassicae* is unable to transmit the virus (90).

SOUTHERN BEAN MOSAIC VIRUS (SBMV)

This virus occurs mainly in warm temperate and tropical regions of the Americas and Africa, but has also been reported from France and India (153). It is transmitted through seed of French bean and cowpea and by leaf beetles. The virus is released into the soil from the roots of infected bean plants (122), and when bean seeds were germinated in such virus-contaminated soil, many bean seedlings were found to be infected (141). Recent research by Teakle (138a) has demonstrated that bean plants became infected when transplanted into or above soil containing either root or top debris of infected bean plants. Although soil transmission in nature has not yet been reported, it seems likely that it occurs.

SOWBANE MOSAIC VIRUS (SoMV)

This virus is probably of worldwide distribution in species of the Chenopodiaceae, and has also been recovered occasionally from plants in other families, such as grapevine (8). It is seed-borne and readily transmitted in *Chenopodium* spp. by handling. It was transmitted when *Chenopodium* bait seedlings were grown in soil containing virus-infested debris (141). There is no information on natural transmission of this virus in soil.

TOBACCO NECROSIS VIRUS (TNV)

Usually this virus causes localized infection in plants and the occasional mechanical transmission is probably insufficient to ensure survival of the virus in nature in the absence of the fungus vector *Olpidium brassicae*. However, strains that infect hosts systemically, such as the CN strain in *Chenopodium quinoa,* may be released from roots in large amounts and be able to infect their systemic host mechanically when grown in soil contaminated with infective leachate (151).

TOMATO BUSHY STUNT VIRUS (TBSV)

This virus was first isolated from tomato in England (121), and because of the high stability of its particles and its ease of mechanical transmission, it was quickly used in classical studies on the nature of plant viruses. Other strains distinct from the tomato strain have been found in other hosts, including petunia in Italy (89), sweet cherry (*Prunus avium* L.) in North America (2b), and pelargonium in Britain (54).

Although the pelargonium leaf curl strain is spread in vegetatively propagate cuttings (54), transmission of many other strains is through the soil. Several workers have observed the recovery of TBSV in bait plants such as petunia or *Nicotiana clevelandii* planted in virus-containing natural soil (89, 56; R. Stace-Smith, personal communication).

Evidence implicating a fungal vector in soil is based mainly on the observation that application of Captan fungicide at 0.2% active to infective soil greatly reduced its infectivity to French bean and zucchini squash bait plants (88), although not inactivating the virus *in vitro*. However, at the 0.2% active dosage Captan may stunt the roots and tops of tomato seedlings (D. S. Teakle, unpublished data), so it is possible that the Captan acted by altering the susceptibility of the French bean and squash roots to virus infection, rather than acting on a vector. Further, attempts to find a fungal vector, such as *Olpidium brassicae* (20, 139) or *Lagenocystis radicis* (23), have so far been unsuccessful.

Although it is possible that TBSV has an unidentified soil-borne vector, it is more likely that the virus is transmitted only mechanically in soil. This conclusion seems feasible, because the virus is released in large amounts from infected plants (122), the virus is highly stable (Table 7.1), and bait plants are often readily infected when grown in soil infested with either sap extracts or root or top debris of plants thought unlikely to contain a natural vector (76, 141).

Some of the TBSV released from the roots of living plants or from plant debris will be washed out of soil in drainage water and will be detectable in river or lake water (150, 152). Another possible explanation of its occurrence in tomatoes grown in the presence of sewage water (150, 91; A. J. Galindo, personal communication) is its ability to survive and infect after passage through the human alimentary tract. It is possible that infective TBSV can move between countries inside man or other migratory animals, or in tomato or other crop plants that are the subject of international commerce.

Factors Affecting Transmission

Although mechanical transmission in soil is probably much more common than has been generally recognized, it is still an inefficient process when compared with transmission by either soil-borne or aerial vectors. It may be important because it results in virus being available for transmission by other means. For instance, TMV, ToMV, and GMV can be transmitted between tops of plants, and at least in the case of TMV and ToMV this may be more efficient than soil transmission (16).

What are the factors influencing the extent of mechanical transmission of viruses in soil? Although little research has been done in this area, it is known, or can be predicted, that many different factors are involved. These include:

1. *Amount of free virus (inoculum) in the soil.* A large, vigorous, systemically infected plant with a high concentration of virus will contribute substantially to the virus content of the soil through virus release from both living roots and root and top debris. Release of GMV and SBMV as

determined by assaying drainage water from leaf-inoculated host plants is greatest between flowering and early senescence, as shown by Shukla *et al.* (120) and Teakle (138a).

The particle stability of the virus will also be important in determining the amount of free virus in the soil. Stable viruses will tend to accumulate in soil, whereas labile viruses will be lost rapidly. It is significant that all of the viruses that are regularly transmitted mechanically in soil are highly stable with respect to tests of longevity *in vitro* and thermal inactivation point (Table 7.5). Sugarcane mosaic virus, which is only moderately stable, was not transmitted when sorghum seed was planted in soil infested with root debris, but only when roots or drainage water or infected plants contacted healthy seedlings (10).

Another factor affecting the amount of free virus in the soil will be adsorption to soil particles. If a virus is adsorbed, it will lose infectivity. Apparently viruses differ widely in their adsorption, since a tobacco necrosis virus was not adsorbed to any appreciable extent by three soils differing widely in their mineral and organic contents (151), whereas TRV adsorbed strongly to illite or montmorillonite clays but not kaolinite, and TMV adsorbed slightly or poorly to these clays (160). In contrast to TMV, the Rotterdam-B strain of TMV adsorbed strongly to montmorillonite or illite clays in suspension (146). The amount of leaching is also important, since it will remove free virus from the soil (59).

2. *Susceptibility of the bait plant.* Use of a highly susceptible bait plant that is systemically infected will aid in detection of any infection that occurs, e.g., *Nicotiana clevelandii* has been used for detection of TBSV in soil (56). It is probable that most viruses that can infect via the tops of plants can also infect via the roots (86), but efficiency of infection of roots may be relatively low.

3. *Suitability of the environment.* This will affect both the amount of inoculum in the soil and the susceptibility of the bait plant. It has been shown that air-drying soil greatly reduced its content of extractable TMV (2a), SBMV (122), TBSV (122), and GMV (120) compared with similar soil kept moist. Temperature undoubtedly affects both the release of virus from a living plant and the susceptibility of the bait plant. In this latter connection, recent research by Teakle (138a) has demonstrated that French bean was more readily infected when grown in SBMV-infested soil at 26°C than at either 21° or 32°C, which are less suitable for plant growth.

4. *Duration and degree of exposure of the bait plant to infective soil.* A plant growing in infective soil may be exposed to infection during the greater part of its life. This is in contrast to the situation occurring with sap inoculation of leaves, when exposure to virus is relatively short and most infection occurs within seconds or minutes (173). Presumably infection of roots occurs via the natural wounds (172), which are being continually produced as roots grow, and are more numerous the greater the extent and vigor of the root system.

The origin of natural wounds leading to virus infection is not known, but since roots change in size with changes in water content brought about by changes in soil moisture and atmospheric humidity, this could be important. Growth of old or new roots could also result in natural wounds.

Fungus-Transmitted Viruses

Properties

Based on particle morphology, the viruses with fungus vectors are of three major types (Table 7.2). The viruses with isometric particles are cucumber necrosis virus (CNV) (27), the serotypes of tobacco necrosis virus (TNV), and their satellite viruses. In natural infections, they are found in the roots of many annual and perennial species. Virus particles of CNV and TNV are small, 26 and 31 nm in diameter, respectively, and contain a unipartite single-stranded RNA genome. Cucumber necrosis virus, found naturally only in greenhouse cucumbers, is readily sap-transmissible and has a wide host range. The properties and host range of CNV and TNV, in particular the cucumber strains, are similar; however, they are not serologically related.

Tobacco necrosis virus is known to have a wide host range and is occasionally associated with diseases of economically important crops, namely the Augusta disease of tulip (98, 164), stipple streak of bean (4, 5), necrosis disease of cucumber (144, 163), and rusty root complex disease of carrot (75). Satellite virus multiplies only in the presence of TNV and considerable specificity exists between serologically different strains of satellite virus and TNV strains on which a given satellite virus depends for replication (73). The virions are about 17 nm, containing single-stranded RNA of molecular weight 0.3×10^6.

The viruses with straight, tubular particles may be placed into two groups, based on the ranges of molecular weight of coat proteins and serological relationships (Table 7.2). Type members of the two groups of viruses with straight, tubular particles are soil-borne wheat mosaic virus (SBWMV) and tobacco stunt virus (TSV). Understanding of the genomic RNA compositions of these viruses is fragmentary, but increasing. Evidence for a bipartite genome has been presented for SBWMV (118, 119, 156), beet necrotic yellow vein virus (BNYVV) (125), and peanut clump virus (145). Generally speaking, viruses of the SBWMV group share the following characteristics. They produce two types of rigid tubular particles approximately 18–25 nm wide and 250–300 nm long (long particles, virion-1) and 110–190 nm long (short particles, virion-2), with a single species coat protein of $20–25 \times 10^3$ daltons. The BNYVV contains an additional particle component 390 nm long. Long-particle RNA (RNA-1) and short-particle RNA (RNA-2) are required for infectivity. They are persistently transmitted by plasmodiophorid fungi in soil. The two viruses

TABLE 7.2. Properties of viruses reported to be transmitted or potentially transmitted by soil-borne lower fungi.

Particle shape and virus type	Virus[a]	Particle size (nm)	Genomic components (mol. wt. × 10³)	Coat protein (mol. wt. × 10³)	Vector	Virus–vector relation	References[b]
Isometric	Cucumber necrosis	31	—	—	Olpidium radicale	Nonpersistent	27 (82)
	Satellite	17	0.3	23.0	O. brassicae	Nonpersistent	69c (15)
	Tobacco necrosis	26	1.3–1.6	22.6–33.5	O. brassicae	Nonpersistent	69b (14)
Straight tubular	Soil-borne wheat mosaic*	300 + 110–160 × 20	2.3 + 0.8–1.2	19.7	Polymyxa graminis	Persistent	14a (77)
	Potato mop-top*	300 + 150 × 20	—	19.8	Spongospora subterranea	Persistent	37c (138)
Soil-borne wheat mosaic virus	Broadbean necrosis*	250 + 150 × 25	—	21.4	?(Polymyxa spp.)	? Persistent	61b (223)
	Beet necrotic yellow vein	390 + 270 + 65–105 × 20	2.3 + 1.8 + 0.7 + 0.6	21.0	P. betae	Persistent	125 (144)
	Oat golden stripe	300 + 150 × 20	—	—	P. graminis	? Persistent	105 (235)
	Rice stripe necrosis	300 − 110 × 20	—	—	P. graminis	? Persistent	29
	Peanut clump	245 + 190 × 20	2.1 + 1.7	23.0	P. graminis	? Persistent	145
	Hypochoeris mosaic	240–260 + 120–140 × 21	—	24.5	—	—	17a (273)
	Nicotiana velutina mosaic	125–150 × 18	—	—	—	—	107c (189)
Tobacco stunt virus	Tobacco stunt†	200–375 × 22	—	50–52	O. brassicae	Persistent	80, 81, 92, 93
	Lettuce big vein¹	200–375 × 22	—	50–52	O. brassicae	Persistent	81, 93
Filamentous	Barley yellow mosaic‡	550 + 275 × 13	—	—	P. graminis	Persistent	61c (143)
Barley yellow mosaic virus	Oat mosaic	600–750 × 13	—	—	P. graminis	Persistent	45a (145)
	Rice necrosis mosaic‡	550 + 275 × 13	—	—	P. graminis	Persistent	61a (172)
	Wheat spindle streak mosaic	190–1975 × 13	—	—	P. graminis	Persistent	120a (167)
	Wheat yellow mosaic‡	550–275 × 13	—	—	? P. graminis	? Persistent	157

[a]Those viruses having the same designation (*, †, ‡) are reportedly serologically related.
[b]Numbers in parentheses refer to CMI/AAB Descriptions of Plant Viruses.

belonging to the tobacco stunt virus group, tobacco stunt (TSV) and lettuce big-vein viruses (LBVV), are rigid tubular particles approximately 22 nm wide and 200–375 nm long, containing a single species coat protein of 50– 52×10^3 daltons (93). They appear to contain RNA that behaves like a double-stranded RNA of 1.2–3.4×10^6 daltons (S. A. Masri and C. Hiruki, unpublished data).

The viruses with filamentous particles have been placed in the barley yellow mosaic virus group. They have been little studied, but appear to be similar to the viruses of the SBWMV group in having a persistent relationship with their plasmodiophorid vector. All induce the formation of pinwheel inclusions in infected cells.

Mode of Transmission

Certain species in two groups of root-infecting fungi serve as vectors of the above-mentioned viruses. In the Chytridiales, *Olpidium* spp. transmit TNV, satellite virus, CNV, TSV, and LBVV; in the Plasmodiophorales, *Polymyxa* spp. transmit the viruses of soil-borne wheat mosaic, beet necrotic yellow vein, peanut clump, and several others, and *Spongospora subterranea* transmits potato mop-top virus (PMTV). Other viruses that are strongly suspected to be soil-borne with a possibility of having fungus vectors also are listed in Table 7.2. This prediction is based on their close similarities in basic properties of virus particles to the known fungus-transmitted viruses.

The life cycle of *Olpidium* spp. is simple: the zoosporangia, formed mostly in the epidermal cells of infected roots, release the uniflagellate zoospores through exit tubes into soil water outside host cells. The zoospores move to another root, attach to its surface, and encyst (1). Each zoospore incites an infection canal, which becomes visible about 2–3 hr later. The protoplasm of the zoospore moves through the canal into the host cell, leaving the cyst wall outside. In 2–3 days a thallus grows in the cell and develops into a new zoosporangium, replacing host cytoplasm. Newly formed zoospores emerge through one or more exit tubes. Some zoospores fuse in pairs before encystment takes place. Thick-walled resting sporangia (50), which are formed in the host cells and later released into the soil, can withstand drying and other adverse conditions, and may germinate to produce zoospores. The plasmodiophorid vectors have a similar life cycle to that of *Olpidium* (135, 137).

Olpidium transmits different viruses in two ways, which may be termed nonpersistent and persistent. In the nonpersistent mode the particles of TNV (134), satellite virus (72), or CNV (26) are acquired by zoospores *in vitro* and carried into root cells. The particles first attach specifically to the surface of the zoospore (124, 143), and those on the surface of the flagellum are quickly taken into the zoospore protoplasm and then transferred with the protoplasm to a root cell. Subsequent growth of the *Olpidium* in the root cell is not essential for virus multiplication; heating roots for 10 sec at 50°C, 2–24 hr after inoculation with virus-carrying zoospores kills *O. brassicae* without preventing infection and replication of TNV (70). Apparently these viruses are not retained in resting spores, and the association with *Olpidium* is short-lived (nonpersistent).

By contrast, TSV (46, 48, 51) and LBVV (22) are apparently held internally, and are retained by resting spores of *Olpidium* for many years. They are acquired by the fungus while it is growing in plant cells and are not acquired by zoospores *in vitro*. Similar behavior, although less well studied, was observed with SBWMV (28, 110), BNYVV (125, 126), and PMTV (67) in their plasmodiophorid vectors. Resting spores remain viable under air-dried conditions and can probably retain the viruses for many years and the viruses are transmitted to roots or tubers by the zoospores after germination. However, evidence for multiplication of these viruses in the fungal vectors is lacking.

In an ultrastructure study of various stages in the life cycles, *P. graminis* and *P. betae* were examined in SBWMV-infected wheat and BNYVV-infected sugar-beet roots (84). In spite of an earlier report of detection of viruslike particles in *P. betae* zoospores (125), virus particles were not seen within fungal plasmodia, zoosporangia, or cystosori. On the other hand, particles of both viruses were observed in the host tissue in close contact with the fungal vector (84).

Nematode-Transmitted Viruses

Properties

Two groups of viruses, the nepoviruses with isometric particles and the tobraviruses with tubular particles, are transmitted by nematodes, in particular, root-feeding ectoparasites (38). Tobacco ringspot virus (TRSV) is the type member of the Nepovirus group, which has 18 member viruses and an additional five tentative members (Table 7.3). Generally speaking, the nepoviruses have isometric virions of 25–30 nm diameter with three components, 49–56S (top), 86–128S (middle), and 115–134S (bottom), which contain respectively, 0, 27–40, and 42–46% single-stranded RNA. The virions with a single-species coat protein of molecular weight 53–60 $\times$ 10^3 contain either of two RNA components of molecular weight 2.0–2.6 (2.8) $\times$ 10^6 (RNA-1) and 1.4–2.2 (2.4) $\times$ 10^6 (RNA-2) (101, 108). Both RNA genome components are required for infectivity. The nepoviruses have wide host ranges, inducing ringspot symptoms and causing economically significant losses to fruit trees, grapevines, and other diverse groups of plants such as vegetables, ornamentals, small fruits, and special crops. Of 18 nepoviruses known, four viruses occur in biologically diverse but serologically related strains that may be placed in subgroups (Table 7.3). The five viruses listed as the tentative members of the Nepovirus group are aberrant in that they have anomalous coat protein compositions al-

TABLE 7.3. Properties of viruses reported to be transmitted or potentially transmitted by nematodes.[a]

Particle shape and virus group	Virus	Particle size (nm)	Genome components[b] (mol. wt. × 10^6)	Coat protein (mol. wt. × 10^3)	Vector	References[c]
Isometric						
Nepovirus						
	Tobacco ringspot (TRSV) subgroup					
	TRSV, type strain	29	2.4 (2.8) + 1.4 (1.3)	57	*Xiphinema americanum*	123a (17)
	TRSV, eucharis mottle strain	29	2.4 + 1.4	—		123a (17)
	Potato black ringspot	25	2.5 + 1.5	59	—	111a (206)
	Olive latent ringspot	28	2.7 + 1.4	58	—	114
	Arabis mosaic (AMV) subgroup					
	AMV, type strain	30	2.4 (2.8) + 1.4 (1.3)	54	*X. diversicaudatum*	99b (16)
	AMV (H), hop strain	30	2.4 + 1.4	54	*X. diversicaudatum*	99b (16)
	Grapevine fanleaf	30	2.4 + 1.4	54	*X. index, X. italiae*	45b (28), 107
	Raspberry ringspot (RRV) subgroup					
	RRV, type (Scottish) strain	30	2.4 (2.8) + 1.4 (1.4)	54	*Longidorus elongatus*	99a (6), (198)
	RRV, English strain	30	2.4 (2.8) + 1.4 (1.4)	54	*L. macrosoma*	99a (6), 99e (198)
	Arracacha virus A	26	2.5 + 1.4	53	—	68, 68 (216)
	Mulberry ringspot	25	—, — (1.5)	—	*L. martini*	154a (142)
	Artichoke Italian latent	30	2.4 + 1.5 (1.68)	54	*L. apulus*	90c (176)
	Tomato black ring (TBRV) subgroup					
	TBRV, potato bouquet strain	30	2.5 (2.8) + 1.5 (1.7)	57	*L. attenuatus*	99c (38)
	TBRV, beet ringspot strain	30	2.5 (2.8) + 1.5 (1.7)	57	*L. elongatus*	99c (38)
	Cocoa necrosis	30	2.4 + 1.5	60	—	75a (173)
	Grapevine chrome mosaic	30	2.4 + 1.5 (1.63)	—	—	90b (103)
	Artichoke yellow ringspot	30	2.4 + 2.1	54	—	107b (271)
	Myrobalan latent ringspot	28	2.6 (2.8) + 1.9 (2.0)	53	—	27a (160)
	Hibiscus latent ringspot	28	2.6 + 2.0	54	—	18, 19 (233)
	Blueberry leaf mottle	28	2.4 + 2.2	54	—	107a (267)
	Cherry leaf roll (CLRV) subgroup					
	CLRV, cherry strain	30	2.4 (2.8) + 2.1 (2.3)	54	(*X. diversicaudatum, X. coxi*)	24a (80)

	CLRV, many serotypes from birch, blackberry, dogwood, elderberry, elm, raspberry, rhubarb, and walnut	—	—, —	—	64, 65, 113, 166
	Peach rosette mosaic	28	2.5 + 2.2	57	*X. americanum*	26a (150)
	Chicory yellow mottle	30	2.4 + 2.0	54	—	104, 106a (132)
	Grapevine Bulgarian latent	30	2.2 + 2.1	54	—	90d (186)
	Lucerne Australian latent	25	2.4 + 2.1	55	—	66
	Tomato ringspot (TomRV) subgroup				
	TomRV, type strain	28	2.3 (2.8) + 2.2 (2.4)	58	*X. americanum*	123b (18)
	TomRV, grape yellow vein strain	28	—, —	—	*X. americanum*	123b (18)
	Tentative members				
	Strawberry latent ringspot	30	2.6 (2.9) + 1.6 (1.4)	44, 29	*X. diversicaudatum, X. coxi*	99d (126)
	Cherry raspleaf	30	2.0 + 1.5	24, 22.5	*X. americanum*	123c (159)
	Artichoke vein-banding	—	—, —	27.5, 24.5, 22	—	33
	Tomato top necrosis	26		—	—	3
	Satsuma dwarf	26	1.9 + 1.7	42, 21	—	157a (208)
Bromovirus	Brome mosaic	25	1.1 + 1.0 + 0.8 + 0.3	20	*L. macrosoma, X. diversicaudatum, X. coxi*	82a (180), 116
Dianthovirus	Carnation ringspot	30	1.5 + 0.5	38	*X. diversicaudatum, L. elongatus, L. macrosoma*	55a (21)
Ilarvirus	Prunus necrotic ringspot	23	—, —	25	*L. macrosoma*	31a (5)
Tubular					
Tobravirus	Tobacco rattle	190 + 45–115 × 23	2.5 + 0.7–1.3	24	*Paratrichodorus* and *Trichodorus* spp.	37a (12)
	Pea early-browning	210 + 105 × 21	2.5 + 1.3	24	*Paratrichodorus* and *Trichodorus* spp.	37b (120)

[a] From: Murant (99), with some modifications.
[b] From determinations in polyacrylamide gels under nondenaturing conditions; values in parentheses are determined in denaturing conditions.
[c] Numbers in parentheses refer to CMI/AAB Descriptions of Plant Viruses.

though all of them resemble nepoviruses in having three sedimenting virion components, bipartite genomes, and wide host ranges, while at least some have nematode vectors [strawberry latent ringspot virus (SLRV) and cherry raspleaf virus (CRLV)] and are transmitted through seed (SLRV, CRLV, and Satsuma dwarf virus).

There are three additional isometric viruses, brome mosaic virus (BMV), carnation ringspot virus (CaRSV), and prunus necrotic ringspot virus, which have been reported to be transmitted by nematodes under laboratory conditions only. They are not related to nepoviruses, but belong to the Bromovirus, Dianthovirus, and Ilarvirus groups, respectively. These viruses are listed separately in Table 7.3.

The tobraviruses, the second major group of nematode-transmitted viruses, have rigid, tubular, bipartite particles 21–23 nm in diameter, with two predominant lengths, characteristic of the particular isolates. Virion-1 of different isolates has similar modal lengths; 185–197 nm for TRV (42), and about 210 nm for pea early browning virus (PEBV) (12). Virion-2 of different isolates ranges from 45 to 115 nm. Their single-stranded RNA genomes occur in two parts (RNA-1 and RNA-2), each having distinct genetic functions, which will be discussed later. The molecular weights of RNA-1 and RNA-2 are 2.5×10^6 and 0.7–1.4×10^6, respectively (24). Virion-1 and virion-2 are serologically indistinguishable. Moderate to distant serological relationships exist between European, North American, and Brazilian strains of TRV and they may be separated into three serotypes (42). Intergrading isolates also occur (40).

Mode of Transmission

Two families of nematodes are responsible for transmission of viruses belonging to the Nepovirus and Tobravirus groups: in the order Dorylaimida, which includes relatively few species known to be plant parasites, the family Longidoridae includes two genera, *Xiphinema* (with six vector species) and *Longidorus* (five vector species), which can act as efficient vectors of nepoviruses (82). The family Trichodoridae contains seven species of *Paratrichodorus* and five species of *Trichodorus* as vectors of tobraviruses (82).

Longidorids are large nematodes (2–12 mm long) having a hollow axial stylet with a slitlike aperture throughout its length, about 100–300 μm. The stylet is supported by the odontophore, which also serves as an ejecting structure. The basal bulb contains saliva-producing gland cells (171).

Xiphinema spp. (169, 170) feed in the region of cell elongation behind the root cap of seedling roots. Following initial lip rubbing, the nematode thrusts its stylet vigorously into the cell wall and completes perforation within a few seconds. The stylet continues to attack the cell until it reaches and perforates the walls of the underlying cell. After penetrating a few cells, it starts ingesting the cytoplasmic contents of the attacked cells by a rapid pumping action of the bulb, which follows a definite sequence of events lasting a few seconds. Food ingestion by *X. index* is intermittent; it feeds on an individual cell for several minutes only, then its stylet penetrates the deeper layer of cells far below the epidermis. Consequently, root growth is retarded and the root starts to swell, usually at the region of elongation. Several days' continuous feeding transforms the swelling into a terminal gall, which in turn attracts feeding nematodes (170). *Longidorus* spp. show quite a different feeding behavior compared to *Xiphinema* spp. The stylet is thrust into root tips without odontophore rotation and continues to penetrate until it is fully protracted in the root tissue. Salivation is thought to occur within 30–60 min between final stylet penetration and ingestion proper (171). The nematode may feed at one site for several hours continuously with its long stylet penetrating inner tissues, probably reaching vascular elements, and is more sedentary in feeding habit.

The nepoviruses transmitted by *Xiphinema* spp. appear to persist longer in their vectors than those transmitted by *Longidorus* spp. For example, raspberry ringspot (RRV) and tomato black ring viruses (TBRV) rarely persisted in *L. elongatus* for 3 months (100, 127), whereas TRSV persisted for almost 1 year in *X. americanum* at 10°C (9). However, such extended virus longevity in nematodes does not imply that viruses multiply in their vectors. Furthermore, infectious virus particles are not retained through a molt (41, 130). Following ingestion from virus-infected plants, virus particles accumulate as a monolayer on surfaces of the nematode stomodaeum: typical examples include RRV and TRSV on the stylet guiding sheath of *L. elongatus* (131) and certain viruses transmitted by *Xiphinema* spp. on the cuticular lining of the esophageal lumen (94, 133). TRV particles become associated with the cuticular lining of the pharynx and esophagus (132) and persist for an extended period of time (162). At molting, however, these surfaces are shed. Nonvector longidorids do not retain particles of viruses that are transmitted by related vector species (45).

Trichodorids most commonly attack epidermal cells and root hairs, moving from cell to cell and staying only a few minutes at each cell (170). They feed at or just behind the zone of elongation of young roots, avoiding the apical meristem. A feeding cycle is composed of the following phases: cell wall exploration, wall perforation, salivation, ingestion, and departure from the feeding site.

In trichodorids the stylet often measures up to 200 μm. While both groups of vectors often feed on the young tissue near the root tips, *Trichodorus* spp. usually feed on epidermal cells and move about actively. Host specificity varies from species to species of nematode. Both larvae and adult nematodes can acquire, retain, and transmit viruses, but there is no evidence of transovarial transmission or virus multiplication in the vectors. Viruses are apparently not retained through the molt.

The reversible association of virus particles with specific sites in the food canal (131) may play an important role in virus transmission. Virus particles transmitted by *Longidorus* spp. become associated with the esophageal wall, and particles of TRV attach to these sites when plant sap containing virus particles passes down the food canal during acquisition feeding, and dissociate when saliva passes in the reverse direction into host cells during inoculation feeding. A hypothesis and supporting evidence suggest that the protein surface of the virus particles is involved in determining vector specificity. Comparative tests showed a close correlation between serological specificity and vector specificity among strains of nepoviruses (36, 44). Work with virus pseudorecombinants demonstrated that the vector specificity of RRV is determined by a genome that carries the gene for virus coat protein (44).

Insect-Transmitted Viruses

Remarkably few viruses have been reported to be insect-transmitted in soil, and apparently roots are of little importance as a route for virus infection by this means. There are two exceptions. First, lettuce mosaic virus (LMV), a potyvirus, can be transmitted by the root aphid *Pemphigus bursarius* (95). However, this mode of transmission of LMV is less common than transmission by leaf-feeding aphids, and the virus may be common in localities where root aphids are absent.

Second, maize chlorotic mottle virus, an isometric virus about 30 nm in diameter, survives in infested maize debris from which it is apparently acquired by larvae of the western rootworm, *Diabrotica virgifera*. The larvae transmit the virus when they feed on developing maize roots. Later, adults of several chrysomelid species can transmit the virus efficiently between the tops of maize plants (158). Control involves rotating maize with a nonhost, such as *Sorghum bicolor* or *Glycine max*, or drying infested roots for 3–4 weeks, which greatly reduces or eliminates their virus content.

Other Soil-Borne Viruses

Although most soil-borne viruses have natural means of transmission that are known, there are a few for which little evidence is yet available. Until convincing evidence is forthcoming, they are placed in a fifth group.

Maize white line mosaic virus infects maize in the United States causing stunting and loss in yield. Particles of the virus are isometric and have a diameter of 35 nm (174). Infection occurs most commonly in low-lying wet areas of fields, and transmission through soil to maize seedlings has been demonstrated repeatedly. Mechanical transmission to leaves of maize or other test plants has not been reported. Evidence implicating a fungal vector is the presence of *Polymyxa graminis* and *Olpidium* sp. in roots (11) and the reduction in transmission achieved by applying the fungicide Benomyl to the soil (87).

Ecological Significance of Properties of Viruses

Several properties of viruses are important or potentially of significant value in relation to ecological consideration of soil-borne virus diseases. Among them are (1) the stability of viruses in soil, (2) the nature of viral genetic elements, in particular divided genome systems, and (3) the affinity of viruses with their vectors, namely the ability to be acquired, retained, and transmitted by their vectors, which altogether form a basis for virus–vector specificity.

Stability in Soil

With regard to stability in soil, two important aspects must be considered, namely the stability of the virus itself *in vitro* and the stability of the association of a given virus with its vector. Since the former aspect has been dealt with in the preceding section, here more attention will be paid to the latter. Virus stability in soil in terms of transmissibility can be treated by categorizing viruses into those that are resistant and those that are sensitive to air-drying of infested soil. Whereas the transmissibility of fungus-associated viruses is relatively resistant to air-drying, that of nematode-associated viruses is sensitive. The virus–fungal vector relationship can be further characterized by the type of association, such as the location of the virus in relation to the vector fungus. Two types, those associated externally with vector spores (nonpersistent type) and those associated internally (persistent type), are known to occur. There is no parallel between the stability of virus infectivity *in vitro* and the stability of virus association with its vector. Rather, unstable viruses are uniquely stabilized for significantly extended periods of time by associating with particular vector fungi in the soil. For example, TSV remains infectious in sap at room temperature only a few hours at most (49). However, its viability is extended up to several years upon association with *O. brassicae* (46) and, moreover, TSV longevity in association with *O. brassicae* has recently been extended to at least 20 years (C. Hiruki, unpublished data). SBWMV remains infective at 15°C up to 3 months (155), whereas longevity of the same virus is extended to 3 years by association with *P. graminis* (109). BNYVV, with an *in vitro* longevity of 5 days at 20°C, can persist at least 4 years upon association with *P. betae* (125). Actual mechanism(s) of stabilization of virus infectivity by association with a specific vector fungus in any of the above-mentioned examples are not yet known. TNV, which is transmitted by *O. brassicae* in a nonpersistent manner, appears to survive in living roots, root debris, or soil water. It may be adsorbed to clay particles, and infective virus particles can be released into buffer solution when properly suspended (122). The virus spreads in splashing water and is acquired by zoospores of *O. brassicae* for further transmission, creating serious problems in managing greenhouse- and field-grown crops.

Survival of nematode-transmitted viruses is influenced largely by ecological factors, such as the presence of infected weeds and weed seeds.
Woody perennials play a significant role in the survival of nematode-
transmitted viruses by serving as a stable reservoir. Virus-infected weed
seed offers an important and often decisive means of survival in the ecosystem (100). The rate of spread of nematode-transmitted viruses is rather
slow. However, when a noncultivated elm root system spreads into an
adjoining pasture land, transmission of arabis mosaic virus by nematodes
may increase to a rate of 11 m/year (96).

Divided Genome

Possession of a divided genome is common among soil-borne viruses, particularly those transmitted by fungi or nematodes (Tables 7.2–7.4). The
divided-genome viruses are those that require more than one nucleic acid
component for infection and those in which these nucleic acids are packed
separately (83). This genetic system allows such viruses independent

TABLE 7.4. Genetic determinants for biological properties of some soil-borne
viruses.

Virus[a]	Genomic RNA	Properties	References
SBWMV	RNA-1	Infectivity	156
		Virus concentration (requires RNA-2)	156
	RNA-2	Coat protein (serotype)	156
		Particle length	156
		Inclusion type	156
RRV	RNA-1	Host range in raspberry	39
		Invasion type	39
		Symptom suppression in *Petunia hybrida*	39
		Seed transmission	35, 39
	RNA-2	Coat protein (serotype)	43, 44
		Vector specificity	44
		Symptom type (*P. hybrida*)	43
TRV	RNA-1	Proteins of molecular weight 17×10^4 and 14×10^4	40
		RNA replicase (?)	40
		Lesion type	40
		Invasion type	40
	RNA-2	Coat protein	40
		Protein of molecular weight 3.1×10^4	40
		Particle length (virion-2)	40
		Invasion type (Solanaceae)	40

[a]SBWMV, Soil-borne wheat mosaic virus (vector, *Polymyxa graminis*); RRV, raspberry ringspot virus
(vector, *Longidorus elongatus*); TRV, tobacco rattle virus (vector, *Paratrichodorus* and *Trichodorus*
spp.).

translation of a number of different proteins from the whole genome and high-frequency recombination between related virus strains.

SBWMV, a *Polymyxa*-transmitted tubular virus, is a bipartite virus consisting of virion-1 and virion-2 of various lengths that are characteristic for each isolate of different origin. The combination of RNA-1 and RNA-2 is essential for infection and replication (Table 7.4). Repeated subculturing by sap inoculation results in RNA-2 mutants that are smaller than normal but still functional (except for the smallest RNA-2) upon being combined with RNA-1. Plants infected with RNA-1 and normal RNA-2 showed symptoms milder than those with increased amounts of smaller RNA-2 (118). SBWMV virions shorter than virion-2 occur in wheat plants after an infection period of several months, probably as a result of spontaneous deletion mutation of RNA-2 (119, 119a). Virion-2, which contains RNA-2, controls particle length, serotype, and type of inclusion bodies (156).

The proportion of divided-genome viruses transmitted by nematode vectors among soil-borne viruses seems to be unusually high (Table 7.3). The obvious advantage of having divided genomes is that there will be increased possibilities of reassortment of genetic material during virus replication, so adapting to changing environments and new host conditions. Presumably a nematode may be able to mediate and increase the possibility of reassortment by acquiring a heterogeneous virus population.

The genome of RRV, a nepovirus, is composed of two RNA species (Table 7.4). Serological specificity and transmissibility by the nematode *L. elongatus* were found to be determined by RNA-2. This fact suggests that the coat protein of virus particles is involved in the acquisition and transmission of the virus particles. Genetic determinants for the ability to infect various plants systemically are located in RNA-1. Severity of systemic infections and ability to infect a raspberry cultivar are also determined by RNA-1. Determinants on RNA-1 and RNA-2 play a part in determining certain other biological features in virus–host interactions (38). It was also shown that the reassortment of genetic determinants permits novel combinations of newly acquired properties, such as vector specificity and ability to infect a new host (108). However, some of the pseudorecombinants thus produced are less stable than their parental viruses (63).

In tobraviruses, RNA-1 is infective and replicates without forming coat protein. On the other hand, virion-2 is noninfective. Virion-2 needs the presence of RNA-1 for the production of complete virions including both virion-1 and virion-2 (Table 7.4). Repeated subculturing of certain tobraviruses often results in a change in length of virion-2 (42). Earlier reviews document the evolution of the divided-genome concept with this virus (32, 40). RNA-1 controls replication, symptoms, and virion length. RNA-2 controls the production of coat protein, virion length, and symptoms. Some recognition specificity between RNA-1 and RNA-2 appears to exist, since some degree of compatibility for pseudohybridization between different combinations of strains occurs. The interstrain compatibility has probably not been investigated sufficiently to provide a clear-cut explanation of possible mechanism(s) of pseudohybridization. New combinations of these characteristics in the pseudorecombinants would expand the range of adaptability to different sets of conditions. Characteristics connected with divided genomes that might confer on a virus new capability to survive under selective pressure would include altered host range, improved adaptation to vector transmission, speedy invasion of host tissue, and symptom variants, which lead to tolerant host–virus interaction with a high virus concentration (32).

Virus–Vector Specificity

The most common organism feeding upon a particular plant species is not necessarily capable of serving as a vector of viruses that can affect that particular plant. For example, although there are several reports on the influence of endotrophic mycorrhizae on virus disease development (25, 62, 115, 117), these observations do not implicate the mycorrhizal fungi as virus vectors. Therefore, to serve as efficient fungal or nematode vectors of plant viruses, they must be able to form certain specific relationships with viruses. Likewise, it is also conceivable that there are some properties with which viruses must be equipped in order to achieve specific association with their respective vector organisms.

Electron microscopy of zoospores from virus-transmitting zoospore suspensions of TNV and CNV, both transmitted in a nonpersistent manner, revealed numerous virus particles adhering to the surface of the zoospore plasmalemma and axonemal sheath of vector isolates, but not of nonvector isolates (124, 143, 142). This adsorption specificity could be based on differences in surface charge (97) or surface structure of zoospores and viruses. Attachment of virus particles to the zoospore surface is therefore the first requirement for fungal transmission to occur. Further phases in the transmission process involve movement of virus by endocytosis into the zoospore protoplast during or after encystment and its release after the zoospore protoplast has entered the host cell (142). There is a wide variation in the vector efficiency of different isolates of *O. brassicae*. For example, crucifer isolates do not transmit any of several strains of TNV (71, 97, 140) apparently because of their inability to acquire the virus (143), and the lack of virus transmission is not correlated with the host's reaction to the *Olpidium* isolates used. Furthermore, at the other extreme, the lettuce isolates used in different laboratories consistently acquire abundant virus particles and efficiently transmit TNV strains. However, studies with a group of *Olpidium* isolates (71, 142, 143) show that there is a range of vector specificities, and the selected isolates of *Olpidium* may transmit certain strains of TNV to a certain host, but not all TNV strains to all host species tested. Vector specificity is complex and may be due to less efficient virus acquisition or penetration of the root, to failure of transmission at any later stages after fungal penetration, or to unfavorable responses of the host cells. In an analysis of a complex of relationships between host, fungus, and virus, a wide range of species was tested for susceptibility to *O. brassicae* and to TSV by sap inoculation or inoculation with viruliferous zoospores (47). The species tested fell into four groups: (a) susceptible to both *O. brassicae* and TSV; (b) susceptible to *O. brassicae,* but not to TSV; (c) susceptible to TSV, but not to *O. brassicae;* (d) resistant to both. Some species were susceptible to TSV introduced by *O. brassicae,* but not by sap inoculation.

With regard to the vector specificity of nematode-transmitted viruses, much progress has been made with nepoviruses (60, 154). Differences in vector specificity of nepovirus strains are based on differences in serological specificity (36, 44, 45). Moreover, experiments with strains and pseudorecombinant isolates of RRV have indicated that vector specificity is determined by the same genome RNA that carries the coat protein cistron (44). The coat protein, therefore, may determine specificity of retention of particles in the nematode, and hence vector specificity. Virus particles that have become detached from their site of retention are thought to be ejected into healthy cells along with saliva. The two parts of nepovirus genomes (RNA-1 and RNA-2) are both needed for infection, but are contained in different virus particles (43, 102, 108), so that a mechanism allowing the inoculation of a cell with a considerable number of particles should favor transmission. This may explain the suitability of nematodes as vectors for this type of virus, because nematodes feed on virus-infected plants and internally accumulate a large number of virus particles prior to transmission to particular host plants.

Various nepoviruses apparently can be transmitted in laboratory tests by longidorids that are not associated with them in the field (159), and various *Longidorus* and *Xiphinema* spp. are reported to transmit brome mosaic virus (BMV) (116) and carnation ringspot virus (CRSV) (31). It remains to be seen whether these combinations bear broader ecological significance, as additional evidence has been reported recently regarding transmission of CRSV by *L. elongatus* in orchards (77) and BMV by aphids and rust fungus (165). In transmission of tobraviruses, there is also some evidence of vector specificity, although few comparisons have been made of the transmission of different isolates by the same vector species. A British isolate of PEBV was transmitted by *Paratrichodorus anemones,* whereas a Dutch isolate of PEBV was not (37), and only one of five Dutch isolates of TRV was transmitted by *P. pachydermus* (161). It would be of interest to know whether vector specificity depends on the virus coat protein as it may in the nepoviruses (36, 45).

Conclusion

Soil-borne viruses are a rather heterogeneous collection of viruses, which may be placed in at least 13 plant virus groups (Table 7.5). They may be transmitted either mechanically or by means of a soil-inhabiting fungus,

TABLE 7.5. Particle stability in sap and number of genome components of groups of soil-borne viruses.[a]

Method of transmission and virus group	Number of genome components	Thermal inactivation point (°C)	Longevity in vitro (days)	Dilution end-point
Mechanical				
Dianthovirus	2	85–90	70	10^{-5}
Sobemovirus	1	85–95	28–140	10^{-5}–10^{-6}
Tobacco necrosis virus	1	85–95	7–28	10^{-6}
Tombusvirus	1	90	490	10^{-6}
Tobamovirus	1	95	365 +	10^{-6}–10^{-7}
Tobravirus	2	78–85	42	10^{-6}
Potyvirus[b]	1	55–65	4–109	10^{-2}–10^{-4}
Fungi, nonpersistent				
Tobacco necrosis virus	1	85–95	7–28	10^{-6}
TNV satellite virus	1	95	7–84	10^{-5}
Fungi, persistent				
Soil-borne wheat mosaic virus	2	50–65	1–4	10^{-1}–10^{-3}
Tobacco stunt virus	?	75–80	<1	10^{-2}–10^{-3}
Barley yellow mosaic virus	?	45–65	1–7	10^{-2}–10^{-4}
Nematodes				
Dianthovirus	2	85–90	70	10^{-5}
Nepovirus	2	55–70	7–70	10^{-3}–10^{-4}
Tobravirus	2	78–85	42	10^{-6}
Insects				
Maize chlorotic mottle virus (ungrouped)	?	?	?	?
Potyvirus[b]	1	55–65	4–10	10^{-2}–10^{-4}

[a]Information from Boswell and Gibbs (13), and Hiruki (49) CMI/AAB Descriptions of Plant Viruses, cocited in Tables 7.1–7.3.
[b]Rarely soil-borne.

nematode, or insect vector. Occasionally a virus that is vector-borne, such as TNV transmitted by *Olpidium brassicae* or TRV transmitted by trichodorid nematodes, is also mechanically transmitted in soil (136), but usually there is only one route by which a virus is soil-borne.

Some of the properties of the viruses are closely correlated with their mode of transmission in soil. Viruses that are mechanically transmitted are generally highly stable, as indicated by a high thermal inactivation point and longevity *in vitro,* and are concentrated in sap, as indicated by a high dilution end-point (Table 7.5). These properties are advantageous because of the low efficiency of mechanical transmisson in soil, the adverse effects of the soil environment on survival of free virus, and the diluting effect of soil water on free viruses. Nonpersistent transmission of viruses by fungi resembles that of mechanically transmitted viruses, in that the viruses have a phase when they are free in the soil, but their carriage into the plant is efficient following an association with the zoospore surface.

In contrast, the viruses that are transmitted by fungi in a persistent manner have particles that are relatively unstable and occur in low concentration in sap (Table 7.5). These viruses are highly dependent for natural transmission on their specific vectors, and therefore they have a need for close compatibility with their vector fungi, in which they can survive for many years. These viruses lead a protected, intracellular life inside their vector or host plant, and are not exposed to the soil environment directly.

The viruses that are transmitted by nematodes occupy an intermediate position regarding the need for high particle stability and a high concentration in host sap. The nematode-transmitted viruses are always acquired by their vectors directly from the host plant, and therefore avoid exposure to the outside soil environment. However, within the nematode they are located extracellularly in a carbohydrate matrix along the walls of the upper alimentary tract (129), in which position they are exposed to a somewhat unfavorable environment of plant sap and nematode saliva. Therefore, as well as having a specific interaction with the surface of the alimentary tract, they need moderately high particle stability (Table 7.5).

The possession of a divided genome is a property of all the nematode-transmitted viruses and of at least one of the groups of viruses transmitted in a persistent manner by fungi (Table 7.5). However, a divided genome is absent from five of the seven groups of viruses that are mechanically transmitted and both of the groups of viruses that are nonpersistently transmitted by fungi (Table 7.5). Possibly the viruses that are internally borne in fungi or nematodes have evolved along the lines of genetic flexibility provided by a divided genome, whereas the viruses that are externally borne in fungi or are mechanically transmitted in soil have evolved along the lines of transmission efficiency provided by an undivided genome.

Vector specificity is a character of all fungus-transmitted viruses (Table 7.2) and all nematode-transmitted viruses, although in the latter case a virus may be transmitted by two closely related nematodes (Table 7.3). In general the fungus- and nematode-transmitted viruses have no close counterparts among viruses regularly transmitted by aerial vectors. Exceptions to this rule are viruses in the barley yellow mosaic virus group, which have particles and pinwheel inclusion bodies morphologically similar to those of the aphid-borne potyviruses.

None of the soil-borne viruses is known to multiply in its vector. It seems likely, therefore, that the viruses are plant viruses rather than fungal, nematode, or insect viruses. It is possible that the ancestors of some of the viruses, such as those in the tobacco necrosis virus, soil-borne wheat mosaic virus (SBWMV), tobravirus, and dianthovirus groups, were mechanically transmitted in soil, but later acquired soil-borne vectors and thereafter evolved in association with these vectors, which greatly increased their efficiency of transmission. If the distant serological relationship between certain viruses in the SBWMV and tobamovirus groups (74, 103, 106) is an indication of a common origin, then the tobamoviruses might be descendants of viruses that have failed to acquire a vector and hence have evolved toward efficient mechanical transmission in soil and other media, whereas the SBWMV group acquired plasmodiophorid vectors.

No doubt the interest in soil-borne viruses will continue to increase, with new viruses and vectors discovered and additional information on the ecology and control of soil-borne viruses. Perhaps the area most likely to receive increased attention in the near future is the occurrence of soil-borne viruses in waters, including natural streams, rivers, and lakes, sewage, and nutrient solutions used for growing crops such as lettuce, cucumbers, and tomatoes. It is now clear that some viruses, such as tomato bushy stunt, tobacco necrosis, and tobacco mosaic viruses, are washed from soil and/or sewage into waters and can survive either free or within their vector for a considerable time (150, 152). Plants irrigated with such virus-containing waters could contract infection, and new methods of control, such as the addition of nonionic detergents to water to destroy zoospores transmitting lettuce big-vein virus (149), may be necessary.

Acknowledgments. We wish to thank those people who contributed to this review by personal communications and advice. Portions of the work described here were supported by grants from the Natural Sciences and Engineering Research Council of Canada (A3843, IC0125). C.H. was a holder of a CSFP senior visiting professorship from the Australian Government during 1984–1985.

References

1. Alderson, P. G., and Hiruki, C., 1977, Scanning electron microscopy of zoospores of *Olpidium brassicae,* free or attached to tobacco roots, *Phytopathol. Z.* **90**:123–131.
2. Allen, W. R., 1981, Dissemination of tobacco mosaic virus from soil to plant leaves under greenhouse conditions, *Can. J. Plant Pathol.* **3**:163–168.
2a. Allen, W. R., 1984, Mode of inactivation of TMV in soil under dehydrating conditions, *Can. J. Plant Pathol.* **6**:9–16.
2b. Allen, W. R., and Davidson, T. R., 1967, Tomato bushy stunt virus from *Prunus avium* L. I. Field studies and virus characterization, *Can. J. Bot.* **45**:2375–2383.
3. Bancroft, J. B., 1968, Tomato top necrosis virus, *Phytopathology* **58**:1360–1363.
4. Bawden, F.C., and van der Want, J. P. H., 1949, Bean stipple streak caused by a tobacco necrosis virus, *Tijdschr. Plantenziekten* **55**:142–150.
5. Behncken, G. M., 1968, Stipple streak disease of French bean caused by a tobacco necrosis virus in Queensland, *Aust. J. Agric. Res.* **19**:731–738.
6. Behncken, G. M., 1970, Some properties of a virus from *Galinsoga parviflora, Aust. J. Biol. Sci.* **23**:497–501.
6a. Behncken, G. M., Francki, R. I. B., and Gibbs, A. J., 1982, Galinsoga mosaic virus, CMI/AAB Descriptions of Plant Viruses, No. 252, Com-

monwealth Mycological Institute/Association of Applied Biologists, Kew, Surrey, England.

7. Beijerinck, M. S., 1898, Uber ein Contagium vivum fluidum als Ursache der Fleckenkrankheit der Tabaksblatter, *Verh. Akad. Wet. (Amsterdam)* **65:**3–21 [translation in: *Phytopathological Classics,* No. 7, American Phytopathological Society, St. Paul, Minnesota, pp. 33–52].

8. Bercks, R., and Querfurth, G., 1969, Uber den Nachweis des Sowbane Mosaic Virus in Reben, *Phytopathol. Z.* **66:**365–373.

9. Bergeson, G. B., Athow, K. L., Laviolette, F. A., and Thomasine, M., 1964, Transmission, movement and vector relationships of tobacco ringspot virus in soybean, *Phytopathology* **54:**723–728.

10. Bond, W. P., and Pirone, T. P., 1970, Evidence for soil transmission of sugarcane mosaic virus, *Phytopathology* **60:**437–440.

11. Boothroyd, C. W., and Zitter, T. A., 1981, Maize white line mosaic—1980, in: *Proceedings 36th Northeastern Corn Improvement Conference,* New York, New York.

12. Bos, L., and van der Want, J. P. H., 1962, Early browning of pea, a disease caused by a soil- and seed-borne virus, *Tijdschr. Plantenziekten* **68:**368–390.

13. Boswell, K. F., and Gibbs, A. J., 1983, *Viruses of Legumes 1983. Descriptions and Keys from Virus Identification Data Exchange,* Australian National University, Canberra. 139 pp.

14. Bowen, R., and Plumb, R. T., 1979, The occurrence and effects of red clover necrotic mosaic virus in red clover (*Trifolium pratense*), *Ann. Appl. Biol.* **91:**227–236.

14a. Brakke, M. K., 1971, Soil-borne wheat mosaic virus, CMI/AAB Descriptions of Plant Viruses, No. 77, Commonwealth Mycological Institute/Association of Applied Biologists, Kew, Surrey, England.

15. Broadbent, L., 1965, The epidemiology of tomato mosaic. 8. Virus infection through tomato roots, *Ann. Appl. Biol.* **55:**57–66.

16. Broadbent, L., 1976, Epidemiology and control of tomato mosaic virus, *Annu. Rev. Phytopathol.* **14:**75–96.

17. Broadbent, L., Read, W. H., and Last, F. T., 1965, The epidemiology of tomato mosaic. X. Persistence of TMV-infected debris in soil, and the effects of soil partial sterilization, *Ann. Appl. Biol.* **55:**471–483.

17a. Brunt, A. A., and Stace-Smith, R., 1983, Hypochoeris mosaic virus, CMI/AAB Descriptions of Plant Viruses, No. 273, Commonwealth Mycological Institute/Association of Applied Biologists, Kew, Surrey, England.

18. Brunt, A. A., Barton, R. J., Phillips, S., and Lana, A. O., 1980, Hibiscus latent ringspot virus, a newly recognized virus from *Hibiscus rosa-sinensis* (Malvaceae) in western Nigeria, *Ann. Appl. Biol.* **96:**37–43.

19. Brunt, A. A., Barton, R. J., and Phillips, S., 1981, Hibiscus latent ringspot virus, CMI/AAB Descriptions of Plant Viruses, No. 233, Commonwealth Mycological Institute/Association of Applied Biologists, Kew, Surrey, England.

20. Campbell, R. N., 1968, Transmission of tomato bushy stunt virus unsuccessful with Olpidium, *Plant. Dis. Rep.* **52:**379–380.

21. Campbell, R. N., 1979, Fungal vectors of plant viruses, in: H. P. Molitoris, M. Hollings, and H. A. Wood (eds.), *Fungal Viruses,* Springer-Verlag, New York, pp. 8–24.

22. Campbell, R. N., and Fry, P. R., 1966, The nature of the associations be-

tween *Olpidium brassicae* and lettuce big-vein and tobacco necrosis virus, *Virology* **29**:222–233.

23. Campbell, R. N., Lovisolo, O., and Lisa, V., 1975, Soil transmission of petunia asteroid strain of tomato bushy stunt virus, *Phytopathol. Medit.* **14**:82–86.

24. Cooper, J. I., and Mayo, M. A., 1972, Some properties of the particles of three tobravirus isolates, *J. Gen. Virol.* **16**:285–297.

24a. Cropley, R., and Tomlinson, J. A., 1971, Cherry leaf roll virus, CMI/AAB Descriptions of Plant Viruses, No. 80, Commonwealth Mycological Institute/ Association of Applied Biologists, Kew, Surrey, England.

25. Daft, M. J., and Okusanya, B. O., 1973, Effect of endogone mycorrhiza on plant growth. V. Influence of infection on the multiplication of viruses in tomato, petunia, and strawberry, *New Phytol.* **72**:975–983.

26. Dias, H. F., 1970, Transmission of cucumber necrosis virus by *Olpidium cucurbitacearum* Barr & Dias, *Virology* **40**:828–839.

26a. Dias, H. F., 1975, Peach rosette mosaic virus, CMI/AAB Descriptions of Plant Viruses, No. 150, Commonwealth Mycological Institute/Association of Applied Biologists, Kew, Surrey,England.

27. Dias, H. F., and McKeen, C. D., 1972, Cucumber necrosis virus, CMI/ AAB Descriptions of Plant Viruses, No. 82, Commonwealth Mycological Institute/Association of Applied Biologists, Kew, Surrey, England.

27a. Dunez, J., Delbos, R., and Dupont, G., 1976, Myrobalan latent ringspot virus, CMI/AAB Descriptions of Plant Viruses, No. 160, Commonwealth Mycological Institute/Association of Applied Biologists, Kew, Surrey, England.

28. Estes, A. P., and Brakke, M. K., 1966, Correlation of *Polymyxa graminis* with transmission of soil-borne wheat mosaic virus, *Virology* **28**:772–774.

29. Fauquet, C., and Thouvenel, J. C. 1983, Association d'un nouveau virus en batonnet avec la maladie de la necrose a rayures du Riz, en Cote-d'Ivoire, *C. R. Acad. Sci. Paris Ser. III* **296**:575–580.

30. Fletcher, J. T., 1969, Studies on the overwintering of tomato mosaic in root debris, *Plant Pathol.* **18**:97–108.

31. Fritzsche, R., and Kegler, H., 1968, Nematoden als Vektoren von Viruskrankheiten der Obstgewächse, *Tagungsber. Dtsch. Akad. Landwirtschaftswiss. (Berl.)* **97**:289–295.

31a. Fulton, R. W., 1970, Prunus necrotic ringspot virus, CMI/AAB Descriptions of Plant Viruses, No. 5, Commonwealth Mycological Institute/Association of Applied Biologists, Kew, Surrey, England.

32. Fulton, R. W., 1980, Biological significance of multicomponent viruses, *Annu. Rev. Phytopathol.* **18**:131–146.

33. Gallitelli, D., Rana, G. L., and Di Franco, A., 1978, Il virus della scolorazione perinervale del carciofo, *Phytopathol. Medit.* **17**:1–17.

34. Hampton, R. O., and Hancock, C. L., 1981, Soil-related greenhouse spread of bean mild mosaic virus, *Phytopathology* **71**:223 (abstract).

35. Hanada, K., and Harrison, B. D., 1977, Effects of virus genotype and temperature on seed transmission of nepoviruses, *Ann. Appl. Biol.* **85**:79–92.

36. Harrison, B. D., 1964, Specific nematode vectors for serologically distinctive forms of raspberry ringspot and tomato black ring viruses, *Virology* **22**:544–550.

37. Harrison, B. D., 1967, Pea early-browning virus, Report of the Rothamsted Experimental Station for 1966, p. 115.

37a. Harrison, B. D., 1970, Tobacco rattle virus, CMI/AAB Descriptions of Plant Viruses, No. 12, Commonwealth Mycological Institute/Association of Applied Biologists, Kew, Surrey, England.

37b. Harrison, B. D., 1973, Pea early-browning virus, CMI/AAB Descriptions of Plant Viruses, No. 120, Commonwealth Mycological Institute/Association of Applied Biologists, Kew, Surrey, England.

37c. Harrison, B. D., 1974, Potato mop-top virus, CMI/AAB Descriptions of Plant Viruses, No. 138, Commonwealth Mycological Institute/Association of Applied Biologists, Kew, Surrey, England.

38. Harrison, B. D., 1977, Ecology and control of viruses with soil-inhabiting vectors, *Annu. Rev. Phytopathol.* **15**:331–360.

39. Harrison, B. D., and Hanada, K., 1976, Competitiveness between genotypes of raspberry ringspot virus is mainly determined by RNA 1, *J. Gen. Virol.* **31**:455–458.

40. Harrison, B. D., and Robinson, D. J., 1978, The tobraviruses, *Adv. Virus Res.* **23**:25–77.

41. Harrison, B. D., and Winslow, R. D., 1961, Laboratory and field studies on the relation of arabis mosaic virus to its nematode vector *Xiphinema diversicaudatum* (Micoletzky), *Ann. Appl. Biol.* **49**:621–633.

42. Harrison, B. D., and Woods, R. D., 1966, Serotypes and particle dimensions of tobacco rattle viruses from Europe and America, *Virology* **28**:610–620.

43. Harrison, B. D., Murant, A. F., and Mayo, M. A., 1972, Two properties of raspberry ringspot virus determined by its smaller RNA, *J. Gen. Virol.* **17**:137–141.

44. Harrison, B. D., Murant, A. F., Mayo, M. A., and Roberts, I. M., 1974a, Distribution of determinants for symptom production, host range and nematode transmissibility between the two RNA components of raspberry ringspot virus, *J. Gen. Virol.* **22**:233–247.

45. Harrison, B. D., Robertson, W. M., and Taylor, C. E., 1974b, Specificity of retention and transmission of viruses by nematodes, *J. Nematol.* **6**:155–164.

45a. Hebert, T., and Panizo, C. H., 1975, Oat mosaic virus, CMI/AAB Descriptions of Plant Viruses, No. 145, Commonwealth Mycological Institute/Association of Applied Biologists, Kew, Surrey, England.

45b. Hewitt, W. B., Martelli, G., Dias, H. F., and Taylor, R. H., 1970, Grapevine fanleaf virus, CMI/AAB Descriptions of Plant Viruses, No. 28, Commonwealth Mycological Institute/Association of Applied Biologists, Kew, Surrey, England.

46. Hiruki, C., 1965, Transmission of tobacco stunt virus by *Olpidium brassicae,* *Virology* **25**:541–549.

47. Hiruki, C., 1967, Host specificity in transmission of tobacco stunt virus by *Olpidium brassicae, Virology* **33**:131–136.

48. Hiruki, C., 1972, Persistence of tobacco stunt virus in resting sporangia of *Olpidium brassicae, Int. Virol.* **2**:249 (abstract).

49. Hiruki, C., 1975, Host range and properties of tobacco stunt virus, *Can. J. Bot.* **53**:2425–2434.

50. Hiruki, C., and Alderson, P. G., 1976, Morphology and distribution of resting sporangia of *Olpidium brassicae* in tobacco roots, *Can. J. Bot.* **54**:2820–2826.

51. Hiruki, C., Alderson, P. G., Kobayashi, N., and Furusawa, I., 1975, The nature of the infectious agent of tobacco stunt in relation to its vector,

Olpidium brassicae, in: *Proceedings of the 1st Intersectional Congress of International Association of Microbiological Societies,* Vol. 3, pp. 297–302.

52. Hiruki, C., Rao, D. V., Chen, M. H., Okuno, T., and Figueiredo, G., 1984, Characterization of sweet clover necrotic mosaic virus, *Phytopathology* **74**:482–486.

52a. Hollings, M., and Huttinga, H., 1976, Tomato mosaic virus, CMI/AAB Descriptions of Plant Viruses, No. 156, Commonwealth Mycological Institute/ Association of Applied Biologists, Kew, Surrey, England.

53. Hollings, M., and Stone, O. M., 1964, Investigations of carnation viruses. I. Carnation mottle virus, *Ann. Appl. Biol.* **53**:103–118.

54. Hollings, M., and Stone, O. M., 1965, Studies of pelargonium leaf curl virus II. Relationships to tomato bushy stunt and other viruses, *Ann. Appl. Biol.* **56**:87–98.

55. Hollings, M., and Stone, O. M., 1970a, Carnation mottle virus, CMI/AAB Descriptions of Plant Viruses, No. 7, Commonwealth Mycological Institute/ Association of Applied Biologists, Kew, Surrey, England.

55a. Hollings, M., and Stone, O. M., 1970b, Carnation ringspot virus, CMI/ AAB Descriptions of Plant Viruses, No. 21, Commonwealth Mycological Institute/Association of Applied Biologists, Kew, Surrey, England.

56. Hollings, M., and Stone, O. M., 1975, Serological and immunoelectrophoretic relationships among many viruses in the tombusvirus group, *Ann. Appl. Biol.* **80**:37–48.

57. Hollings, M., and Stone, O. M., 1977a, Cymbidium ringspot virus, CMI/ AAB Descriptions of Plant Viruses, No. 178, Commonwealth Mycological Institute/Association of Applied Biologists, Kew, Surrey, England.

57a. Hollings, M., Stone, O. M., 1977b, Red clover necrotic mosaic virus, CMI/ AAB Descriptions of Plant Viruses, No. 181, Commonwealth Mycological Institute/Association of Applied Biologists, Kew, Surrey, England.

58. Hollings, M., Komuro, Y., and Tochihara, H., 1975, Cucumber green mottle mosaic virus, CMI/AAB Descriptions of Plant Viruses, No. 154, Commonwealth Mycological Institute/Association of Applied Biologists, Kew, Surrey, England.

59. Hollings, M., Stone, O. M., and Barton, R. J., 1977, Pathology, soil transmission and characterization of cymbidium ringspot, a virus from cymbidium orchids and white clover (*Trifolium repens*), *Ann. Appl. Biol.* **85**:233–248.

60. Hoy, J. W., Mircetich, S. M., and Lownsbery, B. F., 1984, Differential transmission of Prunus tomato ringspot virus strains by *Xiphinema californicum, Phytopathology* **74**:332–335.

61. Hunter, J. A., Chamberlain, E. E., and Atkinson, J. D., 1958, Note on transmission of apple mosaic virus by natural grafting, *N. Z. J. Agric. Res.* **1**:80–82.

61a. Inouye, T., and Fujii, 1977, Rice necrosis mosaic virus, CMI/AAB Descriptions of Plant Viruses, No. 172, Commonwealth Mycological Institute/ Association of Applied Biologists, Kew, Surrey, England.

61b. Inouye, T., and Nakasone, W., 1980, Broad bean necrosis virus, CMI/AAB Descriptions of Plant Viruses, No. 223, Commonwealth Mycological Institute/Association of Applied Biologists, Kew, Surrey, England.

61c. Inouye, T., and Saito, Y., 1975, Barley yellow mosaic virus, CMI/AAB Descriptions of Plant Viruses, No. 143, Commonwealth Mycological Institute/Association of Applied Biologists, Kew, Surrey, England.

62. Jabajihare, S. H., and Stobbs, L. W., 1984, Electron microscopic examination of tomato roots coinfected with *Glomus* sp. and tobacco mosaic virus, *Phytopathology* **74**:277–279.

63. Jones, A. T., and Duncan, G. H., 1980, The distribution of some genetic determinants in the two nucleoprotein particles of cherry leaf roll virus, *J. Gen. Virol.* **50**:269–277.

64. Jones, A. T., and Murant, A. F., 1971, Serological relationship between cherry leaf roll, elm mosaic and golden elderberry viruses, *Ann. Appl. Biol.* **69**:11–15.

65. Jones, A. T., and Wood, G. A., 1978, The occurrence of cherry leaf roll virus in red raspberry in New Zealand, *Plant Dis. Rep.* **62**:835–838.

66. Jones, A. T., Foster, R. L. S., and Mohamed, N. A., 1979, Purification and properties of Australian lucerne latent virus, a seed-borne virus having affinities with nepoviruses, *Ann. Appl. Biol.* **92**:49–59.

67. Jones, R. A. C., and Harrison, B. D., 1969, The behaviour of potato mop-top virus in soil and evidence for its transmission by *Spongospora subterranea* (Wallr.) Lagerh, *Ann. Appl. Biol.* **63**:1–17.

68. Jones, R. A. C., and Kenten, R. H., 1978, Arracacha virus A, a newly recognized virus infecting arracacha (*Arracacia xanthorrhiza;* Umbelliferae) in the Peruvian Andes, *Ann. Appl. Biol.* **90**:85–91.

68a. Jones, R. A. C., and Kenten, R. H., 1980, Arracacha virus A, CMI/AAB Descriptions of Plant Viruses, No. 216, Commonwealth Mycological Institute/Association of Applied Biologists, Kew, Surrey, England.

69. Johnson, J., 1937, Factors relating to the control of ordinary tobacco mosaic, *J. Agric. Res.* **54**:239–273.

69a. Kado, C. I., 1971, Sowbane mosaic virus, CMI/AAB Descriptions of Plant Viruses, No. 64, Commonwealth Mycological Institute/Association of Applied Biologists, Kew, Surrey, England.

69b. Kassanis, B., 1970a, Tobacco necrosis virus, CMI/AAB Descriptions of Plant Viruses, No. 14, Commonwealth Mycological Institute/Association of Applied Biologists, Kew, Surrey, England.

69c. Kassanis, B., 1970b, Satellite virus, CMI/AAB Descriptions of Plant Viruses, No. 15, Commonwealth Mycological Institute/Association of Applied Biologists, Kew, Surrey, England.

70. Kassanis, B., and Macfarlane, I., 1964, Transmission of tobacco necrosis virus by zoospores of *Olpidium brassicae, J. Gen. Microbiol.* **36**:79–93.

71. Kassanis, B., and Macfarlane, I., 1965, Interaction of virus strain, fungus isolate, and host species in the transmission of tobacco necrosis virus, *Virology* **26**:603–612.

72. Kassanis, B., and Macfarlane, I., 1968, The transmission of satellite viruses of tobacco necrosis virus by *Olpidium brassicae, J. Gen. Virol.* **3**:227–232.

73. Kassanis, B., and Phillips, M. P., 1970, Serological relationship of strains of tobacco necrosis virus and their ability to activate strains of satellite virus, *J. Gen. Virol.* **9**:119–126.

74. Kassanis, B., Woods, R. D., and White, R. F., 1972, Some properties of potato mop-top virus and its serological relationship to tobacco mosaic virus, *J. Gen. Virol.* **14**:123–132.

75. Kemp, W. G., and Barr, D. J. S., 1978, Natural occurrence of tobacco necrosis virus in a rusty-root disease complex of *Daucus carota* in Ontario, *Phytopathol. Z.* **91**:203–217.

75a. Kenten, R. H., 1977, Cacao necrosis virus, CMI/AAB Descriptions of Plant Viruses, No. 173, Commonwealth Mycological Institute/Association of Applied Biologists, Kew, Surrey, England.

76. Kleinhempel, H., and Kegler, G., 1982, Transmission of tomato bushy stunt virus without vectors, *Acta Phytopathol. Hung.* **17**:17–21.

77. Kleinhempel, H., Gruber, G., and Kegler, G., 1980, Investigations on carnation ringspot virus in fruit trees, *Acta Phytopathol. Hung.* **15**:107–111.

78. Komuro, Y., 1971, Cucumber green mottle mosaic virus on cucumber and watermelon and melon necrotic spot virus on muskmelon, *Japan Agric. Res. Q.* **6**:41–45.

79. Komuro, Y., and Iwaki, M., 1969, Presence of tobacco mosaic virus in roots of field-grown tomato plants healthy in appearance, *Ann. Phytopathol. Soc. Japan* **35**:294–298.

80. Kuwata, S., and Kubo, S., 1981, Rod-shaped particles found in tobacco plants infected with tobacco stunt agent, *Ann. Phytopathol. Soc. Japan* **47**:264–268.

81. Kuwata, S., Kubo, S., Yamashita, S., and Doi, Y., 1983, Rod-shaped particles, a probable entity of lettuce big vein virus, *Ann. Phytopathol. Soc. Japan* **49**:246–251.

82. Lamberti, F., 1981, Combating nematode vectors of plant viruses, *Plant Dis.* **65**:113–117.

82a. Lane, L. C., 1977, Brome mosaic virus, CMI/AAB Descriptions of Plant Viruses, No. 180, Commonwealth Mycological Institute/Association of Applied Biologists, Kew, Surrey, England.

83. Lane, L. C., 1979, The nucleic acids of multipartite, defective and satellite plant viruses, in: T. C. Hall and J. W. Davies (eds.), *Nucleic Acids in Plants,* Vol. II, CRC Press, Boca Raton, Florida, pp. 65–110.

84. Langenberg, W. G., and Giunchedi, L., 1982, Ultrastructure of fungal plant virus vectors *Polymyxa graminis* in soilborne wheat mosaic-infected wheat and *P. betae* in beet necrotic yellow vein virus-infected sugar beet, *Phytopathology* **72**:1152–1158.

85. Lanter, J. M., McGuire, J. M., and Goode, M. J., 1982, Persistence of tomato mosaic virus in tomato debris and soil under field conditions, *Plant Dis.* **66**:552–555.

86. Leggat, F. W., and Teakle, D. S., 1976, Symptoms in roots rub-inoculated with ten different viruses, *Z. Pflkrank. Pflschutz* **88**:195–203.

87. Louie, R., Gordon, D. T., Knoke, J. K., Gingery, R. E., Bradfute, O. E., and Lipps, P. E., 1982, Maize white line mosaic virus in Ohio, *Plant Dis.* **66**:167–170.

88. Lovisolo, O., 1966, Indagini so virosi di plante ornamentale, in: *Atti del Primo Congresso dell'Unione Fitopatalogia Mediterranea,* Bari. Part I & II, 574–584.

89. Lovisolo, O., Bode, O., and Volk, J., 1965, Preliminary studies of the soil transmission of petunia asteroid mosaic virus (= "petunia" strain of tomato bushy stunt virus), *Phytopathol. Z.* **53**:324–342.

90. Lyness, E. W., Teakle, D. S., and Smith, P. R., 1981, Red clover necrotic mosaic virus isolated from *Trifolium repens* and *Medicago sativa* in Victoria, *Aust. Plant Pathol.* **10**:6–7.

90a. Martelli, G., Quacquerelli, A., and Russo, M., 1971, Tomato bushy stunt virus, CMI/AAB Descriptions of Plant Viruses, No. 69, Commonwealth

Mycological Institute/Association of Applied Biologists, Kew, Surrey, England.

90b. Martelli, G., Quacquerelli, A., and Russo, M., 1972, Grapevine chrome mosaic virus, CMI/AAB Descriptions of Plant Viruses, No. 103, Commonwealth Mycological Institute/Association of Applied Biologists, Kew, Surrey, England.

90c. Martelli, G. P., Rana, G. L., and Savino, V., 1977, Artichoke Italian latent virus, CMI/AAB Descriptions of Plant Viruses, No. 176, Commonwealth Mycological Institute/Association of Applied Biologists, Kew, Surrey, England.

90d. Martelli, G. P., Quacquerelli, A., and Gallitelli, D., 1978, Grapevine Bulgarian latent virus, CMI/AAB Descriptions of Plant Viruses, No. 186, Commonwealth Mycological Institute/Association of Applied Biologists, Kew, Surrey, England.

91. Martinez, A. J., Galindo, A. J., and Rodriguez, M. R. 1974, Estudio sobre la enfermedad del 'Pinto' del jitomata (*Lycopersicon esculentum* Mill.) en la región de Actopan, Hgo. *Agrociencia* **18**:71–78 [*Rev. Plant Pathol.* **55**:627, 1976.].

92. Masri, S. A., and Hiruki, C., 1982, Purification and characterization of tobacco stunt virus, in: *IV International Conference on Comparative Virology,* Banff, Canada, p. 59.

93. Masri, S. A., and Hiruki, C., 1983, A new group of elongated plant viruses having a capsid protein of an unusually high molecular weight, *Can. J. Plant Pathol.* **5**:208 (abstract).

94. McGuire, J. M., Kim, K. S., and Douthit, L. B., 1970, Tobacco ringspot virus in the nematode *Xiphinema americanum, Virology* **42**:212–216.

95. McLean, D. L., 1962, Transmission of lettuce mosaic virus by a new vector, *Pemphiqus bursarius, J. Econ. Entamol.* **55**:580–583.

96. McNamara, D. G., 1980, The spread of arabis mosaic virus through noncultivated vegetation, *Plant Pathol.* **29**:173–176.

97. Mowat, W. P., 1968, *Olpidium brassicae:* Electrophoretic mobility of zoospores associated with their ability to transmit tobacco necrosis virus, *Virology* **34**:565–568.

98. Mowat, W. P., 1970, Augusta disease in tulip—A reassessment, *Ann. Appl. Biol.* **66**:17–28.

99. Murant, A. F., 1981, Nepoviruses, in: E. Kurstak (ed.), *Handbook of Plant Virus Infections and Comparative Diagnosis,* Elsevier/North-Holland Biomedical Press, Amsterdam, pp. 197–238.

99a. Murant, A. F., 1970a, Raspberry ringspot virus, CMI/AAB Descriptions of Plant Viruses, No. 6, Commonwealth Mycological Institute/Association of Applied Biologists, Kew, Surrey, England.

99b. Murant, A. F., 1970b, Arabis mosaic virus, CMI/AAB Descriptions of Plant Viruses, No. 16, Commonwealth Mycological Institute/Association of Applied Biologists, Kew, Surrey, England.

99c. Murant, A. F., 1970c, Tobacco black ring virus, CMI/AAB Descriptions of Plant Viruses, No. 38, Commonwealth Mycological Institute/Association of Applied Biologists, Kew, Surrey, England.

99d. Murant, A. F., 1974, Strawberry latent ringspot virus, CMI/AAB Descriptions of Plant Viruses, No. 126, Commonwealth Mycological Institute/Association of Applied Biologists, Kew, Surrey, England.

99e. Murant, A. F., 1978, Raspberry ringspot virus, CMI/AAB Descriptions of Plant Viruses, No. 198, Commonwealth Mycological Institute/Association of Applied Biologists, Kew, Surrey, England.

100. Murant, A. F., and Lister, R. M., 1967, Seed transmission in the ecology of nematode-borne viruses, *Ann. Appl. Biol.* **59:**63–76.

101. Murant, A. F., and Taylor, M., 1978, Estimates of molecular weights of nepovirus RNA species by polyacrylamide gel electrophoresis under denaturing conditions, *J. Gen. Virol.* **41:**53–61.

102. Murant, A. F., Mayo, M. A., Harrison, B. D., and Goold, R. A., 1972, Properties of virus and RNA components of raspberry ringspot virus, *J. Gen. Virol.* **16:**327–338.

103. Nakasone, W., and Inouye, T., 1978, Serological relationships between broad bean necrosis virus and viruses of the tobamovirus group, *Ann. Phytopathol. Soc. Japan* **44:**97 (abstract).

104. Piazzolla, P., Gallitelli, D., Vovlas, C., and Quacquarelli, A., 1979, Nuovidati sui virus della maculatura gialla della cicoria, *Phytopathol. Medit.* **17:**149–152.

104a. Pirone, T. P., 1972, Sugarcane mosaic virus, CMI/AAB Descriptions of Plant Viruses, No. 88, Commonwealth Mycological Institute/Association of Applied Biologists, Kew, Surrey, England.

105. Plumb, R. T., and Macfarlane, I., 1977, Cereal diseases, Rothamsted Experimental Station Report for 1976, Part 1, Harpenden, England.

106. Powell, C. A., 1976, The relationship between soil-borne wheat mosaic virus and tobacco mosaic virus, *Virology* **71:**453–462.

106a. Quacquerelli, A., Martelli, G. P., and Vovlas, C., 1974, Chicory yellow mottle virus, CMI/AAB Descriptions of Plant Viruses, No. 132, Commonwealth Mycological Institute/Association of Applied Biologists, Kew, Surrey, England.

107. Quacquarelli, A., Gallitelli, D., Savino, V., and Martelli, G. P., 1976, Properties of grapevine fanleaf virus, *J. Gen. Virol.* **32:**349–360.

107a. Ramsdell, D. C., and Stace-Smith, R., 1983, Blueberry leaf mottle virus, CMI/AAB Descriptions of Plant Viruses, No. 267, Commonwealth Mycological Institute/Association of Applied Biologists, Kew, Surrey, England.

107b. Rana, G. L., Kyriakopoulou, P. E., and Martelli, G. P., 1983, Artichoke yellow ringspot virus, CMI/AAB Descriptions of Plant Viruses, No. 271, Commonwealth Mycological Institute/Association of Applied Biologists, Kew, Surrey, England.

107c. Randles, J. W., 1978, Nicotiana velutina mosaic virus, CMI/AAB Descriptions of Plant Viruses, No. 189, Commonwealth Mycological Institute/Association of Applied Biologists, Kew, Surrey, England.

108. Randles, J. W., Harrison, B. D., Murant, A. F., and Mayo, M. A., 1977, Packaging and biological activity of the two essential RNA species of tomato black ring virus, *J. Gen. Virol.* **36:**187–193.

108a. Rao, A. L. N., and Hiruki, C., 1985, Clover primary leaf necrosis virus, a strain of red clover necrotic mosaic virus, *Plant Dis.* **69:**959–961.

109. Rao, A. S., 1968, Biology of *Polymyxa graminis* in relation to soil-borne wheat mosaic virus, *Phytopathology* **58:**1516–1521.

110. Rao, A. S., and Brakke, M. K., 1969, Relation of soil-borne wheat mosaic virus and its fungal vector, *Polymyxa graminis, Phytopathology* **59:**581–587.

111. Roberts, F. M., 1948, Experiments on the spread of potato virus X between plants in contact, *Ann. Appl. Biol.* **35**:266–278.

111a. Salazar, L. F., and Harrison, B. D., 1979, Potato black ringspot virus, CMI/AAB Descriptions of Plant Viruses, No. 206, Commonwealth Mycological Institute/Association of Applied Biologists, Kew, Surrey, England.

112. Salomao, T. A., Silberschmidt, K., Weigl, D. R., and Chagas, C. M., 1977, A transmissao dos virus do grupo "rattle" do fu mo atraves do solo, sem a interferencia do vetor, *Cientifica* **5**:45–54.

113. Savino, V., Quacquarelli, A., Gallitelli, D., Piazzolla, P., and Martelli, G. P., 1977, Il virus dell' accartocciamento fogliare del ciliegio nel noce. I. Identificazione e carratterizzazione, *Phytopathol. Medit.* **16**:96–102.

114. Savino, V., Gallitelli, D., and Barba, M., 1983, Olive latent ringspot virus, a newly recognized virus infecting olive in Italy, *Ann. Appl. Biol.* **103**:243–249.

115. Schenck, N. C., and Kellam, M. K., 1978, The influence of vesicular arbuscular mycorrhiza on disease development, Florida Agricultural Experimental Station Bulletin, No. 799.

116. Schmidt, H. B., Fritzsche, R., and Lehmann, W., 1963, Die Übertragung des Weidelgrasmosaik Virus durch Nematoden, *Naturwissenschaften* **50**:386.

117. Schonbeck, F., and Spengler, G., 1979, Detection of TMV in mycorrhizal cells of tomato by immunofluorescence, *Phytopathol. Z.* **94**:84–86.

118. Shirako, Y., and Brakke, M. K., 1984a, Two purified RNAs of soil-borne wheat mosaic virus are needed for infection, *J. Gen. Virol.* **65**:119–127.

119. Shirako, Y., and Brakke, M. K., 1984b, Spontaneous deletion mutation of soil-borne wheat mosaic virus RNA II, *J. Gen. Virol.* **65**:855–858.

119a. Shirako, Y., and Ehara, Y., 1986, Comparison of the *in vitro* translation products of wild-type and a deletion mutant of soil-borne wheat mosaic virus, *J. Gen. Virol.* **67**:1237–1245.

120. Shukla, D. D., Shanks, G. J., Teakle, D. S., and Behncken, G. M., 1979, Mechanical transmission of galinsoga mosaic virus in soil, *Aust. J. Biol. Sci.* **32**:267–176.

120a. Slykhuis, J. T., 1976, Wheat spindle streak mosaic virus, CMI/AAB Descriptions of Plant Viruses, No. 167, Commonwealth Mycological Institute/Association of Applied Biologists, Kew, Surrey, England.

121. Smith, K. M., 1935, A new virus of the tomato, *Ann. Appl. Biol.* **22**:731–741.

122. Smith, P. R., Campbell, R. N., and Fry, P. R., 1969, Root discharge and soil survival of viruses, *Phytopathology* **59**:1678–1687.

123. Sol, H. H., 1963, Some data on the occurrence of rattle virus at various depths in the soil and on its transmission, *Neth. J. Plant. Pathol.* **69**:208–214.

123a. Stace-Smith, R., 1970a, Tobacco ringspot virus, CMI/AAB Descriptions of Plant Viruses, No. 17, Commonwealth Mycological Institute/Association of Applied Biologists, Kew, Surrey, England.

123b. Stace-Smith, R., 1970b, Tomato ringspot virus, CMI/AAB Descriptions of Plant Viruses, No. 18, Commonwealth Mycological Institute/Association of Applied Biologists, Kew, Surrey, England.

123c. Stace-Smith, R., and Hansen, A. J., 1976, Cherry rasp leaf virus, CMI/

AAB Descriptions of Plant Viruses, No. 159, Commonwealth Mycological Institute/Association of Applied Biologists, Kew, Surrey, England.

124. Stobbs, L. W., Cross, G. W., and Manocha, M. S., 1982, Specificity and methods of transmission of cucumber necrosis virus by *Olpidium radicale* zoospores, *Can. J. Plant Pathol.* **4:**134–142.

125. Tamada, T., 1975, Beet necrotic yellow vein virus, CMI/AAB Descriptions of Plant Viruses, No. 144, Commonwealth Mycological Institute/Association of Applied Biologists, Kew, Surrey, England.

126. Tamada, T., Abe, H., and Baba, T., 1975, Beet necrotic yellow vein virus and its relation to the fungus *Polymyxa betae,* in: *Proceedings of the First Intersectional Congress of International Association of Microbiological Societies,* Vol. 3, pp. 313–320.

127. Taylor, C. E., 1970, The association of *Longidorus elongatus* with raspberry ringspot and tomato black ring viruses, *Zesz. Probl. Postep. Nauk Roln.* **92:**283–289.

128. Taylor, C. E., 1980, Nematodes, in: K. F. Harris and K. Maramorosch (eds.), *Vectors of Plant Pathogens,* Academic Press, New York, pp. 375–416.

129. Taylor, C. E., 1983, Nematode transmission of plant viruses, in: *Abstracts of Papers 4th International Congress of Plant Pathology, Melbourne,* p. 10.

130. Taylor, C. E., and Raski, D. J., 1964, On the transmission of grape fanleaf by *Xiphinema index, Nematologia* **10:**489–495.

131. Taylor, C. E., and Robertson, W. M., 1969, The location of raspberry ringspot and tomato black ring viruses in the nematode vector, *Longidorus elongatus* (de Man), *Ann. Appl. Biol.* **64:**233–237.

132. Taylor, C. E., and Robertson, W. M., 1970a, Location of tobacco rattle virus in the nematode vector, *Trichodorus pachydermus* Seinhorst, *J. Gen. Virol.* **6:**179–182.

133. Taylor, C. E., and Robertson, W. M., 1970b, Sites of virus retention in the alimentary tract of the nematode vectors, *Xiphinema diversicaudatum* (Micol). and *X.* index (Thorne and Allen), *Ann. Appl. Biol.* **66:**375–380.

134. Teakle, D. S., 1962, Transmission of tobacco necrosis virus by a fungus, *Olpidium brassicae, Virology* **18:**224–231.

135. Teakle, D. S., 1972, Transmission of plant viruses by fungi, in: C. I. Kado and H. O. Agrawal (eds.), *Principles and Techniques in Plant Virology,* Van Nostrand-Reinhold, New York, pp. 226–247.

136. Teakle, D. S., 1973, Use of the local lesion method to study the effect of celite and inhibitors on virus infection of roots, *Phytopathol. Z.* **77:**209–215.

137. Teakle, D. S., 1980, Fungi, in: K. F. Harris and K. Maramorosch (eds.), *Vectors of Plant Pathogens,* Academic Press, New York, pp. 417–438.

138. Teakle, D. S., 1983, Zoosporic fungi and viruses—Double trouble, in: S. T. Buczacki (ed.), *Zoosporic Plant Pathogens,* Academic Press, New York, pp. 233–248.

138a. Teakle, D.S., 1986, Abiotic transmission of southern bean mosaic virus in soil. *Aust. J. Biol. Sci.* **39**: 353–359.

139. Teakle, D. S., and Gold, A. H., 1963, Further studies of *Olpidium* as a vector of tobacco necrosis virus, *Virology* **19:**310–315.

140. Teakle, D. S., and Hiruki, C., 1964, Vector specificity in *Olpidium, Virology* **24:**539–544.

141. Teakle, D. S., and Morris, T. J., 1981, Bean seeds acquire southern bean mosaic virus from soil, *Plant Dis.* **65**:599–600.

142. Temmink, J. H. M., 1971, An ultrastructural study of *Olpidium brassicae* and its transmission of tobacco necrosis virus, *Meded. Landbhuwhogesch. Wageningen* **71**:1–135.

143. Temmink, J. H. M., Campbell, R. N., and Smith, P. R., 1970, Specificity and site of *in vitro* acquisition of tobacco necrosis virus by zoospores of *Olpidium brassicae, J. Gen. Virol.* **9**:201–213.

144. Thomas, W., and Fry, P. R., 1972, Cucumber systemic necrosis caused by a strain of tobacco necrosis virus, *N. Z. J. Agric Res.* **15**:857–866.

145. Thouvenel, J. C., and Fauquet, C., 1981, Peanut clump virus, CMI/AAB Descriptions of Plant Viruses, No. 235, Commonwealth Mycological Institute/Association of Applied Biologists, Kew, Surrey, England.

146. Thung, T. H., Hadiwidjaja, T., 1958, Some remarks on Rotterdam-B virus, in: *Proceedings of the 3rd Conference on Potato Virus Diseases,* Lisse-Wageningen, pp. 233–237.

147. Tomaru, K., Maeda, S., and Enomoto, Y., 1971, Infection of tobacco plant with cucumber mosaic virus through roots, *Ann. Phytopathol. Soc. Japan* **37**:63–69.

148. Tomlinson, J. A., and Faithfull, E. M., 1976, Virus disease, 26th Annual Report for 1975, National Vegetable Research Station, Wellesbourne, Warwick, England, pp. 105–112.

149. Tomlinson, J. A., and Faithfull, E. M., 1979, Effects of fungicides and surfactants on the zoospores of *Olpidium brassicae, Ann. Appl. Biol.* **93**:13–19.

150. Tomlinson, J. A., Faithfull, E., Flewett, T. H., and Beards, G., 1982, Isolation of infective tomato bushy stunt virus after passage through the human alimentary tract, *Nature* **300**:637–638.

151. Tomlinson, J. A., Faithfull, E. M., Webb, M. J. W., Fraser, R. S. S., and Seeley, N. D., 1983a. *Chenopodium* necrosis: A distinctive strain of tobacco necrosis virus isolated from river water, *Ann. Appl. Biol.* **102**:135–147.

152. Tomlinson, J. A., Faithfull, E. M., and Fraser, R. S. S., 1983b, Plant viruses in river water, in 33rd Annual Report for 1982, National Vegetable Research Station, Wellesbourne, Warwick, England, pp. 81–82.

153. Tremaine, J. H., and Hamilton, R. I., 1983, Southern bean mosaic virus, CMI/AAB Descriptions of Plant Viruses, No. 274, Commonwealth Mycological Institute/Association of Applied Biologists, Kew, Surrey, England.

154. Trudgill, D. L., Brown, D. J. F., and Robertson, W. M., 1981, A comparison of the effectiveness of the four British virus vector species of *Longidorus* and *Xiphinema, Ann. Appl. Biol.* **99**:63–70.

154a. Tsuchizaki, T., 1975, Mulberry ringspot virus, CMI/AAB Descriptions of Plant Viruses, No. 142, Commonwealth Mycological Institute/Association of Applied Biologists, Kew, Surrey, England.

155. Tsuchizaki, T., Hibino, H., and Saito, Y., 1973, Comparisons of soil-borne wheat mosaic virus isoates from Japan and the United States, *Phytopathology* **63**:634–639.

156. Tsuchizaki, T., Hibino, H., and Saito, Y., 1975, The biological functions of short and long paticles of soil-borne wheat mosaic virus, *Phytopathology* **65**:523–532.

157. Usugi, T., and Saito, Y., 1975, Purification of barley yellow mosaic and wheat yellow mosaic viruses by cesium chloride density equilibrium centrifugation, *Ann. Phytopathol. Soc. Japan* **41**:87 (abstract).
157a. Usugi, T., and Saito, Y., 1979, Satsuma dwarf virus, CMI/AAB Descriptions of Plant Viruses, No. 208, Commonwealth Mycological Institute/Association of Applied Biologists, Kew, Surrey, England.
158. Uyemoto, J. K., 1983, Biology and control of maize chlorotic mottle virus, *Plant Dis.* **67**:7–10.
159. Valdy, R. B., 1972, Transmission of raspberry ringspot virus by *Longidorus caespiticola, L. leptocephalus* and *Xiphinema diversicaudatum* and of arabis mosaic virus by *L. caespiticola* and *X. diversicaudatum, Ann. Appl. Biol.* **71**:229–234.
160. Van der Want, J. P. H., 1952, Some remarks on soil-borne virus disease, *Proceedings Conference on Potato Virus Diseases,* Wageningen-Lisse, the Netherlands, pp. 71–75.
161. Van Hoof, H. A., 1968, Transmission of tobacco rattle by *Trichodorus* species, *Nematologica* **14**:20–24.
162. Van Hoof, H. A., 1970, Some observatons on retention of tobacco rattle virus in nematodes, *Neth. J. Plant Pathol.* **76**:329–330.
163. Van Koot, Y., and van Dorst, J. H. M., 1955, Een nieuwe virusziekte bij komkommers, *Tijdschr. Plantenziekten* **61**:163–164.
164. Van Slogteren, D. H. M., and Visscher, H. F., 1967, Transmission of a tobacco virus, causing "Augusta disease" to the roots of tulip by zoospores of the fungus *Olpidium brassicae* (Wor.) Dang, *Meded. Rijksfac. Landbouwwet. Gent* **32**:927–938.
165. Von Wechmar, M. B., and Rybicki, E. P., 1983, Epidemiology of cereal viruses in South Africa, in: *Abstracts of Papers, 4th International Congress of Plant Pathology, Melbourne,* p. 35.
166. Walkey, D. G. A., Stace-Smith, R., and Tremaine, J. H., 1973, Serological, physical, and chemical properties of strains of cherry leaf roll virus, *Phytopathology* **63**:566–571.
167. Waterworth, H., 1981, Bean mild mosaic virus, CMI/AAB Descriptions of Plant Viruses, No. 231, Commonwealth Mycological Institute/Association of Applied Biologists, Kew, Surrey, England.
168. Waterworth, H. E., Meiners, J. P., Lawson, R. H., and Smith, F. F., 1977, Purification and properties of a virus from El Salvador that causes mild mosaic in bean cultivars, *Phytopathology* **67**:169–173.
169. Wyss, U., 1977, Feeding processes of virus-transmitting nematodes, *Proc. Am. Phytopathol. Soc.* **4**:30–41.
170. Wyss, U., 1981, Ectoparasitic root nematodes: Feeding behavior and plant cell responses, in: B. M. Zuckerman and R. A. Rohde (eds.), *Plant Parasitic Nematodes,* Vol. 3, Academic Press, New York, pp. 325–351.
171. Wyss, U., 1982, Virus-transmitting nematodes: Feeding behavior and effect on root cells, *Plant Dis.* **66**:639–644.
172. Yarwood, C. E., 1960, Release and preservation of virus by roots, *Phytopathology* **50**:111–114.
173. Yarwood, C. E., and Fulton, R. W., 1967, Mechanical transmission of plant viruses, in: K. Maramorosch and H. Koprowski (eds.), *Methods in Virology,* Vol. 1, Academic Press, New York, pp. 237–266.

173a. Zaitlin, M., and Israel, H. W., 1975, Tobacco mosaic virus (type strain), CMI/AAB Descriptions of Plant Viruses, No. 151, Commonwealth Mycological Institute/Association of Applied Biologists, Kew, Surrey, England.

174. de Zoeten, G. A., Arny, D. C., Grau, C. R., Saad, S. M., and Gaard, G., 1980, Properties of the nucleoprotein associated with maize white line mosaic in Wisconsin, *Phytopathology* **70:**1019–1022.

8
Immunoelectron Microscopy of Plant Viruses and Mycoplasma

Yogesh C. Paliwal

Introduction

Immunoelectron microscopy (IEM) is a general term that covers a group of techniques that utilize the specificity of an antigen–antibody reaction in electron microscopic investigations of biological specimens. More specifically, IEM may be defined as electron microscope (EM) viewing of an immunological reaction where antigen in a liquid suspension or *in situ* is allowed to react with antibody followed by certain treatments of the adduct to obtain specific qualitative or quantitative information. Although IEM was first introduced to plant viruses by Anderson and Stanley (3), Ball and Brakke (4) provided the first IEM procedure, named "leaf dip serology," as an improvement over negative staining for plant virus studies. In this classical IEM method, a purified virus preparation, sap from fresh leaf cuts, or a crushed piece of infected leaf is mixed with a suitable dilution of homologous antiserum. After incubation the preparation is viewed in the EM to detect virus particles in the form of clumps; hence the method has also been known as "clumping" (66). Because of the problems of nonspecific clumping observed under some conditions and unreliability in cases of antibody excess or low virus concentrations, this method is not commonly used now.

Currently, two different approaches to the utilization of IEM are available. In the first, IEM techniques are employed to study antigens present in liquid suspensions, while in the second antigens are studied *in situ* usually in fixed and ultrathin sectioned biological materials with the help of antibodies labeled with suitable markers detectable in the EM. The IEM techniques most useful in plant virus work are those allowing study of antigens in liquid suspensions. Immunosorbent electron microscopy (ISEM) is one such IEM technique, and was first introduced to plant vi-

Yogesh C. Paliwal, Chemistry and Biology Research Institute, Agriculture Canada, Ottawa, Ontario K1A OC6, Canada.
© 1987 by Springer-Verlag New York Inc. *Current Topics in Vector Research*, Vol. 3.

rology by Derrick (21). Essentially, ISEM consists of adsorption of antibodies to a support film on the EM grid, use of this antigen-receptive surface to trap particles of antigenically related viruses from a suspension, and viewing of the trapped particles in the microscope after suitable contrasting. ISEM has been very useful in studies of plant viruses in suspensions and is widely employed for their specific detection in crude plant juice especially when virus concentration is too low to be detected by routine negative staining. The greatest advantage of ISEM is its extremely high sensitivity combined with the direct viewing of the virus or mycoplasma itself and not a reaction product. Other, more recently introduced IEM techniques, such as "decoration" (66) and colloidal gold-labeled antibodies (31, 77, 106), by themselves or in combination with ISEM, have a variety of applications and are becoming increasingly popular. In ultrastructural studies, use of IEM with labeled antibodies allows determination of subcellular sites of localization and possibly replication of viruses that are difficult to discern, and of other virus-related antigens in the plant cells and vector tissues. IEM with labeled antibodies has not been widely used in plant virus and mycoplasma studies, hence only recent, promising developments and potential applications in this area will be discussed. Since IEM methods for detection and study of antigens in suspensions are more commonly used and have a greater variety of valuable applications in plant virus and mycoplasma research, a discussion of these methods and their many applications is the main subject of this chapter. For more detailed information on techniques and procedures of IEM of virus suspensions the reader is referred to two recent reviews (63, 65).

Techniques

Immunoelectron Microscopy of Antigens in Suspensions

The classical "leaf dip serology" (4), its improved version provided by Milne and Luisoni (66) under the name "clumping," the more sensitive and versatile ISEM (21), and "decoration" (66) essentially constitute the IEM techniques for the study of virus or mycoplasma in suspensions. Only ISEM and decoration will be discussed here.

IMMUNOSORBENT ELECTRON MICROSCOPY

Electron microscope grids of copper are most commonly used in ISEM, but when long (several hours to overnight) treatment with crude juice or buffers is required, nickel grids may be preferred, as copper reacts to form harmful salts that degrade virus particles (73). Also, certain treatments (detailed later in this section) to enhance trapping of virus particles require nickel grids. Films of parlodion or formvar coated with carbon have been widely used on the grids. Butvar B-98 (Monsanto Co., St. Louis,

MO.) films without carbon also have been reported to possess stability and virus trapping ability as good as those of formvar or parlodion (96). Films of carbon alone, although somewhat questionable in routine use due to brittleness, seem to allow trapping of more particles than other films (86). Heavy carbon films cast on formvar, used after dissolving away the formvar from the grid squares by treatment with chloroform, have also provided good results (75). Furthermore, it is now generally agreed that for consistent results grids freshly carboned should be used (17, 65, 75). Carbon surfaces are quite hydrophobic, and making them hydrophilic by glow discharge ion etching (80, 110) resulted in increased entrapment of virus particles especially when a "protein A" treatment was used before coating of the grids with antibodies (49, 75, 110). Routinely, if protein A is not to be used, heavy carbon films can be positively charged simply with an ethidium bromide treatment to increase trapping efficiency (110). Alternatively, trapping efficiency of parlodion–carbon grids and perhaps of carbon films on other plastics as well can be increased by a 30-min UV-light (1700 μW/cm^2) treatment prior to antibody coating (79).

Antisera of low or high titers from early or later bleedings and unfractionated antiserum or the gamma-globulin fraction all can be successfully used in ISEM. Before antibody coating of grids, a precoating with protein A can be applied (95). Protein A, a commercially available protein obtained from the cell walls of the bacterium *Staphylococcus aureus,* has great capacity to specifically bind the Fc portion of the IgG molecule. One molecule of protein A binds two of IgG (24), thus increasing the density of IgG molecules adsorbed on the grids during treatment with antiserum to coat with antibodies. Use of protein A enhances virus trapping efficiency of grids, with reported increases varying from only two times (62) to as high as 300 times (95). However, careful comparison has shown that the increases are usually high only when high concentrations of virus are present in the test samples (48). When only a low-titered antiserum is available, use of protein A alleviates the inhibition of trapping of virus particles or mycoplasma cells due to the use of antiserum at low dilutions or without dilution (12, 47, 48, 65). Also, use of protein A minimizes nonspecific trapping of virus particles on the grids (58). A 5- to 10-min treatment of grids with 10 μg/ml protein A at room temperature gives excellent results (75, 96).

Suitably diluted antisera are used for antibody coating of grids and 0.05 M Tris–HCl buffer, pH 7.2, has been the buffer of choice for diluting sera in a number of ISEM studies (10, 21, 22, 73, 74). Other buffers that have given equally good results are 0.1 M phosphate, pH 7.0 (49, 57, 66), 0.06 M phosphate buffer, pH 6.5 (85, 86), and 0.05 M sodium carbonate buffer, pH 9.6 (105). In a study (17) of the effect of pH on antisera diluted in 0.1 M phosphate buffer (pH ranging from 5.0 to 8.0), the number of particles of different viruses trapped varied only from 0.6 to 2.6 times, a relatively small difference compared to the magnitude of the effect some other factors (discussed later) can have. Antisera with titers of 1/500 or higher generally

give best results at dilutions of 1/1000 or higher, but low dilutions (usually
< 1/100) or undiluted sera cause inhibition of virus particle trapping, pre-
sumably because serum proteins other than gamma globulins compete and
occupy adsorption sites on the support film (49, 64, 73, 74). Such inhibition
is independent of the titers of antisera (49) and the degree of inhibition
varies with different antisera and viruses (69). However, gamma globulins
purified from the antiserum can be used at high concentrations with no
adverse effects on particle counts (48). Although serum coating periods
of 30 min (21, 23, 73) or 1–3 hr (30, 89) have been used, there is evidence
that in most situations 5 min at room temperature is enough to allow all
the antibody adsorption that will occur on the grids (8, 66, 84). Grids
precoated with protein A seem to require at least 10 min to saturate with
IgG from the antiserum (96; Y.C. Paliwal, unpublished). As for the temperature at which coating is done, grids coated at 36–37°C seem to attach
virus particles more firmly than those coated at room temperature (84;
Y.C. Paliwal, unpublished). Antibody-coated grids are thoroughly rinsed,
usually with the same buffer in which antiserum was diluted and drained,
but not dried prior to reacting them with the preparation containing virus.

Methods of making the virus preparations from field or laboratory samples of plants as well as vectors and reaction of the preparations with
antibody-coated grids are among the critical steps for satisfactory ISEM.
Although a variety of buffers have been employed for preparing extracts
of infected tissue or to suspend purified virus, 0.05 M Tris–HCl, pH 7.2
(21), 0.1 M phosphate, pH 7.0 (66), and 0.06 M phosphate, pH 6.5 (85,
86), have been more commonly used. The most appropriate buffer for
preparing tissue extracts for ISEM is the one that preserves the integrity
and morphological details of the virus particles or the mycoplasma cells.
The extraction buffer found most suitable for use in purification of the
virus or mycoplasma from infected plants may work well for ISEM as
well; if not, it will be a good starting point in finding the right extraction
medium. Cohen *et al.* (17) tested the effect of pH of extraction buffer for
four viruses involving five different host plants using 0.1 M phosphate
buffer from pH 5.0 to 8.0. ISEM particle counts varied up to 20 times,
each virus–host system requiring a specific pH. Crude sap from infected
leaves, sap diluted with buffer, clarified tissue extracts, partially or highly
purified virus in sucrose or cesium chloride solutions, and extracts of insect
vectors all can be successfully used in ISEM. Saps of several plant species,
notably grapevine, cocoa, members of Rosaceae, and certain grasses, interfere with virus detection by ISEM (71, 73, 115), presumably due to
high content of phenolic substances and/or tanins and mucilages. Such
substances interact with virus particles either to disrupt them or precipitate
them by forming complexes. Addition of 2.5% nicotine during preparation
of extracts from grapevine and rosaceous plants allowed satisfactory results in ISEM of plum pox virus and several grapevine viruses (8, 71, 89).
In a comparison of additives, polyvinylpyrolidone (PVP) and polyethylene
glycol 6000 were much more effective than nicotine and sodium sulfite

for fan leaf virus-infected grapevine tissue (1). For grasses, use of 2-mer-captoethanol (0.2%) in the extraction buffer allowed improved ISEM detection of barley yellow dwarf virus (73; Y.C. Paliwal, unpublished).

In some host–virus systems, plant proteases, presumed to be mainly carboxypeptidases, are highly active in the sap and cause degradation of virus particles, making electron microscopy impossible or highly unsatisfactory. For example, alfalfa mosaic and bean yellow mosaic (BYMV) viruses in the sap of *Solanum lacineatum* and freezia mosaic and BYMV in freezia sap could only be detected in the EM when tissue was ground in buffer containing 0.1–0.2 mM phenylmethylsulfonyl fluoride (13, 14). Many viruses can be detected by ISEM in crude sap of their host plants prepared by simply grinding the tissue in a pestle and mortar, but for phloem-limited viruses (e.g., barley yellow dwarf and potato leaf roll viruses) special procedures are required for preparing tissue extracts so that ISEM-detectable amounts of virus are released from the fiberous vascular tissue (73, 86). Since mycoplasma cells are easily disruptable, preparation of extracts from infected plants or insect vectors requires gentle trituration and clarification treatments and a medium that protects the integrity of mycoplasma cells (12, 99). In ISEM for diagnostic purposes, incubation of antibody-coated grids with the virus preparation for 15 min at room temperature or 37°C allows trapping of enough particles in most cases (47, 62). With viruses that occur in low concentration in plants or vectors, incubation of grids with the virus preparation for a few hours to overnight or even several days may give significantly increased particle counts (8, 75, 86, 89). Generally, the number of particles trapped on the antibody-coated grids is significantly affected by the period of incubation of grids with the virus preparation, the relationship being linear at least for the first few hours (8, 86). However, with some unstable viruses, increasing incubation periods beyond a certain duration leads to disruption of particles and decrease in numbers of recognizable particles (47). Incubation at low temperatures (4–6°C) may lessen the adverse effects of long incubation periods (75, 86). Vibrating the drops of virus preparation on grids during incubation have been reported to trap up to twice as much virus, but the results were inconsistent (66). Periodic agitation of grids on the drops with a needle during incubation improved trapping of BYDV particles from aphid extracts (75). Also, gentle rotation on a magnetic stirrer of antiserum-coated nickel grids submerged in virus preparations significantly increased trapping of virus particles (102). A thorough washing, rather rigorous after use of crude juice or vector extracts, of the grids with buffer followed by double-distilled water is essential before applying contrast in order to minimize debris on the grids. Use of sucrose (0.4 M) in the washing buffers was reported to reduce greatly the debris on grids (10, 23).

Before examination in the EM, suitable contrast must be applied to the preparations. In early ISEM work, shadowing with platinum/palladium (21) and positive staining (10, 23, 73) were employed with good results. However, negative staining is preferred now because it allows high-resolution visualization of virus particle morphology. Aqueous uranyl acetate and phosphotungstic acid (PTA) are widely used. Other stains, such as ammonium molybdate for blueberry shoestring, carrot red leaf, and several nepoviruses (30, 85, 116) and uranyl formate–NaOH for several luteoviruses and other nepoviruses (75, 85–87), have given good results. Since mycoplasma cells are easily disintegrated and particles of some viruses are unstable in various stains, brief incubation of grids in aldehyde fixatives prior to staining is required (12, 85). Also, alfalfa mosaic virus, ilarviruses, cucumoviruses, tomato spotted wilt virus, some rhabdoviruses, and some geminiviruses are labile in PTA (65), thus requiring other stains. In ISEM for disease diagnostic purposes, visualization of a few characteristic virus particles or mycoplasma cells on the grids is sufficient, but in quantitative work particle counting must be carried out. A method used by Roberts and Harrison (86), Roberts (82), and Roberts and Brown (85) that involves counting particles in one or more randomly chosen fields in each of ten grid squares on each of two duplicate grids and converting the counts to a $1000_\mu m^2$ area basis provides a reliable index for comparison of different virus preparations. Although the limit of detection of plant viruses with ISEM varies with various factors related to the antiserum and the virus preparation used, concentrations in the range of 0.1–10 ng/ml, which represent 10^7–10^9 particles/ml of a small spherical virus, are routinely detectable (5, 21, 30, 65, 74).

Two types of controls should be routinely included in ISEM work. First, grids coated with preimmune serum or antiserum to a virus serologically unrelated to one being studied should be included. The second control can be the untreated grids or those treated with buffer in place of a serum. The degree of particle trapping on the homologous antiserum-coated grids should be judged against the particle counts on serum-coated control grids. Although untreated or buffer-treated grids trap more particles than the serum-coated controls especially from purified virus preparations, it is important to know the particle count of untreated or buffer-treated grids to ascertain the performance of normal serum controls (83).

It has been reported that with a few viruses ISEM either did not increase or gave less than threefold increase in particle counts over the controls (49, 62). Also, the number of particles attaching to buffer-treated support films varies greatly with different viruses. Certain viruses, particularly from dilute, purified preparations, adsorb to buffer-treated grids in significant numbers (49, 62), and hence ISEM may make little improvement in particle counts in such cases (49, 65). Another explanation, put forward by Milne and Lesemann (65), for unsatisfactory ISEM with certain virus preparations is that such purified preparations or crude saps may contain sufficient amounts of free subunits or oligomers of viral coat protein to compete with virus particles and block binding sites of the antibody layer on support films. Indeed, poor ISEM results were obtained with frozen–thawed or long-stored virus preparations, presumably due to the presence of free protein subunits released from virus particles, whereas virus was effectively trapped from fresh preparations (65).

An interesting modification that holds promise of increasing the sensitivity and perhaps provide new applications of ISEM for plant viruses and mycoplasmas has been reported. Katz *et al.* (40), working with animal viruses, demonstrated that protein A-containing bacterial (*S. aureus*) cells could be coated with viral antibodies and that such coated bacteria could "fish out" homologous virus particles from a suspension. Later, Nicolaieff *et al.* (70) showed that antibody-coated bacteria could be used to trap isometric (Fig. 8.1) as well as rod-shaped plant viruses from suspensions. In a comparison of antibody-coated bacteria and support films coated with protein A and antibody it was shown that bacteria were extremely efficient

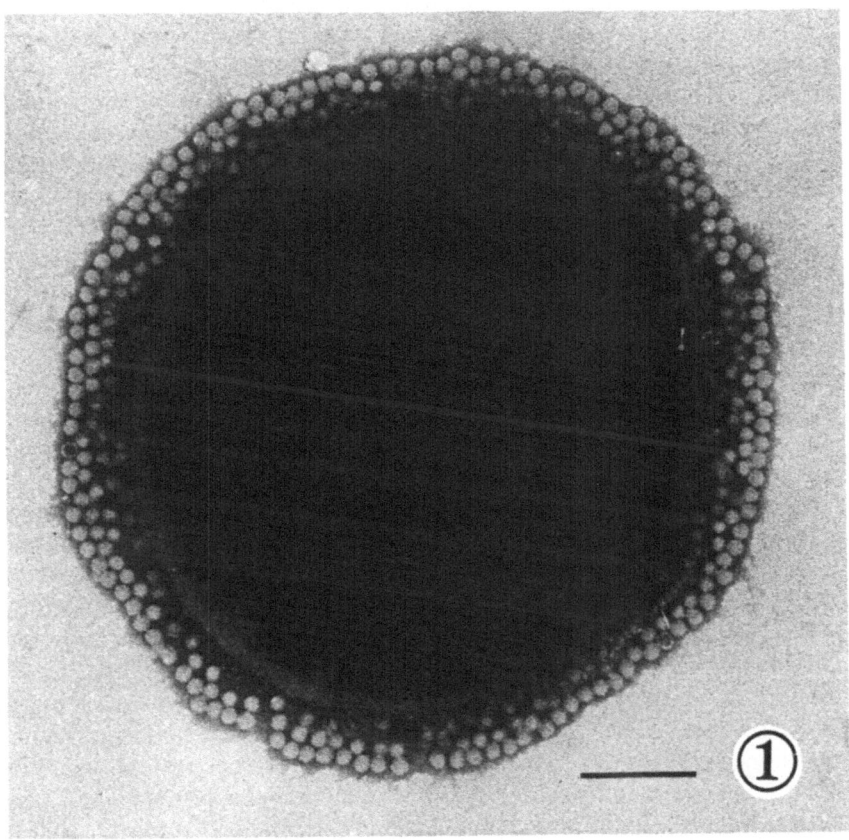

FIGURE 8.1. A bacterial (*Staphylococcus aureus*) cell treated with specific antiserum and subsequently incubated with 500 ng/ml of turnip yellow mosaic virus, showing virus particles trapped on the cell surface. Bar represents 0.5 μm. [Nicolaieff *et al.* (70).]

in trapping isometric virus particles and that their trapping efficiency was slightly superior to that of support films. However, the sensitivity for virus detection of the two methods was about the same, since virions present only at the periphery of the bacterial cells can be seen in the transmission electron microscopy (70). Virus particles attached to most of the surface area of the bacterial cell can be seen by scanning electron microscopy (SEM) (40). Therefore, an ISEM modification providing even higher sensitivity of detection is feasible if a routinely usable, high-resolution SEM method for virus particle visualization can be combined with the use of *S. aureus* cells.

DECORATION

When virus particles applied to a support film for routine negative staining or those trapped on the grids with the help of an antibody layer are treated with suitably diluted antiserum, the particles are coated with a layer of globulin molecules that is clearly visible after negative staining. Since this specific antibody–antigen reaction, known as "decoration" (66), occurs only with the homologous antiserum or with an antiserum to a serologically related virus, it identifies the virus particles with certainty. Decorated particles are more easily recognized in the microscope than the undecorated ones, due to their increased size and because the darkly staining antibody layer around the particles greatly increases contrast (65). Although decoration has been used on virus particles loosely adsorbed to the support films not coated with antiserum (4, 56, 66), virus particles in such preparations often do not remain immobile and are found in clumps after the decoration and staining steps are completed (2; Y.C. Paliwal, unpublished). The decoration method is best utilized when virus particles are first immobilized on the support films by using antibody-coated grids, and following the ISEM protocol, a decoration step is added before negative staining. For this step, grids incubated with the virus preparation are rinsed with buffer and then incubated for 15 min with the antiserum at a dilution 5–10 times less (65, 96) than the titer of the antiserum, washed appropriately, and negatively stained. When only a low-titered antiserum is available, a very low dilution or undiluted serum may be necessary for decoration. This may lead to nonspecific decoration (65), and partial or total degradation of virus particles may occur due to proteolytic activity in the sera (65; Y.C. Paliwal, unpublished). Heating the antisera to 60°C for 10 min prior to use or using only sera that have been stored liquid at 4°C for several months can eliminate or minimize this problem (65). Use of IgG fraction can also circumvent the problem.

In some applications of decoration, where a series of dilutions of the homologous antiserum or a heterologous antiserum (related virus) is used, weak decoration of particles is likely to occur on certain grids. Such decoration is difficult to score with certainty in the microscope. Kerlan *et al.* (42, 43) applied a second decoration to the virus particles, previously subjected to decoration, with sheep anti-rabbit IgG antiserum and then staining negatively. The apparent diameter of virus particles, which is nearly doubled by one decoration, increased to about five times their actual size (42) due to a large, electron-dense halo around the particles. Thus, the detectability of decorated particles in general and of marginally or invisibly decorated ones in particular was further increased by double decoration, since the particles are more conspicuously visible at lower magnifications. This allowed speedier examination of larger fields of view in the microscope. Detectability of the first antibody is also increased because many more particles are scored as decorated after treatment with anti-rabbit IgG antibody.

An alternative to decoration is to treat the virus particles trapped on the grids with antibodies labeled with an electron-dense marker and to look for specific labeling of the particles in the microscope. Pares and Whitecross (77) used such an approach employing TMV antibodies labeled with colloidal gold particles complexed with protein A, and demonstrated that virus particles can be identified as TMV due to specific attachment of TMV antibody visualized as gold particles attached to virus particles. The technique needs some improvement with regard to resolution and specificity, but has potential to be a sensitive alternative to decoration for both virus identification and for quantitative work using gold particle counts.

Immunoelectron Microscopy of Antigens *in Situ*

IEM has had limited application in ultrastructural studies of plant viruses in plant and vector tissues because particles of many viruses can be easily visualized in host cells in ultrathin sections with proper fixation and staining. However, IEM methods have been useful for localization of free viral protein, small spherical viruses that are difficult to distinguish *in situ* from small cell organelles, and cellular inclusions induced by viruses (28, 29, 37, 92, 94).

Detection of viral antigens at the cellular level in tissue sections is achieved by the use of labeled antibodies as an immunologically specific stain. Two labels have been commonly used: ferritin, an electron-dense molecule of about 12 nm diameter, and enzymes such as horseradish peroxidase. In case of antibodies conjugated with ferritin, the antigen–antibody reaction is detected *in situ* by visualizing ferritin molecules tagged to virus particles or other nonparticulate antigens that react with the antibody–ferritin conjugate. Reaction of peroxidase-conjugated antibodies with antigens is detected by treating the tissue with a suitable enzyme substrate, fixing the reaction product to make it electron-dense, and examining the tissue sections in the EM. Peroxidase-labeled antibodies have been extensively used for immunocytochemistry of animal viruses and various soluble antigens of viral or tissue origin in a variety of animal tissues (45). Although infiltration into the tissue and diffusion into the cells of enzyme-labeled IgG is less of a problem than that of ferritin–IgG conjugates because of the large size (12–14 nm) of the latter, enzyme-labeled antibodies have not been used in electron microscope immuno-cytology of plant viruses. The problem of abundance of endogenous per-oxidases in plant cells could be one reason for this.

For preparation of ferritin-conjugated antiviral antibody, antisera of reasonably high titers (e.g., at least 1/512–1/1024) against the whole virus or isolated coat protein (94) should be used and a high degree of specificity should be ensured (81). One way to ensure specificity is to isolate IgG from the antiserum, absorb it with the purified virus (homologous antigen), and separate the IgG from the virus–IgG complex (92). A standard method (98) of conjugating ferritin to the antibodies with the help of toluene 2,4-diisocyanate as a coupling agent is widely used (37, 92–94). The ferritin–IgG complex (FIC) is a relatively large entity that does not readily penetrate into the tissue and cannot pass through the intact plasma membrane of cells. To overcome this problem, plant tissue pieces are first fixed in a suitable fixative (e.g., glutaraldehyde) and then various disruptive treat-ments are applied to open the cells to FIC. Pieces of leaf or petiole tissue are frozen and cut open from all four sides in a cryostat, thawed, and incubated in the FIC solution with continuous shaking for a few hours (92, 93). Alternatively, leaf tissue (plasmolysed or unplasmolysed) can be cut into small pieces in a powered chopper and then ultrasonicated in cold before incubation with FIC (37). An indirect method (two-step procedure) like that commonly employed in fluorescent antibody procedures has been used for ferritin as well (28, 29, 94). Disrupted tissue is first incubated with rabbit antiviral IgG to allow its diffusion into the cells for binding with the antigens; sheep anti-rabbit IgG conjugated with ferritin is then applied to bind with rabbit IgG–antigen complexes present in the cells. The indirect method is often more sensitive and offers other advantages (94). The sheep anti-rabbit IgG antibody can be purchased or prepared in large amounts. Large volumes of ferritin conjugates can be prepared and then used for a whole range of different viruses as long as small amounts of specific antiviral sera are available. Some of the viruses or their anti-bodies may be difficult to obtain in larger amounts required in the direct method for individual ferritin conjugation and standardization of the pro-cedure. High nonspecific binding of ferritin has been reported as a problem in the indirect method (94). Shepard et al. (94) found that use of rabbit antiviral antibody at as high a dilution as possible reduced this problem, but incubation of tissue preparation with sheep normal immunoglobulin followed by addition of rabbit antiviral antibody directly to this mixture was most efficient in preventing nonspecific tagging. After the tissue is treated with FIC through the direct or indirect method it is fixed with osmium tetroxide and processed for ultrathin sectioning and electron mi-croscopy using conventional methods.

Colloidal gold as an antibody label for antigens *in situ* offers a promising alternative to ferritin and has recently been employed for animal viruses

(27, 59), plant viruses (45a, 52, 106), and mycoplasmas (32). The method involves fixation of small tissue pieces in 1–3% glutaraldehyde (52, 106) or 4% paraformaldehyde at 0–4°C, dehydration in the usual manner, embedding in Lowicryl (a methacrylic-acrylic polymer) at -20 to $-30°C$ (52, 106), ultrathin sectioning, and then staining with gold–IgG in a direct or indirect procedure. As with the immunoferritin technique, the antiviral IgG of high specificity must be used. Colloidal gold is best prepared by the method of Slot and Geuze (101) and for the indirect method anti-rabbit IgG can be conjugated to colloidal gold by protocols devised by De May *et al.* (20) or Lin and Langenberg (52). The postsectioning staining procedure, as opposed to the preembedding staining employed commonly for immunoferritin and immunoperoxidase methods, has been used for the gold antibody (52, 106). This minimizes the redistribution of intracellular antigens that is likely to occur during the tissue disruption required to achieve diffusion of ferritin antibody. Also, ultrastructural morphology of the tissue is better preserved in the postsectioning staining method (52). Before antibody treatment and gold staining, etching of sections with alcoholic sodium hydroxide is advisable to enhance accessibility of antibody to the antigens in the cells (52). For the indirect method, etched sections are briefly treated with dilute normal goat serum, incubated in rabbit antiviral antiserum, washed thoroughly with normal goat serum, stained with gold–goat anti-rabbit IgG complex, washed again with normal goat serum, and rinsed in distilled water (52). Sections can then be stained with the usual stains, such as uranyl acetate and lead citrate, before examination in the EM. Use of colloidal gold (5 or 8 nm diameter)-conjugated antibodies for localization of plant viruses is claimed to provide higher resolution and lower nonspecific staining than the ferritin antibody (52).

Polystyrene latex particles have nonspecific affinity for gamma globulins (55) and antibody-sensitized latex is widely used in agglutination tests of plant virus preparations and in clinical medicine. Immunolatex methods have been successfully used in conjunction with scanning electron microscopy to detect plant viruses carried adsorbed on surfaces of vectors (34, 50). In the indirect immunolatex labeling technique (50), latex particles (176 nm diameter) are sensitized with specific antiviral IgG following the standard method (6) and mixed with a concentrated, purified virus preparation. Large antigen excess prevents flocculation of latex, but the virus particles bind around the individual latex particles. The surfaces of insect vector mouth parts or other vector surfaces are treated with specific antiviral IgG followed by virus-sensitized latex particles. Specimens are examined in the SEM after coating with gold or gold–palladium. Binding of latex particles as large clumps indicates sites of occurrence of virus particles. Surfaces of nonviruliferous vectors processed with antiviral antibody and viruliferous specimens treated with rabbit normal serum serve as controls. Only a few scattered, single particles of latex are observed on controls (50). In the direct immunolatex technique antibody-sensitized latex particles are used directly to bind to virus particles present on surfaces to be examined, but this method is not as specific as the indirect method (50).

Applications

Rapid Diagnosis of Virus/Mycoplasma Diseases

Suitability of ISEM for diagnostic applications with plant viruses and mycoplasmas is based on many advantages the method offers over other methods of serodiagnosis: (1) ISEM has been considered to be the most sensitive of all serological tests (107). Its sensitivity is 80–1000 times higher than that of the latex flocculation test (105) and often equal to (112), or in a number of cases up to three times better than, that of ELISA (30, 33, 57, 99, 105); (2) the results are unequivocal, as the virus particles or mycoplasma cells themselves are visualized, not a reaction product; (3) false positives are absent, making the technique more reliable; (4) ISEM is rapid; results are obtainable in most cases within 20 min (65) to 6 hr and within 24–36 hr in others; (5) small volumes of antigen preparations (1–2 µl) can be successfully used; (6) ISEM is carried out with small volumes of diluted antisera, thus economizing on precious antisera; (7) low-titered antisera, sera from early bleedings, and antisera without fractionation or conjugation can be used, making the technique widely applicable; (8) it allows detection of a broad range of serologically related strains of a virus or other related viruses within a taxonomic group; (9) antibodies to host plant proteins in the antiserum are not a problem, as virus particles can be distinguished even if some background debris is present on the grids; (10) the method is inexpensive, as there are relatively few steps and less preparation, and small amounts of only few chemicals are required in routine ISEM.

IEM methods (predominantly ISEM) have been successfully used in disease diagnosis and/or pathogen identification of at least 60 different plant virus or mycoplasma diseases (113; author's compilation). Among others, these have included low-titered, difficult to detect luteoviruses and viruses of woody perennials. Symptoms of barley yellow dwarf luteovirus (BYDV) are indistinguishable from those induced by some abiotic stresses and by aster yellows mycoplasma (73). Also, symptomless BYDV infection is characteristic of rye, some wheat cultivars, and many grasses, and some BYDV strains remain symptomless in barley (73, 76). The virus was diagnosed reliably by ISEM in all of these situations in a variety of host plant species, which also included wheat, barley, and oats doubly infected with BYDV and aster yellows (73). ISEM provided a sensitive diagnosis for potato leaf roll virus in pieces of tuber tissue, tuber sprouts, and potato leaves (44, 86). Viruses in woody rosaceous plants, grapevine, and citrus plants often present difficulties in serodiagnosis due to the presence of interfering substances in their tissue extracts and the low virus titers. ISEM has given good results in the diagnosis of a range of different viruses in these plant species (1, 8, 39, 42, 71, 89, 105). In some of these cases, ISEM was more reliable than ELISA, since spurious reactions or insufficient sensitivity prevented detection of viruses in some samples with the latter method (105). Serologically related strains of a virus and other related viruses within the same taxonomic group can be diagnosed by ISEM using one antiserum. The specificity of the method using dilute (e.g., 1/1000 or more) antiserum was reported to be broad enough to detect virus strains or heterologous viruses that differ from the homologous one by a serological differentiation index (SDI) of up to about 4 (114). Broader specificity can be obtained using protein A, low dilution of antiserum, and, if needed, longer incubation of grids with the tissue extract. Using broad-specificity ISEM, four tombusviruses with different degrees of serological relatedness were detected with the help of one antiserum, but standard ELISA, due to narrow specificity, reliably detected only two of these viruses (58). Similarly, three potyviruses of different degrees of relatedness were readily detected with antiserum of one virus (47). Thus, ISEM offers a reliable method for screening of nursery stocks of various fruit crops, ornamentals, and other perennials where symptom-based diagnosis is unreliable.

ISEM can be adapted to detect several serologically unrelated viruses simultaneously. It was used to screen field clovers for five viruses, using grids coated with a mixture of antisera to two morphologically different viruses at a time (36). Similarly, three viruses affecting roses were detected in leaf and petal extracts of field-grown roses using grids coated with a mixture of antisera to the three viruses (105).

The ISEM trapping of virus particles on grids coated with a specific antiserum in itself provides at least a tentative identification of the virus. When the method is set up for high specificity by using a high dilution of antiserum (preferably 1/1000 or higher) for coating the grids and short incubation (5–30 min) with the virus preparation, the antigen–antibody reaction is essentially homologous and the identity of the particles trapped is established. However, decoration of particles trapped on the support films with antibody against the virus suspected to be present identifies the virus (Fig. 8.2) unequivocally (1, 42, 66, 86, 89). An example of the application of the decoration technique is the identification of two serologically distantly related viruses, beet ring spot strain of tomato black ring virus and the grapevine chrome mosaic virus from a mixed infection of the two in *Chenopodium quinoa* (41). Grids coated with a mixture of antisera of both viruses at a dilution of 1/1000 were used to trap the viruses for 10 min from the tissue extract or a mixture of purified virus particles. Particles were then lightly decorated with one of the two antisera at 1/1000 dilution, followed by second decoration of particles with sheep anti-rabbit IgG. Only a certain proportion of the virus particles on the grids were decorated, thereby distinguishing the two viruses.

Decoration, especially when used on virus particles attached to grids not previously coated with antibodies, may reveal a mixed population (Fig. 8.3) of virus particles (53, 67) from field or laboratory samples, alerting the researcher to the presence of unexpected viruses or contaminated cul-

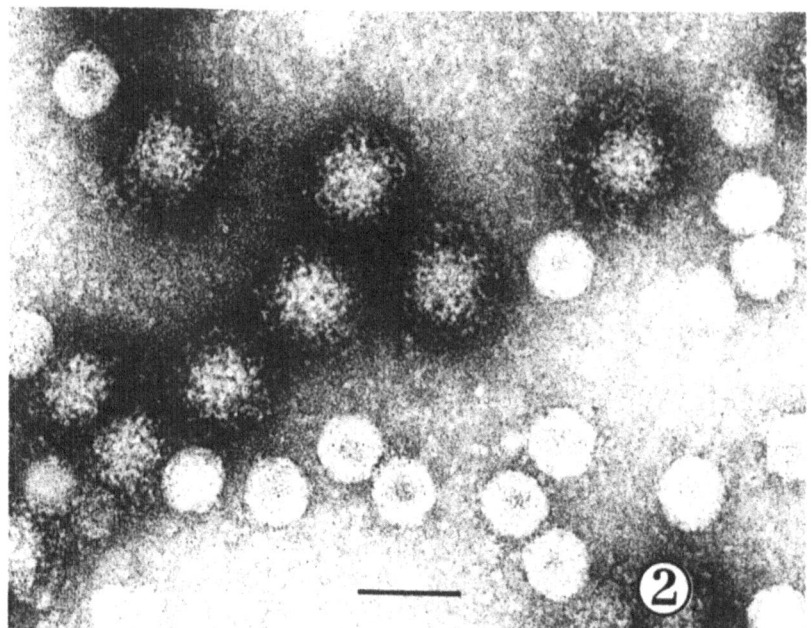

FIGURE 8.2. Decoration of virus particles with specific antibody for identification. A mixture of tobacco necrosis and tomato bushy stunt viruses (TBSV) treated with an antiserum against TBSV. Only the TBSV particles are decorated. Bar equals 50 nm. [Courtesy of R.G. Milne.]

tures. Likewise, contaminant antigens in purified virus preparations or contaminant antibodies in antisera, both either received from elsewhere or prepared in one's own laboratory, can be identified using ISEM decoration (65).

ISEM was first applied to mycoplasma by Derrick and Brlansky (23), who detected helical cells of corn stunt spiroplasma (CSS) in extracts of midvein tissue of infected corn. However, since CSS can be readily diagnosed by light microscopy of juice expressed from leaves (18), ISEM should be more useful in rapid diagnosis of diseases believed to be caused by nonhelical mycoplasma that now require electron microscopy of thin sections of tissue for positive identification. Mycoplasma associated with aster yellows (AY), clover phyllody (CP), and peach-X (PX) diseases were readily detected in clarified juice concentrates from 5 g of leaf tissue using grids coated with respective homologous antisera (99, 100). Although AY and CP are distinct in symptomatology and in the relationships of the pathogen with their leafhopper vectors, they are closely related serologically (99). Together, the two pathogens have an extensive host range that includes many economic and wild plants in almost every angiosperm family. Symptoms of these diseases in some plant hosts can be confused with those induced by viruses, herbicide injury, and sometimes arthropod

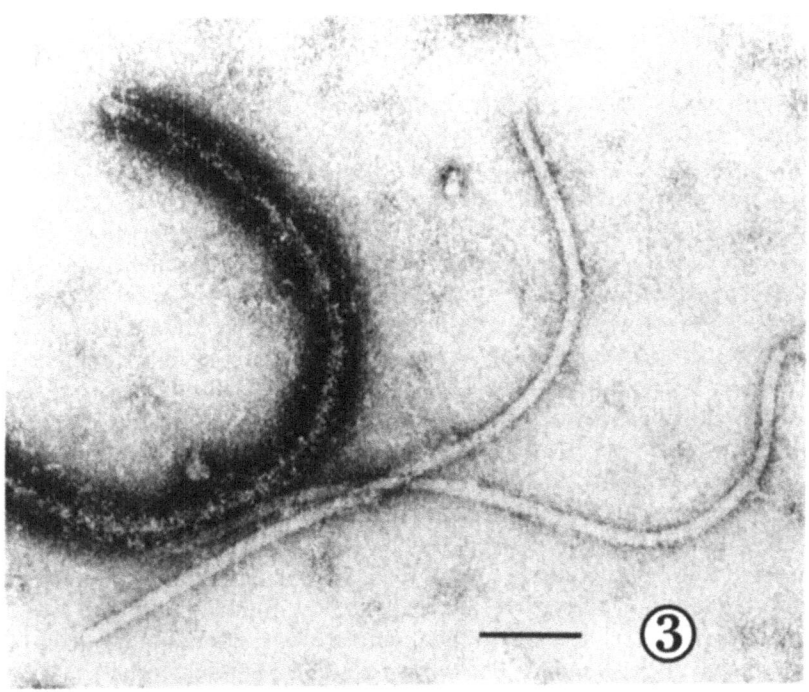

FIGURE 8.3. A naturally occurring mixture of potyvirus and a carlavirus related to carnation latent virus (CLV), partially purified from an orchid. Treatment with CLV antiserum decorated only the carlavirus particles. Bar represents 100 nm. [Courtesy of R.G. Milne.]

pest injury (toxicoses). Due to broad specificity, grids coated with AY antiserum should allow rapid diagnosis of diseased plants to establish possible mycoplasmal etiology if infection of either AY or CP is suspected. Since an antiserum specific to PX mycoplasma and its use in ISEM has been reported (100), it should be possible to develop ISEM protocols designed for rapid identification of PX mycoplasma in peach and herbaceous host plants that may not show characteristic symptoms, but may serve as reservoirs of the pathogen in nature. ISEM also has been employed for the detection of mycoplasma associated with the flavescence dorée disease of grapevine using an antiserum prepared against the mycoplasma extracted from infective leafhopper vectors of the disease. Protein A and undiluted antiserum-coated grids were used and it was necessary to include Tween 20 in the tissue extract and wash buffer, and to fix the mycoplasma cells trapped on the grids with glutaraldehyde before positive staining with ammonium molybdate (12, 60). Many other diseases of great economic importance (e.g., coconut lethal yellowing, paulownia witches broom, citrus likubin, elm phloem necrosis, apple proliferation, maize bushy stunt (corn stunt complex), and stolbur of vegetables) where mycoplasma have been implicated as causal agents lack reliable rapid diagnostic methods, mainly because antisera to these so far nonculturable prokaryotes have not yet been developed and techniques such as ISEM cannot be applied to them.

Although ISEM with its advantages of high sensitivity and reliability is quite suited to meet the challenges of most plant screening and virus identification tasks, the need for costly equipment (EM) and electron microscopy skills and its labor-intensiveness have been cited as reasons for ISEM not being suited for large-scale screening of samples (65, 107). In fact, a fully automated ELISA setup, required for large sample throughout is also costly and unless used continuously may not provide the best return on the investment as compared to the return on investment with an EM. An EM is not an "exclusively for viral diagnosis" piece of equipment and is usually put to a variety of uses 6–7 days/week, 12-months/year. Lack of general availability of the microscope is sometimes considered another handicap. Although EMs are now a very common instrument and access to one can be gained in most cities, immediate availability of one is not a serious impediment to the use of ISEM. Both antibody-coated grids as well as completed ISEM preparations (stained grids) can be stored desiccated for long periods of time (23, 65, 73, 96). Therefore, arrangements for a field station to receive antibody-coated grids from an EM facility, incubate them with the test antigen preparations, stain, and return to the EM center for examination can work well (73). Requirement of specialized (EM) skills for ISEM is no more stringent than the skills required in other highly sensitive tests, such as ELISA or radioimmunoassay, because EM skills are easily learned and relatively little knowledge of immunology and immunochemistry is required for ISEM.

Although ISEM preparations are made rapidly, examination of a large number of grids can be time-consuming and tiring. Thomas (105) reported that about 100 grids, presumably representing 100 samples, can be examined in a day. Examination of grids could be done much faster if virus particles made conspicuous by single or double decoration (42, 66) are being detected. In any case, operator fatigue in examination of a large number of grids does remain a limiting factor for use of ISEM in large-scale testing of plants, and ELISA is the most suitable technique for mass screening. However, ISEM should be preferred over other methods for small- to medium-scale (up to 100 samples) diagnostic work, or when samples have to be screened intermittently and at short notices. The expense of setting up and keeping ELISA in a state of readiness for only intermittent use is hardly justifiable. ISEM can greatly aid in developing ELISA for the mass screening work because the former can go to work a few months before ELISA is set up for use with a new virus (65). Also, during mass testing, constant monitoring of ELISA by ISEM checks is highly desirable.

Detection of Viruses in Seeds

Detection of viruses in seeds by ISEM is relatively simple, unequivocal in its interpretation, and gives consistent results. The method has been successfully used for detection of barley stripe mosaic (BSMV), blackeye cowpea mosaic (BlCMV), broadbean stain (BBSV), lettuce mosaic (LMV), pea seed borne mosaic (PSbMV), soybean mosaic (SMV), and tobacco ringspot (TRSV) viruses in seed of a number of plant species (10, 33, 51, 54, 90). No special modifications to the ISEM method are required for virus detection in seed. If the presence of more than one serologically related strain of a virus is suspected in a seed lot, the method should be set up for broad specificity. Seed extracts are made either by homogenizing in buffer the whole seed presoaked in water for up to 24 hr (33), homogenizing the dissected embryos (90) or hypocotyls (51) of germinated seed, or by suspending the milled seed powder in a buffer (10, 54). Sucrose (0.4 M) was included in the buffer (Tris or phosphate) in the initial work (10, 33), but it does not seem to be essential. Incubation of antibody-coated grids with tissue extracts for 30 min to 1 hr at room temperature or 37°C has given good results with several viruses and different seeds (10, 33, 51, 54), but 24 hr at 4°C was used for BBSV in broadbean seed (90).

Several workers have tested the performance of ISEM in detecting infected seeds in various known mixtures of virus-free and infected seed. PSbMV and LMV were detected in mixtures containing one infected seed per 100 healthy seeds, but BSMV, SMV, and TRSV were detectable in seed extracts representing one infected seed in 1000 healthy seeds (10, 33). ISEM consistently detected PSbMV in seed lots containing 1–5% infected seed, whereas ELISA gave negative results at these levels due to relatively high A_{405} values of healthy pea seed homogenates (33). ISEM was useful in screening broadbean seed lots, revealing 10–40% infected seed in some lots (90); infectivity assays and agar gel diffusion serology on homogenates of embryos and/or seed coats were unsuccessful in detecting the seed-borne BBSV (16). In a seed certification program, Lister et al. (54) successfully used ISEM for testing bulk barley seed lots for levels of BSMV-infected seed and for monitoring results of the SDS disk agar gel diffusion test.

In seed health evaluation, ISEM is a useful tool for screening a limited number of bulk seed lots. However, if virus particles are made more conspicuous by single or double decoration (42, 66), or a colloidal gold-labeled antibody method (31, 77) of improved specificity is used, grids can be examined faster at lower EM magnifications, resulting in much less operator fatigue during examination. ISEM should then be applicable to large-scale testing of bulk seed lots in seed certification programs and quarantine checks.

Virus Particle Size Determination

Particle size is an important criterion in differentiating viruses, especially the elongated types. A large number of particles must be photographed and measured, particularly when dealing with rod-shaped or filamentous viruses, because of the presence of broken or aggregated particles and the possibility of two types (lengths) of particles being characteristic of the virus (e.g., tobraviruses) (35). Since purification and concentration by ultracentrifugation often leads to some breakage and aggregation of elongated virus particles, crude juice or a leaf crush is preferred for making EM preparations to measure particles. Such EM preparations usually contain only a small number of particles, especially in cases of viruses that have low concentration in plants. It therefore becomes a labor-intensive and expensive task to locate large numbers of particles and to take numerous photographs. If an antiserum to the virus is available, ISEM can greatly increase the number of virus particles on the grids, resulting in much reduced amounts of photography.

Milne and Luisoni (66) were first to advocate the value of ISEM in virus particle size determination and employed the method with TMV and a potex virus. The values of "normal length" obtained were the same as those reported in the literature and there was no evidence of increase in the number of broken particles in ISEM preparations as compared to the "dip" preparation. However, Pares and Whitecross (78) reported that ISEM preparations contained a significantly higher proportion of small, presumably broken, particles of two strains of TMV when trapped from crude juice on antiserum-coated grids as compared with the dip preparations. Except for this report, ISEM has been shown to increase the numbers of particles on the grids of several other viruses without any evidence of increased particle breakage. Flexuous rod-shaped particles of oat mosaic virus were concentrated on the grids by ISEM and were found to have a normal length in the range of 650–750 nm in both ISEM and dip preparations, but smaller (250–550 nm) broken particles were predominant in purified preparations (109). Particle size of an isolate of potato virus S (rigid rods) was accurately determined on the basis of particles measured on ten ISEM micrographs, as opposed to 70 micrographs of dip preparations required to provide an equal number of particles for a comparable size determination. The proportion of small (broken) particles was not increased in ISEM preparations relative to those found in dip preparations (88; D.G. Rose, personal communication). One of the much appreciated uses of ISEM is in the measurement of particles of wheat spindle streak mosaic virus (WSSMV), which occurs in an extremely low concentration in wheat plants and has very long, fragile particles. Only occasional virus particles are found with difficulty in dip preparations, but up to 1000 times more particles were made available by ISEM (38). Particles decorated with WSSMV antiserum and those not subjected to decoration both had a modal length of 1800 nm (range 600–3300 nm). During incubation of antibody-coated grids with leaf extracts for 7 hr or overnight (ISEM), the integrity of particles was unaffected and incubation with dilute antiserum for 1–3 hr for decoration of particles did not result in any increase in small particles (38). Similarly, in ISEM measurement of particles of potato mop top virus (PMTV) from potato leaf extracts, good decoration of particles was achieved by incubation with dilute (1/1000) antiserum for 2 hr at 20°C without any disruption of particles, but 3 hr of decoration partly disrupted the particles, and after 4 hr, no particles remained on the grids (86).

In view of the particle disruption during ISEM reported for TMV (78), but not with another tobamovirus (PMTV) or with other viruses in different taxonomic groups, it appears that certain viruses may be inherently more prone to particle disruption under certain experimental conditions. For such viruses special ISEM protocols may have to be devised after determining the buffers, pH, period of incubation with the antibody-coated grids, and negative stain suitable for preserving the integrity of particles.

Assay of Antigen and Antibody Concentration

Virus particle counts from ISEM grids have a linear relationship with virus concentration (mg/ml of suspending medium). The first introduction of ISEM to plant viruses by Derrick (21) demonstrated a linear relationship between the log of the number of particles trapped on the grids and TMV concentration in the range of 0.01–0.74 μg/ml. Linearity in the relationship between particle numbers on the grids and concentration of the virus has since been further established using purified preparations of several other viruses, both spherical and elongated, e.g., BYDV, carnation etched ring (CERV), cowpea mosaic (CpMV), and cauliflower mosaic (CaMV) viruses (5, 46, 74), as well as in extracts of tissues infected with CpMV and citrus tristeza virus (5, 39). When virus concentration has to be accurately assayed and compared, elaborate particle counting is required, since variation in particle numbers between duplicate grids, between different squares of the same grid, and even between different areas of the same grid square can be significant (69). Therefore, a sufficiently large number of fields should be photographed from different parts of each of two duplicate grids and particle counts converted and expressed for a standard area, e.g., 1000 μm^2 (65, 82). If the particles are conspicuous enough (decorated or double decorated), they can be counted directly on the EM screen, which is faster and less expensive than counting on micrograph negatives. Aggregation of particles, particularly of the elongated types, and uneven distribution of virus on the grids are problems that can usually be resolved by experimenting with buffers used in the virus preparation, using high dilutions of antiserum (5) and suitably agitating or magnetically rotating (102) the grids during incubation with the virus preparation. Also, it should be realized that the linear relationship between particle counts and concentration holds within a certain concentration range. With increasing concentrations of virus, saturation of the support film with virus particles will occur beyond a certain concentration and linearity will be lost. Optimum range for each virus–antibody system should be determined before starting the assays. As an example, the optimal range for CERV is 2–10 μg/ml (46).

IEM assays of virus concentration have been used in a variety of applications. Rate of CpMV multiplication in cowpea protoplasts was assayed in clarified extracts of infected protoplasts by ISEM (5). Ishii and Usugi (39) compared relative concentrations of citrus tristeza virus (CTV), a 2000-nm-long closterovirus, in different parts of individual citrus trees and the titers of four strains of the virus in four different citrus species. Monitoring of success of virus purification procedures can be effectively done with ISEM and the method may offer a distinct advantage over other sensitive methods. Closteroviruses and similar viruses with long, slender particles (e.g., WSSMV) are difficult to purify, since their particles break easily, leading to loss of infectivity and reduced yields of purified virus. Improvements in purification techniques are therefore continuously sought. ISEM assays have been successfully used for CTV (9) and WSSMV (K.Z. Haufler and D.J. Fulbright, personal communication) to monitor the condition of virus particles, correlate particle lengths with infectivity, and determine virus concentration at different stages of purification procedures. Although ELISA revealed high virus concentration in virus fractions from density gradients in assays of CTV, infectivity was low, due to particle breakage as revealed by ISEM (9). Also, ISEM assays were useful in screening wheat cultivars for their reaction to WSSMV, as severity of disease symptoms was correlated with higher particle counts (K.Z. Haufler and D.J. Fulbright, personal communication).

Decoration of virus particles with antibodies can be employed to determine titers of antisera rapidly. Decoration titer of an antiserum is the highest dilution that gives consistent positive decoration (63). Virus particles are trapped on the grids from a purified preparation using a rapid ISEM protocol and grids are incubated with different dilutions of the antiserum, stained, and scored in the EM as positive or negative for decoration. The whole operation can be completed in about 40 min because no particle counting is involved and very small amounts of virus are required (63). Decoration titer of an antiserum by the single-decoration method usually comes about one twofold dilution step higher than the conventional gel diffusion titer (63). Double decoration (42, 43) increases the sensitivity of the system because visibility of the antibody halo around the particles is greatly enhanced due to strong staining and manyfold increase in apparent particle diameter (see "Decoration" under "Techniques"). Thus, many more particles on grids treated with high dilutions of antiserum are scored, as decorated and antiserum titers up to four twofold dilutions higher may be recorded (42, 43).

Determination of Serological Relationships

For determination of degree of serological relationship, especially a distant one, ISEM can be set up for broad specificity using protein A, low dilution of antiserum, and relatively longer (few hours or longer) incubation times with the virus. Using no protein A treatment, very dilute antisera (e.g., 1/1000–1/5000) and a short incubation period with virus is more likely to differentiate sharply between serologically related strains of a virus (47, 65). Relationship between two viruses can be assessed in two ways: particle counts of homologous and heterologous reactions on grids can be compared, and/or decoration titer of an antiserum against two viruses can be determined.

In the particle count method, only the counts that are consistently at least three times higher than those on normal serum grids should be considered significant. Meyer (61) compared the factor by which particle counts on antiserum-coated grids were higher than the counts on normal serum-coated grids in homologous and heterologous reactions to assess relationship between several different potyviruses. Several ISEM investigations of relationship between viruses, e.g., between seven geminiviruses (91), between strains of sugarcane mosaic virus (97), between luteoviruses of potato, carrot, legumes, and cereals (87, 116), and between four tombusviruses (58), have compared homologous and heterologous particle counts and graded the relationships as "close" to "distant," depending upon the difference between the two counts. In an evaluation of the degree of relatedness between strains of BYDV the ratio of heterologous to homologous counts was used as an index of relatedness; a ratio of 1 or near 1 denoting total identity or very close relationship, and 0 indicating no relationship (74). However, in none of the above studies were particle counts from reciprocal tests using sera to all viruses or strains being investigated compared. Such a comparison of reciprocal test particle counts provides greater accuracy in assessing relationships, enabling firmer conclusions. This is demonstrated by the results of Tamada et al. (104) for the relationship between potato leaf roll and beet mild yellowing viruses.

Decoration titers of one or more antisera against each of the viruses or strains being studied can be compared in ISEM by using the serological differentiation index (SDI) (111), which denotes the number of twofold dilution steps by which the homologous and heterologous titers differ. Although this concept has been applied to results obtained by mixing each of the antiserum dilutions of a twofold dilution series with virus particles and finding the highest dilution where antibody attachment to particles (clumping) occurs (87), results are more accurate with the ISEM-decoration method. Purified preparations of the viruses being investigated should be mixed to have about equal concentration of each, and particles trapped on serum-coated grids, and then antisera should be titrated individually against the mixture to determine the decoration titers (63). Homologous and heterologous decoration titers of three dianthoviruses (bipartite genome) and the pseudorecombinants using their RNA-1 and RNA-2 were used by Chen et al. (14a) to evaluate serological relationships and to confirm by ISEM that RNA-1 determines the serological specificity of progeny viruses resulting from pseudorecombinations. Kerlan et al. (43) utilized the increased sensitivity of the double-decoration technique and compared SDIs and reciprocal titer SDIs based on double-decoration titers (ISEM-decoration and clumping) of antisera to measure various degrees of relationships among grapevine chrome mosaic virus and strains of tomato black ring virus. ISEM-double decoration amplified and clearly demonstrated relationships that were considered distant (SDI of 7) and very distant (SDI of 7–9) by the gel diffusion method.

The degree of relationship between viruses revealed by ISEM is in a number of cases qualitatively similar but quantitatively different in comparison with results of other serological tests (49, 74, 91, 116). However, this is not unique to ISEM, as similar differences are revealed between results of serological methods other than ISEM (91, 116). Because the attributes of heterologous reactions in ISEM are not thoroughly understood, one may question the significance of weak relationships revealed only by ISEM. On the other hand, since ISEM is probably the most sensitive serological test known, it is expected to amplify relationships revealed as weak to moderate by other tests and elucidate new relationships that have so far remained obscure.

Localization of Antigens on Virus Particles

For viruses with a complex antigenic structure, ISEM-decoration can be employed to identify sites of specific antigens on virus particles: Particles are first trapped on grids using antiserum to the whole virus, and then a decoration treatment with an antiserum specific for one of the antigens allows visualization of the site of the antigen on the virus particles. Maize rough dwarf virus (MRDV) and other viruses of the plant reovirus subgroup 2 have a complex morphology and antigenic structure. Antigenic groups of smooth cores, "B" spikes, and "A" spiked outer shell of MRDV virions were localized by decoration using specific antisera (7, 56). Decoration was also useful in a study of coat protein encapsidation of TMV RNA. The site of initiation of RNA encapsidation was localized by sequential reconstitution of virus rods with proteins of two different TMV strains and examination of the distribution of proteins on reconstituted rods by decoration with strain-specific antisera (26, 72).

Labeled antibodies also can be used to determine sites of antigens on virus particles. Colloidal gold, being very electron-dense and a small enough particle, provides a label of good resolution. The rod-shaped particles of BSMV contain about 3.7% carbohydrate and a lectin of barley plant precipitates the virus particles (31). Using gold-labeled antibody to barley lectin, it was possible to show that the lectin-binding carbohydrate antigen sites were distributed along the entire surface of BSMV particles and not just at the ends as suspected earlier (31).

Gold-labeled antibodies also have been used in the study of surface antigens of mycoplasma cells. On the cell surface of *Acholeplasma laidlawii* K2, antigenic determinants shared with *A. laidlawii* B antigen were demonstrated using colloidal gold conjugated with anti-*A. laidlawii* B antiserum. Localization and monitoring of movements of gold–antibody label on the cell surface elucidated the lateral mobility of antigenic components of the mycoplasma membrane (32).

Detection of Double-Stranded Ribonucleic Acid

Derrick (22) showed that ISEM can be used to assay double-stranded ribonucleic acid (dsRNA). Grids coated with an antiserum against synthetic dsRNA (68) trapped linear dsRNA molecules from crude extracts of TMV-infected plants and circular dsRNA from plants infected with potato spindle tuber viroid (22, 25). Double-stranded RNA is detected by standard ISEM except that after incubation with tissue extract containing RNA, grids are treated with cytochrome to thicken the RNA molecules trapped on the grids. They are then stained with ethanolic uranyl acetate and examined in EM as such or after rotary shadowing with platinum–palladium (22, 25). A dsRNA-containing polyhedral virus was detected in crude extracts of the fungus *Agaricus bisporus* with the help of anti-dsRNA antiserum. Apparently, parts of dsRNA molecules in the virions were exposed or had partly escaped out of virions to allow trapping of virions on antibody-coated grids (19).

The antiserum to dsRNA does not trap single-stranded RNAs or DNA, but it nonspecifically traps all dsRNAs (natural or synthetic) as well as the DNA–RNA hybrids (103). Hence, the technique has limited application, since it can only demonstrate the presence of "a dsRNA" in crude tissue extracts, but this may be of value in diagnosis of some dsRNA viruses and viroids.

Detection of Virus/Mycoplasma in Vectors

Rapid detection of virus/mycoplasma in their vectors is required for epidemiological monitoring and for determination of the proportion of viruliferous insects in a population, in order to enable prediction of disease outbreaks. Transmission tests reveal the presence of infective virus/mycoplasma in the vectors, but results are obtained only after several weeks. Therefore, relatively rapid and sensitive serological tests such as ISEM and ELISA can be of great value. Individual insects can be readily screened by ISEM for the presence of virus/mycoplasma in them and no special modifications to the ISEM procedure are necessary. However, the insect extract should be carefully prepared, ensuring that virus particles localized in internal organ(s) are released in solution. For example, insufficient release of virus particles from heads of nematodes during grinding was believed to be a reason for failure to detect a nepovirus in this nematode tissue by ISEM (85). Microtissue grinders (75) or "minimortars" (86) have been used to grind small insects and whole nematodes in the presence of buffer and washed carborundum (600 mesh). Microtissue grinders consisting of pestles custom-molded in individual centrifuge tube mortars (75) achieve a thorough homogenization of the tissue because of a 'true' fit of the pestle in the mortar. Ultrasonication of the homogenate may further improve the release of virus from the tissues. Extracts of insects or other vectors are incubated with antibody-coated grids at low temperatures (3–5°C) for 1–4 hr (75, 85, 86).

Luteoviruses, e.g., PLRV, in two aphid vector species (86) and three strains of BYDV in their respective aphid vector species (75) were readily identified in single aphids by ISEM-decoration. Similarly, several nepoviruses have been detected in extracts of single nematode vectors (85). An excellent correlation between ISEM scores and infectivity tests of individual aphids exposed to BYDV-infected plants (75) and of single nematodes previously confirmed on diseased plants (85) was observed. Mycoplasma associated with AY disease was also detected by ISEM in extracts of single leafhoppers of the vector species *Macrosteles fascifrons* (R.C. Sinha and N. Benhamou, unpublished).

In ISEM, very small volumes (1–5 µl) of tissue extracts are required, which allows preparation of concentrated extracts of single insects, and the constituents of insect extracts have not been reported to interfere with detection of the virus. For detecting viruses in vectors by standard ELISA, groups of five or more insects were required to obtain sufficient volumes of tissue extract and detectable virus concentrations (11, 15). Also, virus detection was reported to be inconsistent or not fully correlated with infectivity test results in some aphid–virus systems (108; see also 11, 15). However, drastically reduced ISEM sensitivity was also reported in attempts to detect blueberry shoe-string virus (BBSSV) by adding purified virus to extracts of the aphid *Illinoia pepperi* free of BBSSV (30). Reduced sensitivity may have been due to the long (overnight) incubation of purified virus with the extract of *I. pepperi* and the latter may have exerted proteolytic or other detrimental effects on the virus. If the virus was partially degraded during this treatment, incubation at 37°C with diluted antiserum to decorate the particles could compound the earlier detrimental effect. Use of extracts of aphids fed on infected plants rather than purified virus, and shorter time periods for coating the grids with antiserum, incubation with aphid extracts, and for decoration, may alleviate the problem. The presence of an aphid virus in *I. pepperi* also caused confusion in reading ISEM results in this study (30). A similar problem of nonspecific adsorption on the grids of an aphid virus during ISEM of BYDV in the aphid *Rhopalosiphum padi* was minimized by using protein A and higher dilutions of antiserum to coat the grids (Y.C. Paliwal, unpublished).

As with all other serological techniques, ISEM does not indicate that the insects scored viruliferous are all necessarily virus transmitters. However, it does establish that the insects have acquired the virus and the virus is morphologically stable in the insect's metabolic system(s). As reported for BYDV, the ISEM score of the proportion of viruliferous insects in a population may be slightly higher than the proportion of transmitters (75). However, this can be tolerated in view of the rapid availability of the information, which can be useful in prediction of disease outbreaks.

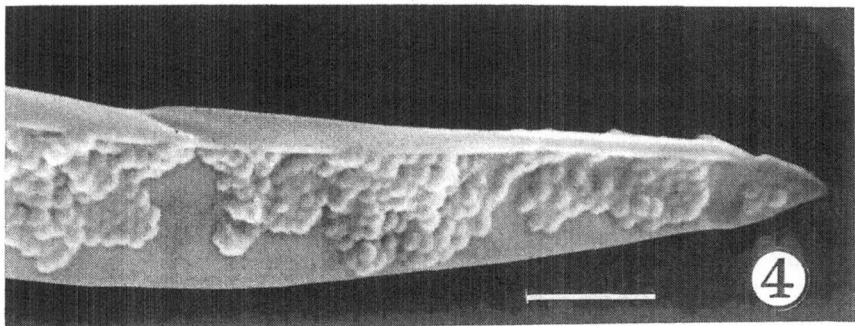

FIGURE 8.4. Immunolatex labeling specific for pea seed-borne mosaic virus (PSbMV) on the tip of mandibular stylets of the potato aphid *Macrosiphum euphorbiae* after a 5-min acquisition feeding on a PSbMV-infected plant. Stylet of the efficient vector biotype of the aphid, showing heavy latex label. Bar represents 2 μm. [Lim *et al.* (50).]

IEM with labeled antibodies can be a useful tool in investigation of vector–virus relationships. In studies of the role of salivary glands of aphid vectors in transmission and vector specificity of three luteoviruses, labeling with ferritin-conjugated antibodies provided unequivocal identification of individual virus particles embedded in basal lamina and invaginations of plasmalemma of the glands (28, 29). The indirect immunolatex method was used to identify PSbMV, a nonpersistantly transmitted potyvirus, on .stylets of its aphid vectors (Figs. 8.4 and 8.5) in scanning EM. The sites of virus attachment on the stylets and the occurrence of different amounts of virus on stylets of efficient and inefficient transmitter aphid biotypes were demonstrated (50). The immunolatex method has also been useful in detecting virus on pollen grains that serve as vectors of seed-borne tree fruit viruses. Latex-conjugated anti-prunus necrotic ring spot virus antibody specifically labeled some areas of the exine of pollen grains from infected sweet cherry trees, demonstrating the sites of occurrence of the

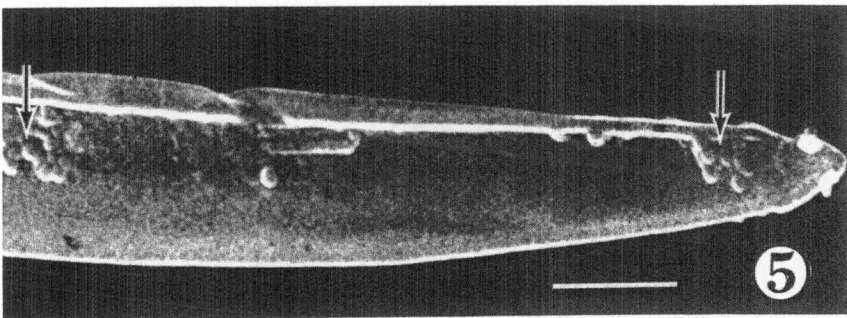

FIGURE 8.5. As for Figure 8.4. Stylet of the inefficient vector biotype of the aphid, showing relatively little labeling (arrows). Bar represents 2 μm. [Lim *et al.* (50).]

virus on the pollen surface (34). Studies of transmission of plant viruses by beetles and mites and through zoospores and resting spores of chytrid fungi are examples of other virus–vector systems where the application of IEM could be very useful.

Localization of Viruses and Other Antigens *in Situ*

Labeled antibodies, including immunoferritin and immunogold, provide a means of localizing and following the development of viral and other antigens in the host plant cells. In some early work, first appearance in tomato leaf cell cytoplasm of TMV antigen and its progressive increase between days 2 and 4 after inoculation was demonstrated by ferritin antibody labeling. Also, viral antigen but not virus particles were consistently detected in the nuclei (92). Immunoferritin tagging experiments with potato virus X-infected tobacco tissues showed that none of the components of laminate inclusions found only in cells of infected plants were antigenically related to the virus or its depolymerized structural protein (93). In similar experiments, crystalline inclusions found in cell nuclei of tobacco etch virus (TEV)-infected plants were found to be antigenically unrelated to TEV protein or to the protein of virus-induced cylindrical inclusions found in cell cytoplasm (94). Hatta and Matthews (37) used ferritin-conjugated antibodies to study sites of synthesis and accumulation of coat protein of tymoviruses in cells of several plant hosts. Coat protein but not nucleoprotein of four tymoviruses was shown to accumulate in the cell nuclei and in close association with clustered vesicles lining the chloroplast outer membrane, providing clues to the sites of assembly of virus particles.

Gold-labeled antiviral antibodies have only recently been utilized for locating virus antigens in cells. They were used to localize red clover mottle virus antigen in pea leaf cells (106), but a need for improvement in resolution and specificity of the gold labeling was evident from the results. Good resolution and specificity of the immunogold labeling was obtained by Lin and Langenberg (52) in elucidation of the sites of localization of BSMV particles in the leaf cells of infected wheat plants. A low-magnification survey of large areas of immunogold-stained thin sections could be done to determine distribution of the virus in the tissues and to assess qualitatively the virus concentration in the cells. In another study, gold labeled antibodies to wheat streak mosaic virus and WSSMV were used to demonstrate that cylindrical inclusions of these viruses in wheat leaf cells are coated with capsid protein (homologous) and that this specific binding of viral protein to the inclusions only in the cells may facilitate cell to cell spread of the virus particles through plasmodesmata (45a). Potentially, the immunogold method can offer greater sensitivity and resolution than the immunoferritin method because of the smaller size and easier detectability of the gold label in the EM. Since immunogold staining works well on ultrathin sections, it may prove useful in identification *in situ* of small spherical viruses that are difficult to distinguish from ribosomes and other small cell organelles in the cell cytoplasm of their plant hosts and vector tissues.

References

1. Accoto, G.P., 1982, Immunosorbent electron microscopy for detection of fan leaf virus in grapevine, *Phytopathol. Medit.* **21**:75–78.
2. Almeida, J.D., Stannard, L.M., and Shersby, A.S.M., 1980, A new phenomenon (SMOG) associated with solid phase immune electron microscopy, *J. Virol. Meth.* **1**:325–330.
3. Anderson, F.A., and Stanley, W.M., 1941, A study by means of the electron microscope of the reaction between tobacco mosaic virus and its antiserum, *J. Biol. Chem.* **139**:339–344.
4. Ball, E.M., and Brakke, M.K., 1968, Leaf-dip serology for electron microscopic identification of plant viruses, *Virology* **36**:152–155.
5. Beier, H., and Shepherd, R.J., 1978, Serologically specific electron microscopy in the quantitative measurement of two isometric viruses, *Phytopathology* **68**:533–538.
6. Bercks, R., and Querfurth, G., 1971, The use of the latex test for the detection of distant serological relationships among plant viruses, *J. Gen. Virol.* **12**:25–32.
7. Boccardo, G., and Milne, R.G., 1981, Enhancement of the immunogenicity of the maize rough dwarf virus outer shell with the cross-linking reagent dithiobis (succinimidyl) propionate, *J. Virol. Meth.* **3**:109–113.
8. Bovey, R., Brugger, J.J., and Gugerli, P., 1982, Detection of fanleaf virus in grapevine tissue extracts by enzyme-linked immunosorbent assay (ELISA) and immune electron microscopy (IEM), in: A.J. McGinnis (ed.), *Proceedings of the 7th Meeting of the International Council for the Study of Viruses and Virus-like diseases of the Grapevine,* Agriculture Canada, Research Branch, pp. 259–275.
9. Brlansky, R.H., 1982, Serologically specified electron microscopy for detection and Identification of plant viruses, Presented at workshop on Ultrasensitive Methods for Virus Detection, University of Puerto Rico, Rio Piedras, Puerto Rico.
10. Brlansky, R.H., and Derrick, K.S., 1979, Detection of seedborne plant viruses using serologically specific electron microscopy, *Phytopathology* **69**:96–100.
11. Carlebach, R., Raccah, B., and Loebenstein, G., 1982, Detection of potato virus Y in the aphid *Myzus persicae* by enzyme-linked immunosorbent assay (ELISA), *Ann. Appl. Biol.* **101**:511–516.
12. Caudwell, A., Meignoz, R., Kuszala, C., Larrue, J., Fleury, A., and Boudon, E., 1982, Purification sérologique et observation ultramicroscopique de l'agent pathogène (MLO) de la flavescence dorée de la vigne dans les extraits liquides de fèves (*Vicia faba* L.) malades, *C. R. Soc. Biol.* **176**:723–729.
13. Cech, M., and Neubauer, S., 1981, Plant proteases as interfering factors in the electron microscopic detection of alfalfa mosaic and bean yellow mosaic viruses in *Solanum laciniatun* Ait, *Biol. Plant. (Prague)* **23**:384–388.
14. Cech, M., Mokra, V., and Branisova, H., 1977, Stabilization of virus particles from the mosaic diseased Freesia by phenylmethyl sulfonylfluoride during purification and storage, *Biol. Plant. (Prague)* **19**:65–70.

14a. Chen, M.H., Hiruki, C., and Okuno, T., 1984, Immunosorbent electron microscopy of dianthoviruses and their pseudorecombinants. *Can. J. Plant Pathol.* **6**:191–195.

15. Clarke, R.G., Converse, R.H., and Kojima, M., 1980, Enzyme-linked immunosorbent assay to detect potato leafroll virus in potato tubers and viruliferous aphids, *Plant Dis.* **64**:43–45.

16. Cockbain, A.J., Bowen, R., and Vorra-Urai, S., 1976, Seed transmission of broad bean stain virus and Echtes Ackerbohnen mosaik-virus in field beans (*Vicia faba*), *Ann. Appl. Biol.* **84**:321–332.

17. Cohen, J., Loebenstein, G., and Milne, R.G., 1982, Effect of pH and other conditions on immunosorbent electron microscopy of several plant viruses, *J. Virol. Meth.* **4**:323–330.

18. Davis, R.E., 1977, Spiroplasma: Role in the diagnosis of corn stunt disease, in: *Proceedings International Maize Virus Disease Colloquium and Workshop,* Ohio Agricultural Research and Development Center, Wooster, Ohio, pp. 92–98.

19. Del Vecchio, V.G., Dixon, C., Lemke, and Paul, A., 1978, Immune electron microscopy of virus-like particles of *Agaricus bisporus, Exp. Mycol.* **2**:138–144.

20. De May, J., Moermans, M., Geuens, G., Nuydens, R., and de Brabander, M., 1981, High resolution light and electron microscopic localization of tubulin with the IGS (immunogold staining method), *Cell Biol. Int. Rep.* **5**:889–899.

21. Derrick, K.S., 1973, Quantitative assay for plant viruses using serologically specific electron microscopy, *Virology* **56**:652–653.

22. Derrick, K.S., 1978, Double-stranded RNA is present in extracts of tobacco plants infected with tobacco mosaic virus, *Science* **199**:538–539.

23. Derrick, K.S., and Brlansky, R.H., 1976, Assay for viruses and mycoplasmas using serologically specific electron microscopy, *Phytopathology* **66**:815–820.

24. Forsgren, A., and Sjoquist, J., 1966, 'Protein A' from *Staphylococcus aureus.* I. Pseudo-immune reaction with human γ-globulin, *J. Immunol.* **97**:822–827.

25. French, R.C., Price, M.A., and Derrick, K.S., 1982, Circular double stranded RNA in potato spindle tuber viroid infected tomatoes, *Nature* **295**:259–260.

26. Fukuda, M., Okada, Y., Otsuki, Y., and Takebe, I., 1980, The site of initiation of rod assembly on the RNA of a tomato and a cowpea strain of tobacco mosaic virus, *Virology* **101**:493–502.

27. Garzon, S., Bendayan, M., and Kurstak, E., 1982, Ultrastructural localization of viral antigens using the protein A-gold technique, *J. Virol. Meth.* **5**:67–73.

28. Gildow, F.E., 1982, Coated-vesicle transport of luteoviruses through salivary glands of *Myzus persicae, Phytopathology* **72**:1289–1296.

29. Gildow, F.E., and Rochow, W.F., 1980, Role of accessory salivary glands in aphid transmission of barley yellow dwarf virus, *Virology* **104**:97–108.

30. Gillett, J.M., Morimoto, K.M., Ramsdell, D.C., Baker, K.K., Chaney, W.G., and Esselman, W.J., 1982, A comparison between the relative abilities of ELISA, RIA and ISEM to detect blueberry shoestring virus in its aphid vector, *Acta Hortic.* **129**:25–29.

31. Giunchedi, L., and Langenberg, W.G., 1982, Efficacy of colloidal gold-labeled antibody as measured in a barley strip mosaic virus-lectin–antilectin system, *Phytopathology* **72**:645–647.

32. Haberer, K., and Frosch, D., 1982, Lateral mobility of membrane-bound

antibodies on the surface of *Acholeplasma laidlawii:* Evidence for virus-induced cell fusion in a procaryote, *J. Bateriol.* **152:**471–478.

33. Hamilton, R.I., and Nichols, C., 1978, Serological methods for detection of pea seed-borne mosaic virus in leaves and seeds of *Pisum sativum, Phytopathology* **68:**539–543.

34. Hamilton, R.I., Nichols, C., and Valentine, B., 1984, Survey for prunus necrotic ringspot and other viruses contaminating the exine of pollen collected by bees, *Can. J. Plant Pathol.* **6:**196–199.

35. Harrison, B.D., and Robinson, D.J., 1981, Tobraviruses, in: E. Kurstak (ed.), *Handbook of Plant Virus Infections: Comparative Diagnosis,* Elsevier/North-Holland Biomedical Press, Amsterdam, pp. 515–540.

36. Harville, B.G., and Derrick, K.S., 1978, Identification and prevalence of white clover viruses in Louisiana, *Plant Dis. Rep.* **62:**290–292.

37. Hatta, T., and Matthews, R.E.F., 1976, Sites of coat protein accumulation in turnip yellow mosaic virus-infected cells, *Virology* **73:**1–16.

38. Haufler, K.Z., and Fulbright, D.W., 1983, Detection of wheat spindle streak mosaic virus by serologically specific electron microscopy, *Plant Dis.* **67:**988–990.

39. Ishii, T., and Usugi, T., 1982, Detection of citrus tristeza virus by serologically specific electron microscopy, *Ann. Phytopathol. Soc. Japan* **48:**231–233.

40. Katz, D., Straussman, Y., Shaher, A., and Kohn, A., 1980, Solid phase immune electron microscopy (SPIEM) for rapid viral diagnosis, *J. Immunol. Meth.* **38:**171–174.

41. Kerlan, C., and Dunez, J., 1983, Application de l'Immunoélectromicroscopie à la détection de deux souches d'un même virus, *Ann. Virol. Inst. Pasteur* **134E:**417–428.

42. Kerlan, C., Mille, B., and Dunez, J., 1981, Immunosorbent electron microscopy for detecting apple chlorotic leaf spot and plum pox viruses, *Phytopathology* **71:**400–404.

43. Kerlan, C., Mille, B., Detienne, G., and Dunez, J., 1982, Comparison of immunoelectronmicroscopy, immunoenzymology (ELISA) and gel diffusion for investigating virus strain relationships, *Ann. Virol.* **133:**3–14.

44. Kojima, M., Chou, T.G., and Shikata, E., 1978, Rapid diagnosis of potato leaf roll virus by immune electron microscopy, *Ann. Phytopathol. Soc. Japan* **44:**585–590.

45. Kurstak, E., Tyssen, P., and Kurstak, C., 1977, Immunoperoxidase technique in diagnostic virology and research: Principles and applications, in: E. Kurstak and C. Kurstak (eds.), *Comparative Diagnosis of Viral Diseases,* Vol. II, Academic Press, New York, pp. 403–448.

45a. Langenberg, W.G., 1986, Virus protein association with cylindrical inclusions of two viruses that infect wheat, *J. Gen. Virol.* **67:**1161–1168.

46. Lawson, R.H., 1982, Quantification of carnation etched ring virus by immunosorbent electron microscopy, *Phytopathology* **72:**708.

47. Lesemann, D.E., 1982, Advances in virus identification using immunosorbent electron microscopy, *Acta Hortic.* **127:**159–173.

48. Lesemann, D.E., and Paul, H.L., 1980, Conditions for the use of protein A in combination with the Derrick method of immuno electron microscopy, *Acta Hortic.* **110:**119–129.

49. Lesemann, D.E., Bozarth, R.F., and Koenig, R., 1980, The trapping of ty-

movirus particles on electron microscope grids by adsorption and serological binding, *J. Gen. Virol.* **48**:257–264.

50. Lim, W.L., De Zoeten, G.A., and Hagedorn, D.J., 1977, Scanning electron-microscopic evidence for attachment of a nonpersistently transmitted virus to its vector's stylets, *Virology* **79**:121–128.

51. Lima, J.A.A., and Purcifull, D.E., 1980, Immunochemical and microscopical techniques for detecting blackeye cowpea mosaic and soybean mosaic viruses in hypocotyls of germinated seeds, *Phytopathology* **70**:142–147.

52. Lin, Na-Sheng, and Langenberg, W.G., 1983, Immunohistochemical localization of barley stripe mosaic virions in infected wheat cells, *J. Ultrastruct. Res.* **84**:16–23.

53. Lisa, V., Luisoni, E., and Milne, R.G., 1981, A possible virus cryptic in carnation, *Ann. Appl. Biol.* **98**:431–437.

54. Lister, R.M., Carroll, T.W., and Zaske, S.K., 1981, Sensitive serologic detection of barley stripe mosaic virus in barley seed, *Plant Dis.* **65**:809–814.

55. Lobuglio, A.F., Rinehart, J.J., and Balcerzak, S.P., 1972, A new immunological marker for scanning electron microscopy, in: *Scanning Electron Microscopy/ 1972*, Part II, IIT Research Institute, Chicago, Illinois, pp 313–320.

56. Luisoni, E., Milne, R.G., and Boccardo, G., 1975, The maize rough dwarf virion. II. Serological analysis, *Virology* **68**:86–96.

57. Luisoni, E., Milne, R.G., and Roggero, P., 1982, Diagnosis of rice ragged stunt virus by enzyme-linked immunosorbent assay and immunosorbent electron microscopy, *Plant Dis.* **66**:929–932.

58. Makkouk, K.M., Koenig, R., and Lesemann, D.E., 1981, Characterization of a tombusvirus isolated from eggplant, *Phytopathology* **71**:572–577.

59. Martin, M.L., and Palmer, E.L., 1983, Electron microscopic identification of rotavirus group antigen with gold-labelled monoclonal IgG, *Arch. Virol.* **78**:279–285.

60. Meignoz, R., Caudwell, A., Kuszala, C., Schneider, C., Larrue, J., Fleury, A., and Boudon, E., 1983, Serological purification and visualization in the electromicroscope of the grapevine flavescence dorée (FD) pathogen (MLO) in diseased plants and infectious vectors extracts, *Yale J. Biol. Med.* **56**:936–937.

61. Meyer, S., 1982, Peanut mottle virus: Purification and serological relationship with other potyviruses, *Phytopathol. Z.* **105**:271–278.

62. Milne, R.G., 1980, Some observations and experiments of immunosorbent electron microscopy of plant viruses, *Acta Hortic.* **110**:129–135.

63. Milne, R.G., 1984, Electron microscopy for the identification of plant viruses in *in vitro* preparations, in: K. Maramorosch and H. Koprowski (eds.), *Methods in Virology*, Vol. 7, Academic Press, New York, pp. 87–120.

64. Milne, R.G., and Lesemann, D.E., 1978, An immunoelectron microscopic investigation of oat sterile dwarf and related viruses, *Virology* **90**:299–304.

65. Milne, R.G., and Lesemann, D.E., 1984, Immunosorbent electron microscopy in plant virus studies, in: K. Maramorosch and H. Koprowski (eds.), *Methods in Virology*, Vol. 8, Academic Press, New York, pp. 85–101.

66. Milne, R.G., and Luisoni, E., 1977, Rapid immune electron microscopy of virus preparations, in: K. Maramorosch and H. Koprowski (eds.), *Methods in Virology*, Vol. 6, Academic Press, New York, pp. 265–281.

67. Milne, R.G., Masenga, V., and Lovisolo, O., 1980, Viruses associated with

white bryony (*Bryonia cretica* L.) mosaic in northern Italy, *Phytopathol. Medit.* **19**:115–120.

68. Moffitt, E.M., and Lister, R.M., 1975, Application of a serological screening test for detecting double-stranded RNA mycoviruses, *Phytopathology* **65**:851–859.

69. Nicolaieff, A., and Regenmortel, M.H.V., 1980, Specificity of trapping of plant viruses on antibody-coated electron microscope grids, *Ann. Virol. Inst. Pasteur* **131E**:95–110.

70. Nicolaieff, A., Katz, D., and Van Regenmortel, M.H.V., 1982, Comparison of two methods of virus detection by immunosorbent electron microscopy (ISEM) using protein A, *J. Virol. Meth.* **4**:155–166.

71. Noel, M.C., Kerlan, C., Garnier, M., and Dunez, J., 1978, Possible use of immunoelectron microscopy (IEM) for the detection of plum pox virus in fruit trees, *Ann. Phytopathol.* **10**:381–386.

72. Otsuki, Y., Takebe, I., Ohno, T., Fukuda, M., and Okada, Y., 1977, Reconstitution of tobacco mosaic virus rods occurs bidirectionally from an internal initiation region: Demonstration by electron microscopic serology, *Proc. Natl. Acad. Sci. USA* **74**:1913–1917.

73. Paliwal, Y.C., 1977, Rapid diagnosis of barley yellow dwarf virus in plants using serologically specific electron microscopy, *Phytopathol. Z.* **89**:25–36.

74. Paliwal, Y.C., 1979, Serological relationships of barley yellow dwarf virus isolates, *Phytopathol. Z.* **94**:8–15.

75. Paliwal, Y.C., 1982a, Detection of barley yellow dwarf virus in aphids by serologically specific electron microscopy, *Can. J. Bot.* **60**:179–185.

76. Paliwal, Y.C., 1982b, Role of perennial grasses, winter wheat, and aphid vectors in the disease cycle and epidemiology of barley yellow dwarf virus, *Can. J. Plant Pathol.* **4**:367–374.

77. Pares, R.D., and Whitecross, M.I., 1982, Gold-labelled antibody decoration (GLAD) in the diagnosis of plant viruses by immuno-electron microscopy, *J. Immunol. Meth.* **51**:23–28.

78. Pares, R.D., and Whitecross, M.I., 1983, A critical examination of the utilization of serum coated grids to increase particle numbers for length determination of rod-shaped plant viruses, *J. Virol. Meth.* **7**:241–250.

79. Pegg-Feige, K., and Doane, F.W., 1983, Effect of specimen support film in solid phase immunoelectron microscopy, *J. Virol. Meth.* **7**:315–319.

80. Reissig, M., and Orwell, S.A., 1970, A technique for the electron microscopy of protein free particle suspensions by the negative staining method, *J. Ultrastruct. Res.* **32**:107–117.

81. Rifkind, R.A., 1976, Ferritin conjugated antibody markers for electron microscopy, in: C.A. Williams and M.W. Chase (eds.), *Methods of Immunology and Immunochemistry,* Vol. 5, Academic Press, New York, pp. 457–463.

82. Roberts, I.M., 1980, A method for providing comparative counts of small particles in electron microscopy, *J. Microsc.* **118**:241–245.

83. Roberts, I.M., 1981a, Factors affecting immunosorbent electron microscopy, *Rep. Scott. Hortic. Res. Inst.* **1980**:106–107.

84. Roberts, I.M., 1981b, Electron microscope serology techniques, in: *Proceedings AAB Workshop on Electron Microscope Serology,* John Innes Institute, Norwich, England, pp. 10–12.

85. Roberts, I.M., and Brown, D.J.F., 1980, Detection of six nepoviruses in

their nematode vectors by immunosorbent electron microscopy, *Ann. Appl. Biol.* **96:**187–192.

86. Roberts, I.M., and Harrison, B.D., 1979, Detection of potato leafroll and potato mop-top viruses by immunosorbent electron microscopy, *Ann. Appl. Biol.* **93:**289–297.

87. Roberts, I.M., Tamada, T., and Harrison, B.D., 1980, Relationship of potato leafroll virus to luteoviruses: Evidence from electron microscope serological tests, *J. Gen. Virol.* **47:**209–213.

88. Rose, D.G., 1983, Some properties of an unusual isolate of potato virus S, *Potato Res.* **26:**49–62.

89. Russo, M., Martelli, G.P., and Savino, V., 1982a, Immunosorbent electron microscopy for detecting sap-transmissible viruses of grapevine, in: A.J. McGinnis (ed.), *Proceedings of the 7th Meeting of the International Council for the Study of Viruses and Virus-like diseases of the Grapevine,* Agriculture Canada, Research Branch, pp. 251–257.

90. Russo, M., Savino, V., and Vovlas, C., 1982b, Virus diseases of vegetable crops in Apulia XXVII. Broad bean stain, *Phytopathol. Z.* **104:**115–123.

91. Sequeira, J.C., and Harrison, B.D., 1982, Serological studies on cassava latent virus, *Ann. Appl. Biol.* **101:**33–42.

92. Shalla, T.A., and Amici, A., 1964, The distribution of viral antigen in cells infected with tobacco mosaic virus as revealed by electron microscopy, *Virology* **31:**78–91.

93. Shalla, T.A., and Shepard, J.F., 1972, The structure and antigenic analysis of amorphous inclusion bodies induced by potato virus X, *Virology* **49:**654–667.

94. Shepard, J.F., Gaard, G., and Purcifull, D.E., 1974, A study of tobacco etch virus-induced inclusions using indirect immunoferritin procedures, *Phytopathology* **64:**418–425.

95. Shukla, D.D., and Gough, K.H., 1979, The use of protein A from *Staphylococcus aureus* in immune electron microscopy for detecting plant virus particles, *J. Gen. Virol.* **45:**533–536.

96. Shukla, D.D., and Gough, K.H., 1983, Characteristics of the protein A-immunosorbent electron microscopic technique (PA-ISEM) for detecting plant virus particles, *Acta Phytopathol. Acad. Sci. Hung.* **18:**173–185.

97. Shukla, D.D., and Gough, K.H., 1984, Serological relationships among four Australian strains of sugarcane mosaic virus as determined by immune electron microscopy, *Plant Dis.* **68:**204–206.

98. Singer, S.J., and Schick, A.F., 1961, The properties of specific stains for electron microscopy prepared by the conjugation of antibody molecules with ferritin, *J. Biophys. Biochem. Cytol.* **9:**519–537.

99. Sinha, R.C., and Benhamou, N., 1983, Detection of mycoplasmalike organism antigens from aster yellows-diseased plants by two serological procedures, *Phytopathology* **73:**1199–1202.

100. Sinha, R.C., and Chiykowski, L.N., 1984, Purification and serological detection of mycoplasmalike organisms from plants affected by peach eastern X-disease, *Can. J. Plant Pathol.* **6:**200–205.

101. Slot, J.W., and Geuze, H.J., 1981, Sizing of protein A colloidal gold probes for immunoelectron microscopy, *J. Cell Biol.* **90:**533–536.

102. Stobbs, L.W., 1984, Effect of grid rotation in a magnetic field on virus adsorption in immunoelectron microscopy, *Phytopathology* **74:**1132–1134.

103. Stollar, B.D., 1973, Nucleic acid antigens, in: M. Sola (ed.), *The Antigens,* Vol. 1, Academic Press, New York, pp. 1–85.
104. Tamada, T., Harrison, B.D., and Roberts, I.M., 1984, Variation among British isolates of potato leafroll virus, *Ann. Appl. Biol.* **104**:107–116.
105. Thomas, B.J., 1980, The detection by serological methods of viruses infecting the rose, *Ann. Appl. Biol.* **94**:91–101.
106. Tomenius, K., Clapham, D., and Oxelfelt, P., 1983, Localization by immunogold cytochemistry of viral antigen in sections of plant cells infected with red clover mottle virus, *J. Gen. Virol.* **64**:2669–2678.
107. Torrance, L., and Jones, R.A.C., 1981, Recent developments in serological methods suited for use in routine testing for plant viruses, *Plant Pathol.* **30**:1–24.
108. Torrance, L., and Jones, R.A.C., 1982, Increased sensitivity of detection of plant viruses obtained by using a fluorogenic substrate in enzyme-linked immunosorbent assay, *Ann. Appl. Biol.* **101**:501–509.
109. Usugi, T., and Saito, Y., 1981, Purification and some properties of oat mosaic virus, *Ann. Phytopathol. Soc. Japan* **47**:581–585.
110. Van Balen, E., 1982, The effect of pretreatments of carbon-coated formvar films on the trapping of potato leafroll virus particles using immunosorbent electron microscopy, *Neth. J. Plant Pathol.* **88**:33–37.
111. Van Regenmortel, M.H.V., 1975, Antigenic relationships between strains of tobacco mosaic virus, *Virology* **64**:415–420.
112. Van Regenmortel, M.H.V., 1982a, *Serology and Immunochemistry of Plant Viruses,* Academic Press, New York, pp. 124–132.
113. Van Regenmortel, M.H.V., 1982b, *Serology and Immunochemistry of Plant Viruses,* Academic Press, New York, pp. 147–169.
114. Van Regenmortel, M.H.V., Nicolaieff, A., and Burckard, J., 1980, Detection of a wide spectrum of virus strains by indirect ELISA and serological trapping electron microscopy (STREM), *Acta Hortic.* **110**:107–115.
115. Vetten, H.J., 1981, Indexing of nepoviruses on *Chenopodium quinoa* after elimination of virus inhibitors in grape leaf extracts, *J. Plant Dis. Protect.* **57**:99–110.
116. Waterhouse, P.M., and Murant, A.F., 1981, Purification of carrot red leaf virus and evidence from four serological tests for its relationship to luteoviruses, *Ann. Appl. Biol.* **97**:191–204.

Index

Current Topics in Vector Research

Edited by Kerry F. Harris

Volume 1   1983

Volume 2   1984